The Emergence of Mathematical Meaning: Interaction in Classroom Cultures

The Studies in Mathematical Thinking and Learning Series
Alan Schoenfeld, Advisory Editor

Carpenter/Fennema/Romberg: *Rational Numbers: An Integration of Research*

Cobb/Bauersfeld: *The Emergence of Mathematical Meaning: Interaction in Classroom Cultures*

Romberg/Fennema/Carpenter: *Integrating Research on the Graphical Representation of Functions*

Schoenfeld: *Mathematical Thinking and Problem Solving*

S. Williams

The Emergence of Mathematical Meaning: Interaction in Classroom Cultures

Edited by

Paul Cobb
Vanderbilt University
Heinrich Bauersfeld
University of Bielefeld

LEA LAWRENCE ERLBAUM ASSOCIATES, PUBLISHERS
1995 Hillsdale, New Jersey					Hove, UK

Copyright © 1995 by Lawrence Erlbaum Associates, Inc.
All rights reserved. No part of this book may be reproduced in any form, by photostat, microform, retrieval system, or any other means, without the prior written permission of the publisher.

Lawrence Erlbaum Associates, Inc., Publishers
365 Broadway
Hillsdale, New Jersey 07642

Cover design by Mairav Salomon-Dekel

Library of Congress Cataloging-in-Publication Data

The emergence of mathematical meaning : interaction in classroom cultures / edited by Paul Cobb, Heinrich Bauersfeld.
    p. cm.
Includes bibliographical references and index.
ISBN 0-8058-1728-X (acid-free). — ISBN 0-8058-1729-8 (pbk. : acid-free)
1. Mathematics—Study and teaching. I. Cobb, Paul. II. Bauersfeld, H.
QA11.E65 1995
372.7'044—dc20   94-39294
                                            CIP

Books published by Lawrence Erlbaum Associates are printed on acid-free paper and their bindings are chosen for strength and durability.

Printed in the United States of America
10 9 8 7 6 5 4 3 2 1

Dedication:
Willie Mack King, Sr.

This volume is dedicated to the memory of Willie Mack King, Sr., second-grade teacher at Ivanhoe Elementary School, Gary, Indiana. Mr. King was the classroom teacher for one of the year-long classroom teaching experiments conducted as part of the Purdue Problem-Centered Mathematics Project. For an entire year, project staff spent an hour or more in his classroom each day. A parade of mathematics education researchers visited his class during that year. Many more have come to know him anonymously as "The Teacher" or "Mr. K" in research reports that document the uniquely creative nature of his mathematics instruction. Without exception, visitors to his classroom were impressed by his overwhelming love for teaching, which found expression in the way he cared for each individual child and by his depth of understanding of and enthusiasm for mathematics that inspired even the least able of the children to make sense of *their* mathematics. He endeared himself to the project staff and to all of the contributors of this book, each of whom had the privilege of experiencing his teaching—both in life, and, even more often, on videotape—and of seeing him "living together" with *his* children. The world is poorer without him.

Contents

	Preface	ix
1	Introduction: The Coordination of Psychological and Sociological Perspectives in Mathematics Education *Paul Cobb and Heinrich Bauersfeld*	1
2	The Teaching Experiment Classroom *Paul Cobb, Erna Yackel, and Terry Wood*	17
3	Mathematical Learning and Small-Group Interaction: Four Case Studies *Paul Cobb*	25
4	Children's Talk in Inquiry Mathematics Classrooms *Erna Yackel*	131
5	Thematic Patterns of Interaction and Sociomathematical Norms *Jörg Voigt*	163
6	An Emerging Practice of Teaching *Terry Wood*	203

7	The Ethnography of Argumentation *Götz Krummheuer*	**229**
8	"Language Games" in the Mathematics Classroom: Their Function and Their Effects *Heinrich Bauersfeld*	**271**
	Glossary	**293**
	Author Index	**299**
	Subject Index	**303**

Preface

The collaboration that gave rise to this book began in 1985 when the two of us shared a room at the Gordon Research Conference on Cybernetics. It was evident from our initial conversations that we were interested in many of the same issues, including that of accounting for the messiness and complexity of mathematical learning as it occurs in classroom situations. However, it also became apparent that we approached these issues from different theoretical perspectives. Bauersfeld stressed the social and interactional aspects of mathematical activity, whereas Cobb treated mathematical development as an individual process of conceptual construction.

In the following years, these initial conversations broadened into a series of meetings that included our colleagues Götz Krummheuer, Jörg Voigt, Terry Wood, and Erna Yackel. These interactions were punctuated by both the exhilaration that comes from a meeting of minds and the perplexity associated with differences in perspectives and languages. Frequently, observations that either the American or German group treated as central appeared peripheral to the other group. It eventually took us 3 years to establish a reasonable basis for communication, and classroom video recordings served as a primary means by which we came to understand each others' positions.

In the course of these discussions, we arrived at the conclusion that psychological and sociological perspectives each tell half of a good story. What was needed was a combined approach that takes individual students' mathematical interpretations seriously while simultaneously seeing their activity as necessarily socially situated. A 3-year project that had as its goal the coordination of psychological and sociological perspectives was subsequently supported by the Spencer Foundation. This book reports the results of that collaboration.

The book is unusual in that it grew out of a systematic research effort and yet reflects the diversity of the individual authors' interests. Thus, it is much more than a compendium of loosely related papers. The process of collaborative argumentations wherein we critiqued and influenced each others' analyses continued throughout the project. Further, video recordings continued to be an invaluable tool, and we agreed to share a single set of recordings and transcripts made in one second-grade classroom in which instruction was generally compatible with current American reform recommendations. Consequently, the reader encounters the same children in multiple situations as they contributed to whole-class discussions, attempted to solve tasks posed by an interviewer, and interacted with another child during small-group problem-solving sessions. Further, the small-group and whole-class episodes sometimes appear in two or more chapters, in which different aspects are brought to the fore depending on the authors' purposes. A rich, complex, and multifaceted view therefore emerges of both the students' and teacher's individual activity, and the inquiry mathematics microculture established in the classroom. At the same time, the reader's familiarity with the classroom provides a grounding for the theoretical constructs developed in the individually authored chapters.

The introductory chapter sets the stage by contrasting the combined constructivist and interactionist approach with several other theoretical traditions, including those of Vygotsky and Piaget. In his chapter, Cobb reports case studies of four pairs of students' small-group activity. This chapter serves to introduce the reader to eight of the children who feature prominently in the subsequent chapters. The relationship between small-group interactions and mathematical learning is dealt with both theoretically and as a pragmatic issue that has implications for instructional improvement. In her chapter, Yackel describes how some of these same eight children varied their mathematical explanations according to the context. Her analysis therefore bears directly on the issue of situated cognition. In addition, she discusses how the classroom teacher supported the students' attempts to explain their mathematical thinking. Wood picks up and broadens this theme in her chapter, in which she analyzes how the teacher gradually reorganized her classroom practices and developed a pedagogical approach compatible with recent reform recommendations.

Voigt's contribution also touches on reform in that he describes how the teacher influenced the students' mathematical activity and leaning without, at the same time, obliging them to use any particular solution method. In addition, he analyzes the classroom discourse to understand how the teacher and students developed the intersubjectivity necessary for mathematically coherent discussions. In his chapter, Krummheuer discusses the related notion of mathematical argumentation. He draws on both small-group and whole-class episodes to document the benefits and limitations of argumentations for conceptual, mathematical learning. Bauersfeld continues this focus in discourse in his chapter, by critiquing traditional approaches to language in mathematics and then proposing a joint constructivist and interactionist alternative.

The key issue that ties the chapters together is that of relating analyses of individual students' thinking to analyses of classroom interactions and discourse, and the classroom microculture. The combined constructivist and interactionist approach taken by the authors draws on symbolic interactionism and ethnomethodology, and thus constitutes an alternative to Vygotskian and Soviet activity theory approaches. The book should therefore be relevant to educators and psychologists interested in situated cognition and the relation between sociocultural processes and individual psychological processes. In addition, the book brings together a range of theoretical constructs and illustrates what they might offer regarding analyses of students' and teachers' actions in complex classroom situations. Further, particular chapters will be of interest to researchers who focus on more specific issues. Those include small-group collaboration and learning (chap. 3); the teacher's activity and growth (chaps. 4 and 6); and language, discourse, and argumentation (chaps. 4, 5, 7, and 8).

It is important to note that the analyses reported in this book grew out of a classroom-based research and development project. Thus, although the primary thrust of the book is theoretical, it grew out of practice and can feed back to guide practice. The theoretical constructs described by the chapter authors are offered as ways of interpreting what might be going on when attempts are made to reform classroom mathematical practices. Analyses of this type are central to developmental or transformational research in that they both lead to the further development of theory and feedback to inform the revision of classroom activities. We therefore hope that this book will contribute to reform in mathematics education.

ACKNOWLEDGMENTS

As with any venture of this type, we must acknowledge the help of a number of people. We are especially grateful to the teachers and students of Tippecanoe County, Indiana, and Gary, Indiana, featured in the video recordings and our discussions. The partnership with these teachers has been enriching beyond measure. A special mention must be given to Graceann Merkel, the second-grade teacher whose instructional practice is discussed in several of the chapters. We also want to express our gratitude to the Spencer Foundation for its support of the analyses reported in this volume. The Foundation's vice president, Marion M. Faldet, and our program officers, Linda May Fitzgerald and Rebecca Barr, have provided encouragement at every step of the project. Alan Schoenfeld wrote a detailed and penetrating review that contributed greatly to the improvement of the book. We are also indebted to Cheryl Burkey for formatting the manuscript. Finally, a special thanks must go to Janet Bowers for her invaluable assistance with the editing process.

Paul Cobb
Heinrich Bauersfeld

1

Introduction: The Coordination of Psychological and Sociological Perspectives in Mathematics Education

Paul Cobb
Vanderbilt University
Heinrich Bauersfeld
University of Bielefeld

The three American and three German contributors to this volume each explore the coordination of cognitive and sociological perspectives in mathematics education. Collectively, the contributions are diverse and yet have an overarching coherence. The diversity reflects differences both in the authors' theoretical backgrounds and in their interests and concerns. Thus, the issues addressed cover the gamut from small-group interactions to students' development of ways of languaging in the mathematics classroom. The coherence reflects the close collaboration of the authors, their use of the same set of video recordings and transcripts, and their compatible epistemological commitments. Each of the authors refers to the basic tenets of both constructivism and social interactionism. Thus, they draw on von Glasersfeld's (1987) characterization of students as active creators of their ways of mathematical knowing, and on the interactionist view that learning involves the interactive constitution of mathematical meanings in a (classroom) culture. Further, the authors assume that this culture is brought forth jointly (by the teacher and students), and that the process of negotiating meanings mediates between cognition and culture (Bauersfeld, Krummheuer, & Voigt, 1988).

The importance of sociological perspectives has long been acknowledged in a variety of disciplines. However, the emergence of this viewpoint in European mathematics education research is a relatively recent development (Balacheff,

1986; Bauersfeld, 1980; Bishop, 1985; Brousseau, 1984). Further, this interpretive stance has only come to the fore in the United States in the last few years. As recently as 1988, Eisenhart wrote with considerable justification that mathematics educators "are accustomed to assuming that the development of cognitive skills is central to human development, [and] that these skills appear in a regular sequence regardless of context or content" (p. 101). The growing trend to go beyond purely cognitive analyses is indicated by an increasing number of texts that question an exclusive focus on the individual learner (Brown, Collins, & Duguid, 1989; Greeno, 1991; Lave, 1988; Newman, Griffin, & Cole, 1989; Nunes, Schliemann, & Carraher, 1993; Saxe, 1991). The analyses presented in this volume contribute to this trend while advancing an alternative position that draws more on symbolic interactionism (Blumer, 1969) and ethnomethodology (Mehan & Wood, 1975) than on the work of Vygotsky and Soviet activity theorists.

We can best clarify the motivations for undertaking the collaborative research reported in this volume by first considering what it ought to mean to know and do mathematics in school, and then by outlining current sociological approaches to mathematical activity.

SCHOOL MATHEMATICS AND INQUIRY MATHEMATICS

The findings of numerous empirical studies indicate that, in the setting of traditional U.S. textbook-based teaching, many students develop conceptions that deviate significantly from those that the teacher intends. Frequently, these conceptions are such that students associate a sequence of symbol manipulations with various notational configurations (Thompson, 1994). Concomitantly, analyses of the social interactions in these classrooms indicate that the construction of such concepts enables students to appear mathematically competent (Gregg, 1993; McNeal, 1992; Schoenfeld, 1987; Yang, 1993). In Much and Shweder's (1978) terms, the mathematical practices established in these classrooms appear to have the quality of instructions to be followed (Cobb, Wood, Yackel, & McNeal, 1992). In other words, the classroom discourse is such that symbol-manipulation acts do not carry the significance of acting mentally on mathematical objects. Further, because there is nothing beyond the symbols to which the teacher and students publicly refer, a mathematical explanation involves reciting a sequence of steps for manipulating symbols. Mathematics as it is constituted in these classrooms, therefore, appears to be a largely self-contained activity that is not directly related to students' out-of-school activities (Cobb, 1987; Confrey, 1990).

These classroom mathematical practices can be contrasted with those constituted in inquiry classrooms, where a radically different microculture has been established and in which the teacher and students together constitute a community of validators (Bussi, 1991; Carpenter, Fennema, Peterson, Chiang, & Loef, 1989; Human, Murray, & Olivier, 1989; Lampert, 1990). The standards of argumentation established in an inquiry classroom are such that the teacher and students typically challenge explanations that merely describe the manipulation of symbols. Further, acceptable explanations appear to carry the significance of acting on taken-as-shared mathematical objects (Cobb et al., 1992). Consequently, from the observer's perspective, the teacher and students seem to be acting in a taken-as-shared mathematical reality, and to be elaborating that reality in the course of their ongoing negotiations of mathematical meanings.

We follow Richards (1991) in calling these two types of classroom microcultures the *school mathematics* and the *inquiry mathematics* microcultures. Davis and Hersh (1981) succinctly captured the distinction between them when they noted that "Mathematicians know they are studying an objective reality. To an outsider, they seem to be engaged in an esoteric communication with themselves and a small group of friends" (pp. 43–44). The public discourse of traditional school mathematics in which explanations involve specifying instructions for manipulating symbols appears to have the characteristics of an esoteric communication. In contrast, the public discourse in an inquiry mathematics classroom is such that the teacher and students appear to act as Platonists who are communicating about a mathematical reality that they experience as objective (Hawkins, 1985). The contributors to this volume all value mathematical activity of this latter type and seek to account for it by coordinating sociological and psychological perspectives. When they take a sociological perspective, they talk of taken-as-shared mathematical objects and describe them as social accomplishments that emerge via a process of interactive constitution. When they take a psychological perspective, they talk of experientially real mathematical objects and describe them as personal constructions that emerge via a process of active conceptual self-organization.

LEARNING MATHEMATICS AND SOCIAL INTERACTION

Two general theoretical positions on the relationship between social processes and psychological development can be identified in the current literature. These positions frequently appear to be in direct opposition in that one gives priority to social and cultural processes, and the other to the individual autonomous learner (Voigt, 1992). The two positions might therefore be termed *collectivism* and *individualism*, respectively. The collectivist position is exemplified by theories developed both in the Vygotskian tradition and in the sociolinguistic tradition. In both cases, mathematical learning is viewed primarily as a process of acculturation. Vygotskian theories, which are currently popular in the United States, locate learning in coparticipation in social practices. Sociolinguistic theories, which are currently more prominent in the United Kingdom, characterize mathematical learning as an initiation into the social tradition of doing mathematics in school. Both types of theories can be contrasted with individualistic theories, which treat mathematical learning almost exclusively as a process of active individual construction. This position is exemplified by neo-Piagetian theories, which view social interaction as a source of cognitive conflicts that facilitate autonomous cognitive development.

The Vygotskian and Activity Theory Tradition

Theorists who work in this tradition tend to assume from the outset that cognitive processes are subsumed by social and cultural processes. Empirical support for this position comes from paradigmatic studies such as those of Carraher, Carraher, and Schliemann (1985), Lave (1988), Saxe (1991), and Scribner (1984), which demonstrate that an individual's arithmetical activity is profoundly influenced by his or her participation in encompassing cultural practices such as completing worksheets in school, shopping in a supermarket, selling candy on the street, and packing crates in a dairy. These findings are consistent with and add credence to the claim that mathematics as it is realized in the microcultures of school mathematics and inquiry mathematics constitutes two different forms of activity. Further, the findings support the view that mathematical practices are negotiated and institutionalized by members of communities.

In making the assumption that priority should be given to social and cultural processes, theorists working in this tradition adhere to Vygotsky's (1979) contention that "The social dimension of consciousness is primary in fact and time. The individual dimension of consciousness is derivative and secondary" (p. 30). From this, it follows that "Thought (cognition) must not be reduced to a subjectively psychological process" (Davydov, 1988, p. 16). Instead, thought should be viewed as "something essentially 'on the surface,' as something located . . . on the borderline between the organism and the outside world. For thought . . . has a life only in an environment of socially constituted meanings" (Bakhurst, 1988, p. 38).

Consequently, the individual in social action is taken as the basic unit of analysis (Minick, 1989). The primary issue to be addressed in this tradition is, then, that of explaining how participation in social interactions and culturally organized activities influences psychological development.

This issue has been formulated in a variety of different ways. For example, Vygotsky (1978) emphasized both social interaction with more knowledgeable others in the zone of proximal development and the use of culturally developed sign systems as psychological tools for thinking. In contrast, Leont'ev (1981) argued that thought develops from practical, object-oriented activity or labor. Several U.S. theorists have elaborated constructs developed by Vygotsky and his students, and speak of cognitive apprenticeship (Brown et al., 1989; Rogoff, 1990), legitimate peripheral participation (Forman, 1992; Lave & Wenger, 1991), or the negotiation of meaning in the construction zone (Newman et al., 1989). In each of these contemporary accounts, learning is located in coparticipation in cultural practices. As a consequence, educational recommendations usually focus on the kinds of social engagements that increasingly enable students to participate in the activities of the expert rather than on the cognitive processes and conceptual structures involved (Hanks, 1991).

An analysis of classroom activity developed within this tradition might both locate it within a broader activity system that takes account of the function of schooling as a social institution and attends to the immediate interactions between the teacher and students (Axel, 1992). This dual focus is explicit in Lave and Wenger's (1991) claim that their "concept of legitimate peripheral participation provides a framework for bringing together theories of situated activity and theories about the production and reproduction of the social order" (p. 47). In general, the individual's participation in culturally organized practices and face-to-face interactions carries the explanatory burden in accounts of cognitive development proposed from this perspective. Thus, the central concern is to delineate the social and cultural basis of personal experience.

Theories developed within this tradition clearly make an important contribution, particularly in accounting for the regeneration of the traditional practices of mathematics instruction. They, therefore, have much to offer in an era of reform in mathematics education that is concerned with the restructuring of the school and takes the issue of ethnic and cultural diversity seriously. However, it can be argued that they are theories of the conditions for the possibility of learning (Krummheuer, 1992). For example, Lave and Wenger (1991), who took a relatively radical position by attempting to avoid any reference to individual psychological processes, said that "A learning curriculum unfolds in *opportunities for engagement* in practice" (p. 93, italics added). Consistent with this formulation, they noted that their analysis of various examples of apprenticeship in terms of legitimate peripheral participation accounts for the occurrence of learning or failure to learn. In contrast, the contributors to this volume seek to understand both what students learn and the processes by which they do so as they participate in a learning curriculum, and to relate these analyses of learning to the processes by which both the curriculum and the encompassing classroom microculture are interactively constituted. As a consequence, theories developed within the Vygotskian and activity theory tradition are not entirely appropriate to the contributors' particular research goals.

The Sociolinguistic Tradition

Walkerdine (1988) and Solomon (1989) are, perhaps, the two most influential proponents of this second collectivist tradition. These theorists, like those working in the Vygotskian tradition, give priority to social and cultural processes. Further, both groups of theorists stress the importance of coparticipation in cultural practices. However, whereas Vygotskian theorists contend that qualitative changes in students' thinking occur as they participate in these practices, Walkerdine and Solomon rejected the view that mathematical development involves the construction of increasingly sophisticated systems of thought. In developing their alternative position, Walkerdine drew on French poststructuralism, whereas Solomon was heavily influenced by, among others, the later Wittgenstein. Despite this difference, both argued that the activity of doing mathematics in school should be viewed as participation in a social or discursive practice. Thus, as Solomon put it, "Understanding is intrinsically social; knowing about numbers entails knowing how and when to use and respond to numbers according to the context in which they appear" (1989, p. 7). Learning mathematics in school is, then, a process of initiation into a pregiven discursive practice and occurs when students act in accord with the normative rules that constitute that practice.

Walkerdine and Solomon made several important contributions while developing their positions. For example, their characterization of traditional school mathematics as a discursive practice adds credibility to the contention that school mathematics and inquiry mathematics constitute two distinct classroom microcultures. Further, Walkerdine in particular illustrated the importance of considering language and conventional sign systems when accounting for mathematical development. In the course of her argument, she explicitly rejected the dualist notion that signifiers refer to or represent features of a preexisting reality, and instead proposed that the function of language is mutual orientation and regulation (cf. Bauersfeld, chap. 8, this volume; Maturana, 1978).

These insights acknowledged, Walkerdine's and Solomon's characterization of mathematical activity is at odds with that advanced by the contributors to this volume. For Solomon, it is an activity in which students learn to act in accord with situated mathematical rules or instructions. Walkerdine, for her part, explicitly argued that the purpose of doing mathematics in school is to produce formal statements that do not signify anything beyond themselves. In these accounts, there is no place for what Davis and Hersh (1981) took to be the key feature of mathematical activity—the creation and mental manipulation of abstract mathematical objects (cf. Schoenfeld, 1991). It would, therefore, seem that Walkerdine's and Solomon's work is tied to the school mathematics microculture. Within the confines of this microculture, their claim that cognitive analyses are irrelevant to accounts of mathematical learning has some merit. It is frequently impossible to infer the mathematical conceptions of any individual student when analyzing video recordings and transcripts of traditional school mathematics lessons (Cobb et al., 1992). However, the claimed irrelevance of cognitive analyses and the contention that doing mathematics involves acting in accord with situated instructions both seem questionable when inquiry mathematics is considered. Theories of mathematical learning developed within the sociolinguistic tradition do not, therefore, appear to be of direct relevance to the research interests of the contributors of this volume.

The Neo-Piagetian Tradition

Perret-Clermont, Doise, and their collaborators attempted to extend Piagetian theory by investigating the role that social interaction plays in individual cognitive development (e.g., Doise & Mugny, 1979; Doise, Mugny, & Perret-Clermont, 1975; Perret-Clermont, 1980; Perret-Clermont, Perret, & Bell, 1989). As part of their rationale, they noted that Piaget, in his earlier writings, stressed that social interaction is necessary for the development of logic, reflectivity, and self-awareness. In an innovative series of studies, they documented the processes by which interpersonal conflicts between learners give rise to individual cognitive conflicts. As learners strive to resolve these conflicts, they reorganize their activity and construct increasingly sophisticated systems of thought.

In contrast to both Vygotskian and sociolinguistic theories, individualistic theories such as those developed in the neo-Piagetian tradition bring the psychological perspective to the fore. The focus is on the individual, autonomous learner as he or she participates in social interactions. Analyses developed in the neo-Piagetian tradition have been extensively critiqued by theorists who attribute a stronger role to social and cultural processes. Thus, Solomon (1989) noted that the role of social interaction in individual development is limited to that of providing a catalyst for development via interpersonal conflict. In her view, this almost exclusive focus on the constructive activity of individual learners gives rise to several difficulties. She argued that it is unreasonable to assume that students' conceptual reorganizations will necessarily constitute a step in their mathematical enculturation. Students could resolve a cognitive conflict in a variety of different ways, only some of which are compatible with the taken-as-shared mathematical practices of the wider community.

Solomon's basic point here was that the development of the ability to participate in such practices cannot be accounted for by autonomous development unless students are attributed foreknowledge of appropriate resolutions to cognitive conflicts. More generally, she contended that the influence of social processes is not limited to the process of learning but instead extends to its products—increasingly sophisticated mathematical ways of knowing. Thus, she argued that students' interpretation of an event as an interpersonal conflict is influenced by the classroom practices in which they participate. For Solomon, both what counts as a problem and as an acceptable solution in the classroom are social through and through. The contributors to this volume accept the central points of Solomon's critique, but question the implication that analyses of individual students' cognitive activity are irrelevant to accounts of mathematical learning in school.

COGNITIVE AND SOCIOLOGICAL PERSPECTIVES

The analyses reported in the following chapters seek to transcend the apparent opposition between collectivism and individualism by coordinating sociological analyses of the microculture established by the classroom community with cognitive analyses of individual students' constructive activities. In this regard, there is full agreement with Saxe and Bermudez's (1992) statement that:

> An understanding of the mathematical environments that emerge in children's everyday activities requires the coordination of two analytic perspectives. The first is a constructivist treatment of children's mathematics: Children's mathematical environments cannot be understood apart from children's own cognizing activities. . . . The

second perspective derives from sociocultural treatments of cognition.... Children's construction of mathematical goals and subgoals is interwoven with the socially organized activities in which they are participants. (pp. 2–3)

This coordination does not, however, produce a seamless theoretical framework. Instead, the resulting orientation is analogous to Heisenberg's uncertainty principle. When the focus is on the individual, the social fades into the background, and vice versa. Further, the emphasis given to one perspective or the other depends on the issues and purposes at hand. Thus, in the view advanced in this volume, there is no simple unification of the perspectives.

The epistemological basis for the psychological perspective elaborated in the following chapters has been developed by von Glasersfeld (1989b). It incorporates both the Piagetian notions of assimilation and accommodation, and the cybernetic concept of viability. The term *knowledge* was used by von Glasersfeld (1992) in "Piaget's *adaptational* sense to refer to those sensory-motor and conceptual operations that have proved viable in the knower's experience" (p. 380). Further, he dispensed with traditional correspondence theories of truth and instead proposed an account that relates truth to the effective or viable organization of activity: "Truths are replaced by viable models—and viability is always relative to a chosen goal" (p. 384). In this model, perturbations that the cognizing subject generates relative to a purpose or goal are posited as the driving force of development. As a consequence, learning is characterized as a process of self-organization in which the subject reorganizes his or her activity in order to eliminate perturbations (von Glasersfeld, 1989b). As von Glasersfeld noted, his instrumentalist approach to knowledge is generally consistent with the views of contemporary neo-pragmatist philosophers such as Bernstein (1983), Putnam (1987), and Rorty (1978).

Although von Glasersfeld (1992) defined learning as self-organization, he acknowledged that this constructive activity occurs as the cognizing individual interacts with other members of a community. Thus, he elaborated that *knowledge* refers to "conceptual structures that epistemic agents, given the range of present experience within their tradition of thought and language, consider *viable*" (p. 381). Further, von Glasersfeld (1989a) contended that "the most frequent source of perturbations for the developing cognitive subject is interaction with others" (p. 136). The interactionist perspective developed by Bauersfeld and his colleagues (Bauersfeld, 1980; Bauersfeld, Krummheuer, & Voigt, 1988) complements von Glasersfeld's cognitive focus by viewing communication as a process of mutual adaptation wherein individuals negotiate meanings by continually modifying their interpretations. However, whereas von Glasersfeld tended to focus on individuals' construction of their ways of knowing, the German group emphasized that:

The descriptive means and the models used in these subjective constructions are not arbitrary or retrievable from unlimited sources, as demonstrated through the unifying bonds of culture and language, through the intersubjectivity of socially shared knowledge among the members of social groups, and through the regulations of their related interactions. (Bauersfeld, 1988, p. 39)

Further, they contended that "Learning is characterized by the subjective reconstruction of societal means and models through negotiation of meaning in social interaction" (Bauersfeld, 1988, p. 39). In accounting for this process of subjective reconstruction, Bauersfeld and his colleagues focused on the teacher's and students' interactive constitution of the classroom microculture. Thus, in their view:

> Participating in the processes of a mathematics classroom is participating in a culture of mathematizing. The many skills, which an observer can identify and will take as the main performance of the culture, form the procedural surface only. These are the bricks of the building, but the design of the house of mathematizing is processed on another level. As it is with culture, the core of what is learned through participation is *when* to do what and *how* to do it. . . . The core part of school mathematics enculturation comes into effect on the meta-level and is "learned" indirectly. (Bauersfeld, chap. 8, this volume)

This discussion of indirect learning clarifies that the occurrence of perturbations is not limited to those occasions when participants in an interaction believe that communication has broken down and explicitly negotiate meanings. Instead, for the German group, communication is a process of often implicit negotiations in which subtle shifts and slides of meaning frequently occur outside the participants' awareness. Newman et al. (1989), speaking within the Vygotskian and activity theory tradition, made a similar point when they said that in an exchange between a teacher and a student, "The interactive process of change depends on . . . the fact that there are two different interpretations of the context and the fact that the utterances themselves serve to change the interpretations" (p. 13). However, it should be noted that Newman et al. used Leont'ev's (1981) sociohistorical metaphor of appropriation to define negotiation as a process of mutual appropriation in which the teacher and students continually coopt or use each others' contributions.

In contrast, Bauersfeld and his colleagues used an interactionist metaphor when they characterized negotiation as a process of mutual adaptation in the course of which the participants interactively constitute obligations for their activity (Voigt, 1985). It can also be noted that in Newman et al.'s account, the teacher is said to appropriate students' actions into the wider system of mathematical practices that he or she understands. The German group, however, took as its primary point of reference the local classroom microculture rather than the mathematical practices institutionalized by wider society. This focus reflects an interest in the process by which the teacher and students constitute the classroom microculture and mathematical practices in the course of their interactions. Further, whereas Vygotskian theorists give priority to social and cultural process, analyses conducted from the interactionist perspective propose that individual students' mathematical activity and the classroom microculture are reflexively related (Cobb, 1989; Voigt, 1992). In this view, individual students are seen as actively contributing to the development of both classroom mathematical practices and the encompassing microculture, and these both enable and constrain their individual mathematical activities. This notion of reflexivity, which is developed in several of the chapters, implies that neither an individual student's mathematical activity nor the classroom microculture can be adequately accounted for without considering the other.

OVERVIEW OF THE CONTRIBUTIONS

The contributions to this volume were developed in the course of a 5-year collaborative research project. At the outset of the project, the authors agreed to share a single set of video recordings and transcripts of small group activities and whole class discussions made in one U.S. second-grade classroom during the 1986–1987 school year. The recordings were ideally suited to the goals of the project because the teacher had succeeded in guiding the development of an inquiry mathematics microculture in her classroom (Cobb, Yackel, & Wood, 1989). The authors were each free to pursue their own research interests when analyzing the recording and transcripts. However, theoretical constructs and empirical analyses were critiqued during week-long meetings held approximately once every 9 months throughout the project. This empirically grounded approach to theory development might be called *collaborative argumentation*.

The chapters are ordered such that the first takes the strongest cognitive perspective and is concerned with individual learning in interaction, whereas the last chapter brings the sociological perspective to the fore and focuses on the relationship between languaging and the classroom microculture. This diversity reflects the researchers' differing theoretical assumptions and focuses. The contributions can therefore be seen to mirror the present situation in mathematics education in which there is a plurality of theoretical models and constructs rather than a generally accepted overarching theory (and both interactionists and constructivists question whether there will ever be one). As a consequence, the contributions exemplify a pluridimensional approach that might offer greater adaptability when analyzing various aspects of mathematical activity in classrooms.

In his chapter, Cobb presents longitudinal case studies of four pairs of second-grade students' small-group activity. Theoretically, the issue that motivated the case studies was that of clarifying the relationship between students' situated conceptual capabilities and their small-group interactions. The view that emerged in the course of the analysis was that of a reflexive relationship between students' mathematical activity and the social relationships they established. On the one hand, the students' cognitive capabilities appeared to constrain the possible forms their small-group interaction could take. On the other hand, the relationships the children actually established constrained the types of learning opportunities that arose, and thus influenced their construction of increasingly sophisticated mathematical ways of knowing. These developing mathematical capabilities, in turn, constrained the ways in which their small-group relationships could evolve, and so on.

Pragmatically, the case studies were conducted to clarify the extent to which small-group collaborative activity facilitates students' mathematical learning. This issue was addressed by relating the types of interaction in which the children engaged to the occurrence of learning opportunities. Two aspects of students' social relationships appear to be crucial for productive small-group activity in an inquiry mathematics classroom. The first is the establishment of a taken-as-shared basis for mathematical communication, and the second is engagement in interactions that involve genuine mathematical argumentation. As these criteria indicate, interactions in which one student explains his or her thinking do not necessarily give rise to learning opportunities for either student. Instead, it seems essential to consider the types of interaction in which students participate when assessing the role that particular activities such as explaining can play in their mathematical development.

In her chapter, Yackel analyzes the ways in which children talk about and explain their mathematical thinking in various social situations. The analysis indicates that there were no systematic differences between the students' explanations in small-group sessions and in whole-class discussions. However, qualitative differences in students' explanations became apparent when the analysis focused on their interpretations of social events, rather than on classroom social arrangements. As an illustration, Yackel accounts for students' explanations to the same task when speaking to the partner, the teacher, and to a researcher during small-group activity, and during the subsequent whole-class discussion, by delineating the student's obligations and expectations in the immediate situation. This relationship between the quality of an explanation and the social situation in which it is developed is reflexive. A student participates in the interactive constitution of the local social situation in the very act of explaining, and that situation constrains the nature of the explanation.

A second issue addressed by Yackel concerns the teacher's role in supporting students' attempts to explain their solutions. The analysis focuses on both the ways in which the teacher facilitated the establishment of situations for explanation, and the ways in which she intervened to help students develop an understanding of what constitutes an acceptable mathematical explanation in an inquiry mathematics classroom. At a more general level, the analysis of students' explanations exemplifies how a sociological perspective can be used to clarify an activity, such as explaining, that is usually characterized almost exclusively in cognitive terms.

The overriding interest that motivates Voigt's contribution is to understand how intersubjectivity is established in inquiry mathematics classrooms. The standard answer, that mathematical thinking involves conceptual necessities, is rejected because (elementary school) students have not yet become members of a mathematical community. The investigation focuses on the teacher's and students' interactions within the classroom microculture while drawing on a cognitive constructivist perspective that takes individual students' subjectivity seriously.

In the course of the analysis, Voigt demonstrates the viability of an interactionist approach in which the teacher and students are seen to mutually influence each other's activity in classroom situations. He elaborates this approach by arguing that the objects and tasks of classroom discourse seem to be ambiguous. The teacher and students develop a taken-as-shared understanding of these objects and events as they negotiate mathematical meanings. As a consequence of this achievement of intersubjectivity, the teacher and students experience their discourse as being thematically coherent. Voigt contends that the mathematical themes that emerge depend on their individual contributions and their negotiation of mathematical meaning, and yet cannot be explained in terms of the thoughts of one person alone. The relationship between mathematical themes and individual contributions therefore appears to be reflexive in nature. As a further point, Voigt reconstructs thematic patterns of interaction that contribute to the stability of classroom mathematical discourse. A central conjecture that emerges from the analysis is that students' participation in the constitution of thematic patterns of interaction and in the evolution of themes supports their construction of increasingly sophisticated conceptual operations.

Voigt clarifies the theoretical constructs developed to account for the achievement of intersubjectivity by applying them to specific classroom episodes. In doing so, he also demonstrates that the teacher did not attempt to specify predetermined ways in which she expected the students to solve tasks. Thus, the teacher did not attempt to influence the students' goals directly, but instead encouraged them to orient their mathematical activity in terms of their own mathematical goals. This did not, however, mean that any mathematical solution was as good as any other. Voigt's analysis indicates that teacher and students mutually constituted sociomathematical norms such as what counts as an insightful solution and, in the process, the teacher indirectly and subtly (in the positive sense) influenced the students' mathematical thinking and development. Voigt argues that because this classroom is representative of several current hopes of mathematics educators, and because crucial aspects of the relationship between teaching and learning mathematics are yet to be explained, the results of his investigation may be seen to justify the intensive, theoretically grounded analysis of relatively short classroom episodes.

The purpose of Wood's chapter is to provide a detailed analysis of the process by which the teacher changed her instructional practice as she interacted with her students. A microanalysis of classroom interactions was conducted to chronicle the shifts the teacher made from the traditional practice of school mathematics to that of teaching for meaning in an inquiry mathematics classroom. This emergence of a new form of practice did not occur in a linear fashion, but instead involved a series of shifts and slides. In particular, the interactional analysis indicates that there was a continuing shift between the traditional pattern of teacher questioning/student response, and a developing inquiry pattern of interaction. Wood argues that the changes the teacher made were underpinned by the interpretations of the researcher's philosophical and theoretical stance that she developed in the course of collaborative conversations. In Wood's view, these interpretations formed the basis from which she constructed an evolving personal theoretical framework. This framework, in turn, guided her actions during classroom interactions in which she encountered a tension between encouraging students' individual constructions and attending to the meanings of mathematics.

The primary purpose of Krummheuer's contribution is to propose a theoretical framework within which to investigate the processes of argumentation that occur in mathematics classrooms. To this end, he discusses several theories of argumentation and conducts a microethnographical reconstruction of a wide range of classroom processes that are subsumed under the notion of argumentation. These include arguing, explaining, justifying, illustrating, exemplifying, and analogizing.

Krummheuer characterizes argumentation as a primarily social process in which cooperating individuals try to adjust their intentions and interpretations by verbally presenting rationales for their actions. Krummheuer notes that participation in this negotiatory aspect of argumentation is frequently assumed to have a strong positive influence on students' conceptual development in mathematics. This widely held assumption can, in fact, be viewed as a social construction accomplished by the community of educators and learners in schools. As such, it is a feature of the "folk psychology" of classroom learning in Bruner's (1983) sense.

Krummheuer argues that, theoretically, the relationship between engaging in argumentation and learning is reflexive. Students' active involvement in a process of argumentation influences their individual learning and, conversely, their cognitive capabilities influence both the course and outcome of a process of argumentation. This reflexive relationship is elaborated in greater detail by both drawing on the concept of format that Bruner (1983) introduced in his studies of early language acquisition, and by considering constructs related to rhetoric and argumentation. Against this background, the limitation of argumentation for conceptual, mathematical learning in regular classroom settings is discussed. In particular, Krummheuer observes that conceptual mathematical learning has as much to do with the development of an everyday-language platform for indicating differences in individual interpretations as it does with the establishment of different formally valid mathematical argumentations. Learning therefore appears to be indirectly related to interactions within a classroom culture of argumentizing.

In his chapter, Bauersfeld offers a theoretical reflection for analyses related to language. He explores language games in the mathematics classroom. He observes that investigations of language in mathematics education typically treat it as an objective entity—as a given body of societal knowledge—and study either the structure or the use of this "body of knowledge." Bauersfeld develops an alternative approach to language by drawing on interactionist and constructivist perspectives, and by building on recent theoretical developments. In doing so, he argues that linguistic processes can be viewed as an accomplishment of language games that are special to each classroom, and in which the teacher and students negotiate taken-as-shared meanings and signs. This theoretical orientation, together with a brief summary of the processes by which students develop "languaging" in the mathematics classroom, constitute a basis for analyzing and interpreting videotapes and transcripts of inquiry mathematics classrooms. The specific issues addressed concern the genetic relationship between language games and the classroom microculture, and the treatment of individual differences. This joint interactionist and constructivist approach to languaging is used to reinterpret several difficulties in the mathematics classroom that stem from illusions of objectively given mathematical realities, from the assumption that a person can unproblematically represent or embody his or her mathematical knowing for someone else, and from the practice of teaching mathematics directly.

REFERENCES

Axel, E. (1992). One developmental line in European activity theories. *Quarterly Newsletter of the Laboratory of Comparative Human Cognition, 14*(1), 8–17.
Bakhurst, D. (1988). Activity, consciousness, and communication. *Quarterly Newsletter of the Laboratory of Comparative Human Cognition, 10,* 31–39.
Balacheff, N. (1986). Cognitive versus situational analysis of problem solving behavior. *For the Learning of Mathematics, 6*(3), 10–12.
Bauersfeld, H. (1980). Hidden dimensions in the so-called reality of a mathematics classroom. *Educational Studies in Mathematics, 11,* 23–41.
Bauersfeld, H. (1988). Interaction, construction and knowledge: Alternative perspectives for mathematics education. In T. Cooney & D. Grouws (Eds.), *Effective mathematics teaching* (pp. 29–46). Reston, VA: National Council of Teachers of Mathematics and Lawrence Erlbaum Associates.
Bauersfeld, H., Krummheuer, G., & Voigt, J. (1988). Interactional theory of learning and teaching mathematics and related microethnographical studies. In H. G. Steiner & A. Vermandel (Eds.), *Foundations and methodology of the discipline of mathematics education* (pp. 174–188). Antwerp: Proceedings of the TME Conference.
Bernstein, R. J. (1983). *Beyond objectivism and relativism: Science, hermeneutics, and praxis.* Philadelphia: University of Pennsylvania Press.
Bishop, A. (1985). The social construction of meaning—A significant development for mathematics education? *For the Learning of Mathematics, 5*(1), 24–28.
Blumer, H. (1969). *Symbolic interactionism: Perspectives and method.* Englewood Cliffs, NJ: Prentice-Hall.
Brousseau, G. (1984). The crucial role of the didactical contract in the analysis and construction of situations in teaching and learning mathematics. In H. G. Steiner (Ed.), *Theory of mathematics education* (Occasional paper 54, pp. 110–119). Bielefeld, Germany: Institut für Didaktik der Mathematik.
Brown, J. S., Collins, A., & Duguid, P. (1989). Situated cognition and the culture of learning. *Educational Researcher, 18*(1), 32–42.
Bruner, J. S. (1983). *Child's talk: Learning to use language.* Oxford, England: Oxford University Press.
Bussi, M. B. (1991, July). Social interaction and mathematical knowledge. In F. Furinghetti (Ed.), *Proceedings of the Fifteenth Conference of the International Group for the Psychology of Mathematics Education* (pp. 1–16). Genoa, Italy: Program Committee of the 15th PME Conference.
Carpenter, T. P., Fennema, E., Peterson, P. L., Chiang, C., & Loef, M. (1989). Using knowledge of children's mathematics thinking in classroom teaching: An experimental study. *American Educational Research Journal, 26,* 499–532.
Carraher, T. N., Carraher, D. W., & Schliemann, A. D. (1985). Mathematics in streets and schools. *British Journal of Developmental Psychology, 3,* 21–29.
Cobb, P. (1987). An investigation of young children's academic arithmetic contexts. *Educational Studies in Mathematics, 18,* 109–124.
Cobb, P. (1989). Experiential, cognitive, and anthropological perspectives in mathematics education. *For the Learning of Mathematics, 9*(2), 32–42.
Cobb, P., Wood, T., Yackel, E., & McNeal, B. (1992). Characteristics of classroom mathematics traditions: An interactional analysis. *American Educational Research Journal, 29,* 573–602.
Cobb, P., Yackel, E., & Wood, T. (1989). Young children's emotional acts while doing mathematical problem solving. In D. B. McLeod & V. M. Adams (Eds.), *Affect and mathematical problem solving: A new perspective* (pp. 117–148). New York: Springer-Verlag.
Confrey, J. (1990). A review of the research on student conceptions in mathematics, science, and programming. In C. B. Cazden (Ed.), *Review of research in education* (Vol. 16, pp. 3–55). Washington, DC: American Educational Research Association.
Davis, P. J., & Hersh, R. (1981). *The mathematical experience.* Boston: Houghton Mifflin.

Davydov, V. V. (1988). Problems of developmental teaching (Part I). *Soviet Education, 30*(8), 6–97.

Doise, W., & Mugny, G. (1979). Individual and collective conflicts of centrations in cognitive development. *European Journal of Psychology, 9,* 105–108.

Doise, W., Mugny, G., & Perret-Clermont, A. N. (1975). Social interaction and the development of cognitive operations. *European Journal of Social Psychology, 5,* 367–383.

Eisenhart, M. A. (1988). The ethnographic research tradition and mathematics education research. *Journal for Reasearch in Mathematics Education, 19,* 99–114.

Forman, E. (1992, August). *Forms of participation in classroom practice.* Paper presented at the International Congress on Mathematical Education, Québec City, Canada.

Greeno, J. G. (1991). Number sense as situated knowing in a conceptual domain. *Journal for Research in Mathematics Education, 22,* 170–218.

Gregg, J. (1993, April). *The interactive constitution of competence in the school mathematics tradition.* Paper presented at the annual meeting of the American Educational Research Association, Atlanta.

Hanks, W. F. (1991). Foreword. In J. Lave & E. Wenger, *Situated learning: Legitimate peripheral participation* (pp. 13–26). Cambridge, England: Cambridge University Press.

Hawkins, D. (1985). The edge of Platonism. *For the Learning of Mathematics, 5*(2), 2–6.

Human, P. G., Murray, J. C., & Olivier, A. I. (1989). *A mathematics curriculum for the junior primary phase.* Stellenbosch, South Africa: University of Stellenbosch, Research Unit for Mathematics Education.

Krummheuer, G. (1992). *Lernen mit "format": Elemente einer interaktionistischen Lerntheorie* [Learning with "formats": Elements of an interactionist learning theory]. Weinhem, Germany: Deutscher Studien Verlag.

Lampert, M. (1990). When the problem is not the question and the solution is not the answer: Mathematical knowing and teaching. *American Educational Research Journal, 27,* 29–63.

Lave, J. (1988). *Cognition in practice: Mind, mathematics and culture in everyday life.* Cambridge, England: Cambridge University Press.

Lave, J., & Wenger, E. (1991). *Situated learning: Legitimate peripheral participation.* Cambridge, England: Cambridge University Press.

Leont'ev, A. N. (1981). Chelovek i kul'tura [Man and culture]. In A. N. Leont'ev, *Problemy razvitiia psikhiki* [Problems of the development of mind] (4th ed., pp. 410–435). Moscow: Moskovskoga Universiteta.

Maturana, H. R. (1978). Biology of language: The epistemology of reality. In G. A. Miller & E. Lennenberg (Eds.), *Psychology and biology of language and thought: Essays in honor of Eric Lennenberg* (pp. 27–63). New York: Academic Press.

McNeal, B. (1992, August). *Mathematical learning in a textbook based classroom.* Paper presented at the International Congress on Mathematical Education, Québec City, Canada.

Mehan, H., & Wood, H. (1975). *The reality of ethnomethodology.* New York: Wiley.

Minick, N. (1989). *L. S. Vygotsky and Soviet activity theory: Perspectives on the relationship between mind and society* (Literacies Institute, Special Monograph Series No. 1). Newton, MA: Educational Development Center, Inc.

Much, N. C., & Shweder, R. A. (1978). Speaking of rules: The analysis of culture in breach. *New Directions for Child Development, 2,* 19–39.

Newman, D., Griffin, P., & Cole, M. (1989). *The construction zone: Working for cognitive change in school.* Cambridge, England: Cambridge University Press.

Nunes, T., Schliemann, A. D., & Carraher, D. W. (1993). *Street mathematics and school mathematics.* New York: Cambridge University Press.

Perret-Clermont, A. N. (1980). *Social interaction and cognitive development in children.* New York: Academic Press.

Perret-Clermont, A. N., Perret, J. F., & Bell, N. (1989, February). *The social construction of meaning and cognitive activity in elementary school children.* Paper presented at the Conference on Socially Shared Cognition, University of Pittsburgh, Pittsburgh, PA.

Putnam, H. (1987). *The many faces of realism.* LaSalle, IL: Open Court.

Richards, J. (1991). Mathematical discussions. In E. von Glasersfeld (Ed.), *Radical constructivism in mathematics education* (pp. 13–52). Dordrecht, Netherlands: Kluwer.

Rogoff, B. (1990). *Apprenticeship in thinking: Cognitive development in social context*. Oxford, England: Oxford University Press.

Rorty, R. (1978). *Philosophy and the mirror of nature*. Princeton, NJ: Princeton University Press.

Saxe, G. B. (1991). *Culture and cognitive development: Studies in mathematical understanding*. Hillsdale, NJ: Lawrence Erlbaum Associates.

Saxe, G. B., & Bermudez, T. (1992, August). *Emergent mathematical environments in children's games*. Paper presented at the International Congress on Mathematical Education, Québec City, Canada.

Schoenfeld, A. H. (1987). What's all the fuss about metacognition? In A. H. Schoenfeld (Ed.), *Cognitive science and mathematics education* (pp. 189–216). Hillsdale, NJ: Lawrence Erlbaum Associates.

Schoenfeld, A. H. (1991). On mathematics as sense-making: An informal attack on the unfortunate divorce of formal and informal mathematics. In J. F. Voss, D. N. Perkins, & J. W. Segal (Eds.), *Informal reasoning and education* (pp. 311–343). Hillsdale, NJ: Lawrence Erlbaum Associates.

Scribner, S. (1984). Studying working intelligence. In B. Rogoff & J. Lave (Eds.), *Everyday cognition: Its development in social context* (pp. 9–40). Cambridge, MA: Harvard University Press.

Solomon, Y. (1989). *The practice of mathematics*. London: Routledge.

Thompson, P. W. (1994). Images of rate and operational understanding of the fundamental theorem of calculus. *Educational Studies in Mathematics, 26*(2–3), 229–274.

Voigt, J. (1985). Patterns and routines in classroom interaction. *Rechereches en Didactique des Mathematiques, 6*, 69–118.

Voigt, J. (1992, August). *Negotiation of mathematical meaning in classroom processes*. Paper presented at the International Congress on Mathematics Education, Québec City, Canada.

von Glasersfeld, E. (1987). Learning as a constructive activity. In C. Janvier (Ed.), *Problems of representation in the teaching and learning of mathematics* (pp. 3–18). Hillsdale, NJ: Lawrence Erlbaum Associates.

von Glasersfeld, E. (1989a). Cognition, construction of knowledge, and teaching. *Synthese, 80*, 121–140.

von Glasersfeld, E. (1989b). Constructivism. In T. Husen & T. N. Postlethwaite (Eds.), *The international encyclopedia of education* (1st Ed., Supplement Vol. 1, pp. 162–163). Oxford, England: Pergamon.

von Glasersfeld, E. (1992). Constructivism reconstructed: A reply to Suchting. *Science and Education, 1*, 379–384.

Vygotsky, L. S. (1978). *Mind and society: The development of higher psychological processes*. Cambridge, MA: Harvard University Press.

Vygotsky, L. S. (1979). Consciousness as a problem in the psychology of behavior. *Soviet Psychology, 17*, 3–35.

Walkerdine, V. (1988). *The mastery of reason: Cognitive development and the production of rationality*. London: Routledge.

Yang, M. T.-L. (1993, April). *A cross-cultural investigation into the development of place value concepts in Taiwan and the United States*. Paper presented at the annual meeting of the American Educational Research Association, Atlanta.

2

The Teaching Experiment Classroom

Paul Cobb
Vanderbilt University
Erna Yackel
Purdue University Calumet
Terry Wood
Purdue University

As noted in the introductory chapter, the contributors agreed to pursue their research interests by analyzing a single set of classroom video recordings and transcripts. These recordings covered both the small-group and whole-class portions of all lessons involving arithmetical activities that were conducted in one second-grade classroom during a 10-week period. The recordings were made in the course of a year-long classroom teaching experiment conducted during the 1986–1987 school year by Cobb, Yackel, and Wood (henceforth known as the American researchers). As originally conceived, this experiment was viewed as a natural extension of the one-on-one constructivist teaching experiment developed by Steffe and his colleagues (Cobb & Steffe, 1983; Steffe, 1983). The decision to conduct the experiment in a second-grade classroom was based on two considerations. First, prior research indicated that the traditional American practice of teaching standard algorithms for adding and subtracting multidigit numbers was detrimental to students' subsequent learning. A majority of the students appeared to construct relatively immature, syntactically based conceptions of place-value numeration and to develop instrumental beliefs about mathematics in school. Second, conceptual models of children's arithmetical learning developed by Steffe and his colleagues (Cobb & Wheatley, 1988; Steffe, Cobb, & von Glaserfeld, 1988; Steffe, von Glaserfeld, Richards, & Cobb, 1983) could be used both to analyze individual children's mathematical activity and to guide the development of instructional activities.

At the outset, the teaching experiment was in fact planned to run for only 3 or 4 months and to focus exclusively on arithmetical concepts. However, the collaborating teacher, who was responsible for all classroom instruction, insisted that it would be impossible to return to textbook-based instruction for the remainder of the school year. At her urging, the American researchers agreed to extend the experiment and to develop instructional activities that addressed all of the cooperating school district's objectives for second-grade mathematics.

The specific instructional activities used in the classroom were developed, modified, and, in some cases, abandoned while the experiment was in progress. To aid this process, two video cameras were used to record every mathematics lesson for the entire school year. Eight of the students faced the cameras when they worked in pairs, thus making it possible to record approximately half of each of the four pairs' mathematical activity throughout the school year. Additional data sources consisted of video-recorded individual interviews conducted with all the students at the beginning, middle, and end of the school year; copies of all the students' written work; and field notes. Initial analyses of these data focused on the quality of the students' mathematical activity and learning. These analyses, together with the classroom teacher's observations, guided the development of the instructional activities.

INSTRUCTIONAL ACTIVITIES

The overall goal when developing the instructional activities was to facilitate the students' construction of increasingly sophisticated conceptual operations. In the case of place-value numeration, for example, the focus was on conceptual operations that enable students to create and coordinate arithmetical units of different ranks. With regard to geometry, the relevant conceptual operations make it possible for students to mentally manipulate spatial images. In general, the instructional activities reflect the cognitive constructivist view that mathematical learning is a process in which students reorganize their activity to resolve situations that they find problematic (Confrey, 1990; Thompson, 1985; von Glasersfeld, 1984). As a consequence, all instructional activities (including those involving arithmetical computation and numeration) were designed to be potentially problematic to students with a wide range of conceptual possibilities (Cobb, Wood, & Yackel, 1991). In addition, the instructional activities were designed to support conceptual and procedural developments simultaneously (Cobb, Yackel, & Wood, 1989; Silver, 1986). The American researchers argued, for example, that a situation in which a student's current computational procedures prove inadequate can give rise to a problem, the resolution of which involves the construction of conceptual knowledge.

In this approach to instructional development, it is the students' inferred experiential realities rather than formal mathematics that constituted the starting point

2. THE TEACHING EXPERIMENT CLASSROOM

for developing instructional activities (Treffers, 1987). Consequently, wherever possible, research-based models of children's mathematical learning were used to guide the development process. This proved to be particularly feasible for arithmetic, where models developed by Steffe et al. (1983, 1988) were used to anticipate both what might be problematic for students when they interpreted specific activities, and what mathematical constructions they might make to resolve these problems. However, in some areas of second-grade mathematics in which research on children's learning was not very extensive, the American researchers relied on intuitions derived from their own prior experiences of interacting with young children.

Because the analyses reported in this volume all deal with lessons in which arithmetical activities were used, only this area of the curriculum is outlined in the following synopsis. The arithmetical activities used at the beginning of the year were designed to facilitate the constructions of thinking strategies (Cobb & Merkel, 1989) and increasingly sophisticated concepts of addition and subtraction. This focus on thinking strategies was justifiable in cognitive terms, because almost all of the beginning second graders in the participating school district could count abstract units of one (Steffe et al., 1983, 1988). The instructional activities included 10-frame activities adapted from Wirtz (1977), and "What's My Rule?" and "War" card games (Kamii, 1985) in both 10-frame and numeral formats. Story problems, balance tasks with single-digit numbers (see Fig. 2.1), and horizontal symbolic sentences sequenced to encourage students to use a previously calculated result to solve a subsequent task were also used. When the teacher used these activities during a 7-week period from the beginning of September to the end of October, she attempted to guide the development of a thinking-strategy orientation, in which students were encouraged to use known sums or differences to solve unknown problems.

Subsequent instructional activities were designed to encourage the development of solutions that involve composite whole-number units (numerical units themselves composed of ones). These activities, which were used during an 11-week period until late December, included a variety of hundreds board and money activities, the "What's My Rule?" whole-class activity, and a whole-class activity conducted on an overhead projector with ten-bars and individual ones. In addition, sequenced story problems, balance tasks with double-digit numbers (see Fig. 2.1), and horizontal sentences with increasingly large two-digit numbers were used. The sequenced story problems included the various problem types identified by research-based classification schemes (e.g., Carpenter & Moser, 1984).

The instructional activities used in the latter half of the school year were designed to give students further opportunities to construct increasingly sophisticated concepts of 10 while completing tasks involving the addition and subtraction of two-digit numbers. Many of these activities included sequences of missing-addend, missing-minuend, and missing-subtrahend tasks presented in both horizontal and vertical column format and an activity called "Four-in-a-Row," which was designed to encourage both computational estimation and the construction of algorithms. In an additional task format called "strips and squares," pictured collections of strips of 10 squares and individual squares were used to present a range of arithmetical tasks (see Fig 2.2). Multiplicative and divisional situations

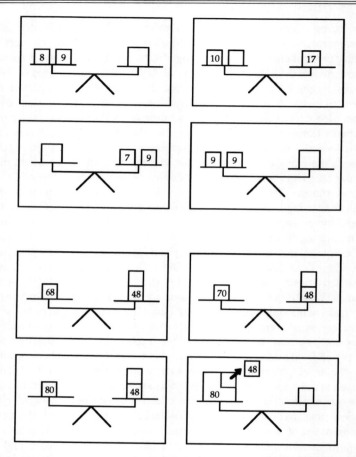

FIG. 2.1. Sample balance activity sheets for thinking strategies in one- and two-digit addition and subtraction.

were also introduced and elaborated throughout the second half of the school year. This interweaving of additive, subtractive, multiplicative, and divisional situations reflected the contention that children's construction of increasingly sophisticated concepts of 10 and of multiplicative units can be accounted for by the progressive construction of a single sequence of conceptual operations (Steffe, 1989; Steffe & Cobb, 1984).

CLASSROOM ORGANIZATION

The instructional activities were used in two types of classroom organizations. In the first, the students typically worked in pairs to solve instructional activities. A

2. THE TEACHING EXPERIMENT CLASSROOM

variety of manipulative materials (e.g., unifix cubes, hundreds boards, plastic coins) were made available, but it was primarily the students' responsibility to decide if the use of a particular manipulative might help them solve their mathematics problems. The teacher, for her part, observed and interacted with the children as they engaged in mathematical activity in pairs. After the students had worked together for perhaps 20 minutes, the teacher led a whole-class discussion of the students' interpretations and solutions. In the second type of classroom organization, the instructional activities were used solely in a whole-class setting. The teacher typically tried to initiate these activities by posing questions, often using an overhead projector and manipulative materials. In both types of classroom organization, the teacher attempted to facilitate discourse in which interpretations and solutions were accepted because they could be explained and justified in terms of actions on mathematical objects. In addition to the complete absence of individual paper-and-pencil seatwork, the teaching experiment classroom differed from

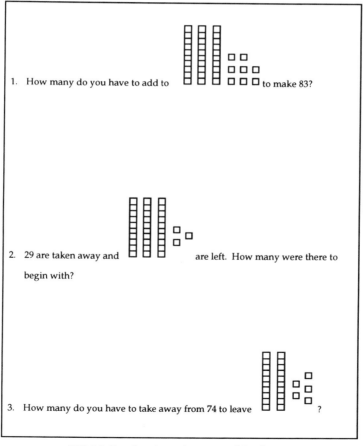

FIG. 2.2. Sample strips-and-squares activity sheet.

traditional classrooms in terms of assessment. The students' written work was not graded and, further, the teacher did not explicitly evaluate their contributions to discussions. Instead, the teacher developed a form of practice in which she framed conflicts between interpretations, solutions, and answers as problems that needed to be resolved.

SOCIAL NORMS

As has been noted, the cognitive goals for the teaching experiment were formulated in terms of the construction of increasingly sophisticated conceptual operations. Three additional goals explicitly articulated at the outset of the experiment addressed aspects of the "hidden curriculum." These concerned the development of intellectual and social autonomy (Kamii, 1985), relational beliefs about mathematics (Skemp, 1976), and task involvement rather than ego involvement as a form of motivation (Nicholls, 1983).

Early in the course of the teaching experiment, it became apparent that the teacher's expectations for the students were in conflict with the students' beliefs about their own role, the teacher's role, and the nature of mathematical activity. Presumably, the students had developed these beliefs as a consequence of their instructional experiences in traditional first-grade classrooms. Largely on her own, the teacher initiated the development of classroom social norms that were compatible with the aforementioned three project goals (Cobb, Yackel, & Wood, 1989).

The social norms for small-group activity that became explicit topics of conversation included persisting to solve personally challenging problems, explaining personal solutions to the partner, listening to and trying to make sense of the partner's explanation, attempting to achieve consensus about an answer, and, ideally, a solution process in situations in which a conflict between interpretations or solutions has become apparent. Social norms for whole-class discussions included explaining and justifying solutions, trying to make sense of explanations given by others, indicating agreement and disagreement, and questioning alternatives in situations in which a conflict between interpretations or solutions has become apparent. It should be stressed that the teacher did not simply list these norms as rules or principles to be followed; instead, she capitalized on specific incidents in which students' activity either instantiated or transgressed a social norm by using them as occasions to discuss her expectations. In the process, most of the students appeared to develop the view that mathematics is an activity in which they were obliged to resolve problematic situations by constructing personally meaningful solutions, and that they should justify these solutions in terms of actions on mathematical objects. Thus, in terminology developed after the teaching experiment was completed, the teacher's attempt to guide the development of an inquiry mathematics microculture seemed to be largely successful.

THE DATA

The classroom video recordings selected for analysis cover the lessons involving arithmetical activities that were conducted between January 13 and March 23, 1987. Transcripts of all whole-class episodes and of two of the four student pairs' small-group activity were developed for the convenience of the German researchers. The decision to restrict the analyses to these lessons was made for four reasons. First, conceptual models of children's arithmetical learning could be used as a resource when conducting the cognitive analyses. Second, video-recorded student interviews conducted in the week prior to January 13 could be used to substantiate the cognitive analyses. Third, previously completed analyses indicated that the classroom social norms negotiated by the teacher and students were relatively stable during the second half of the school year (Cobb et al., 1989). This was an important consideration given that none of the proposed analyses directly focused on the social norms. Fourth, the eight children whose small-group activity was video-recorded worked in the same pairs throughout this 10-week period. This made it easier to analyze the relationship between small-group activity and mathematical learning.

The instructional activities and the number of times that they were used during the 10-week period were:

January: Two-digit addition and subtraction balances (2), multiplication and division balances (1), sequenced two-digit addition and subtraction sentences (2), addition and subtraction story problems (1), strips-and-squares tasks (1).

February: Two-digit addition and subtraction balances (1), multiplication and division balances (1), strips-and-squares tasks (3), multiplication and division story problems (1), sequenced multiplication sentences (1), activities involving the use of play money (6), money story problems (1).

March: Strips-and-squares tasks (2), multiplication and division sentences (1), money story problems (1), balances in which all the numbers are coin values (2).

The following notational conventions are used in subsequent chapters to designate balance and strips-and-squares tasks:

Balance tasks: The symbol "/" is used to separate the two sides of the balance. Thus, the first two tasks in Fig. 2.1 are notated as "8, 9 / _ ," and "10, _ / 17."

Strips-and-squares tasks: The symbol "I" is used to symbolize a pictured strip, and "." to symbolize a pictured square. Thus, the first task in Fig. 2.2 is notated as, "How many do you have to add to III:::: to make 83?"

REFERENCES

Carpenter, T. P., & Moser, J. M. (1984). The acquisition of addition and subtraction concepts in grades one through three. *Journal for Research in Mathematics Education, 15,* 179–202.

Cobb, P., & Merkel, G. (1989). Thinking strategies as an example of teaching arithmetic through problem solving. In P. Traffon (Ed.), *New directions for elementary school mathematics: 1989 yearbook of the National Council of Teachers of Mathematics* (pp. 70–81). Reston, VA: National Council of Teachers of Mathematics.

Cobb, P., & Steffe, L. P. (1983). The constructivist researcher as teacher and model builder. *Journal for Research in Mathematics Education, 14*, 83–94.

Cobb, P., & Wheatley, G. (1988). Children's initial understandings of ten. *Focus on Learning Problems in Mathematics, 10*(3), 1–28.

Cobb, P., Wood, T., & Yackel, E. (1991). A constructivist approach to second grade mathematics. In E. von Glasersfeld (Ed.), *Constructivism in mathematics education* (pp. 157–176). Dordrecht, Netherlands: Kluwer.

Cobb, P., Yackel, E., & Wood, T. (1989). Young children's emotional acts while doing mathematical problem solving. In D. B. McLeod & V. M. Adams (Eds.), *Affect and mathematical problem solving: A new perspective* (pp. 117–148). New York: Springer-Verlag.

Confrey, J. (1990). What constructivism implies for teaching. In R. B. Davis, C. A. Maher, & N. Noddings (Eds.), *Constructivist views on the teaching and learning of mathematics* (Journal for Research in Mathematics Education Monograph No. 4, pp. 107–122). Reston, VA: National Council of Teachers of Mathematics.

Kamii, C. (1985). *Young children reinvent arithmetic: Implications of Piaget's theory*. New York: Teachers College Press.

Nicholls, J. G. (1983). Conceptions of ability and achievement motivation: A theory and its implications for education. In S. G. Paris, G. M. Olson, & W. H. Stevenson (Eds.), *Learning and motivation in the classroom* (pp. 211–237). Hillsdale, NJ: Lawrence Erlbaum Associates.

Silver, E. A. (1986). Using conceptual and procedural knowledge: A focus on relationships. In J. Hiebert (Ed.), *Conceptual and procedural knowledge: The case of mathematics* (pp. 181–198). Hillsdale, NJ: Lawrence Erlbaum Associates.

Skemp, R. R. (1976). Relational understanding and instrumental understanding. *Mathematical Teaching, 77*, 1–7.

Steffe, L. P. (1983). The teaching experiment methodology in a constructivist research program. In M. Zweng, T. Green, J. Kilpatrick, H. Pollak, & M. Suydam (Eds.), *Proceedings of the fourth International Congress on Mathematical Education* (pp. 469–471). Boston: Birkhauser.

Steffe, L. P. (1989, April). *Operations that generate quantity*. Paper presented at the annual meeting of the American Educational Research Association, San Francisco, CA.

Steffe, L. P., & Cobb, P. (1984). Children's construction of multiplicative and divisional concepts. *Focus on Learning Problems in Mathematics, 6*(1 & 2), 11–29.

Steffe, L. P., Cobb, P., & von Glasersfeld, E. (1988). *Construction of arithmetical meanings and strategies*. New York: Springer-Verlag.

Steffe, L. P., von Glasersfeld, E., Richards, J., & Cobb, P. (1983). *Children's counting types: Philosophy, theory and applications*. New York: Praeger Scientific.

Thompson, P. (1985). Experience, problem solving, and learning mathematics: Considerations in developing mathematical curricula. In E. A. Silver (Ed.), *Teaching and learning mathematical problem solving: Multiple research perspectives* (pp. 189–236). Hillsdale, NJ: Lawrence Erlbaum Associates.

Treffers, A. (1987). *Three dimensions: A model of goal and theory description in mathematics instruction—The Wiskobas project*. Dordrecht, Netherlands: Reidel.

von Glasersfeld, E. (1984). An introduction to radical constructivism. In P. Watzlawick (Ed.), *The invented reality* (pp. 17–40). New York: Norton.

Wirtz, R. W. (1977). *Making friends with numbers*. Washington, DC: Curriculum Development Associates.

3

Mathematical Learning and Small-Group Interaction: Four Case Studies

Paul Cobb
Vanderbilt University

The initial motivation for conducting the four case studies of small-group activity presented in this chapter was primarily theoretical. In concert with the general theme of this book, the neo-Piagetian and Vygotskian perspectives were deemed to be inappropriate given the purposes of the investigation. From the neo-Piagetian perspective, social interaction is treated as a catalyst for autonomous cognitive development. Thus, although social interaction is considered to stimulate individual cognitive development, it is not viewed as integral to either this constructive process or to its products, increasingly sophisticated mathematical conceptions. Vygotskian perspectives, on the other hand, tend to subordinate individual cognition to interpersonal or social relations. In the case of adult–child interactions, for example, it is argued that the child learns by internalizing mental functions that are initially social and exist between people. In recent years, several attempts have been made to extend these arguments to small-group interactions between peers (Forman & Cazden, 1985; Forman & McPhail, 1993). It was against the background of these two competing perspectives that an approach was developed that acknowledges the importance of both cognitive and social processes without subordinating one to the other.

In addition to addressing this theoretical issue, the case studies touch on two pragmatic issues. The first concerns the extent to which the children engaged in inquiry mathematics when they worked together in small groups. The second issue concerns the extent to which small-group collaborative activity facilitates children's mathematical learning. At the outset of the teaching experiment, Cobb, Yackel, and Wood's rationale for small-group work was primarily Piagetian and focused on the learning opportunities that arise as children attempt to resolve conflicts in their individual view points. It was only later that the American researchers came to see that the students' constructions have an intrinsically social aspect in that they are both constrained by the group's taken-as-shared basis for communication and contribute to its further development. As a consequence, the rationale for small-group work was modified to take account of the learning opportunities that can arise for children as they mutually adapt to each other's activity and attempt to establish a consensual domain for mathematical activity.

The findings of several previous investigations indicate that small-group interactions can give rise to learning opportunities that do not typically arise in traditional classroom interactions (Barnes & Todd, 1977; Davidson, 1985; Good, Mulryan, & McCaslin, 1992; Noddings, 1985; Shimizu, 1993; Smith & Confrey, 1991; Webb, 1982; Yackel, Cobb, & Wood, 1991). The four case studies extend these analyses by relating the occurrence of learning opportunities to the different types of interaction in which the children engaged. As a consequence, they indicate the extent to which the various types of interaction were productive for mathematical learning.

In the following pages, I set the stage for the case studies by first locating them within the broader context of the classroom microculture. I then discuss the methodology developed to analyze the video recordings and give particular attention to the viability and trustworthiness of the case studies. Here, it is noted that the ways in which the pairs of children interacted frequently varied both within sessions and from one session to the next. However, their social expectations and obligations appeared to be stable once their cognitive construals of tasks and of each others' mathematical actions were taken into account. This made it possible to characterize the social relationships they established in an empirically grounded way. The discussion of the methodology is followed by a description of the cognitive constructs and sociological constructs that proved to be relevant when conducting the analysis. In the case of the cognitive constructs, the focus is on the children's construction of increasingly sophisticated conceptions of 10. The sociological constructs delineate various types of small-group interactions and address relations of authority within a group.

The bulk of the chapter is devoted to the presentation of the four case studies. They are each organized to reflect the reflexivity between the children's mathematical activity and the small-group relationships they established. Thus, each case study first deals with the individual interviews and then with the children's small-group relationship before considering the learning opportunities that arose for the individual children and their mathematical learning. Comparisons across the four case studies are made in the final section of the chapter. There, it is noted that the stability in the children's small-group relationships across the 10-week period covered by the data was matched by stability in each pair of children's cognitive capabilities relative to those of the partner. This suggests that the children's cognitive capabilities and their social relationships may have constrained each other in the sense of limiting possibilities for change.

A further issue addressed in the case studies is the occurrence of learning opportunities as the children engaged in the various types of interaction. The analysis reveals that interactions in which one child routinely attempts to explain his or her thinking are not necessarily productive for either child's learning. In addition, harmony in a group's relationship does not appear to be a good indicator of learning opportunities. In contrast, contentious relationships, in which the children's expectations for each other are in conflict, can be productive. Two features identified in the case studies that appear to be necessary for productive relationships are the development of a taken-as-shared basis for mathematical communication and the routine engagement in interactions in which neither child is an authority. Against the background of a summary and synthesis of the findings, the chapter concludes by first comparing the case study analyses to Piagetian and Vygotskian perspectives, and then by considering their pedagogical implications.

ORIENTATION TO THE CLASSROOM

Because the case studies focus only on the children's small-group interactions, it is important to acknowledge that the small-group and whole-class phases of lessons were interdependent (Bauersfeld, 1992; Schroeder, Gooya, & Lin, 1993). For example, children's small-group activity was influenced by their realization that they would be expected to explain their interpretations and solutions in a subsequent whole-class discussion. Conversely, their small-group activity served as the basis for the whole-class discussions. The relationship between small-group and whole-class activity is addressed more directly elsewhere in this book by Krummheuer's analysis of minicycles (chap. 7, this volume) and Yackel's analysis of explanations in these two social settings (chap. 4, this volume). As their analyses indicate, the classroom microculture both constrained and was sustained by the small-group relationships the children developed.

It is also important to note that the children's participation in the teaching experiment was the first occasion in which they were expected to collaborate to learn. Wood and Yackel (1990) discussed how the teacher intervened during small-group work to help them develop productive relationships. The small-group norms that became explicit topics of conversation included persisting to solve personally challenging problems, explaining personal solutions to the partner, listening to and trying to make sense of the partner's explanations and attempting to achieve consensus about an answer, and, ideally, a solution process in situations where a conflict between interpretations and solutions has become apparent. Classroom observations made during the teaching experiment indicate that the teacher found it less and less necessary to intervene to guide the renegotiation of these norms as the school year progressed (cf. Wood, chap. 6, this volume). By January, when the case studies commence, most of the children appeared able to resolve small-group social conflicts on their own and thus develop relationships that satisfied the teacher's expectations. The case studies do, of course, provide an indication of the extent to which the teacher was able to guide the development of small-group norms. In addition, these analyses allow us to consider the influence of the norms on children's mathematical learning.

DATA CORPUS

In line with the analyses reported in other chapters, the small-group case studies focus on the children's conceptions of place-value numeration and their construction of increasingly sophisticated computational algorithms. The initial data consisted of video recordings of the 27 lessons conducted between January 23 and March 13 of the school year in which the instructional activities involved arithmetic. It should be noted that many of these activities were not explicitly designed to facilitate the children's development of numeration concepts. Thus, the activities included story problems, situations involving money, and tasks designed to support the development of elementary conceptions of multiplication. Because two cameras were used, it was possible to record approximately half of each pair's activity in these 27 sessions. Transcripts were subsequently developed for two of the four pairs' activity, primarily for the convenience of the German members of the research group. Additional data consisted of video-recorded individual interviews conducted with the children in January just prior to the first small-group session.

METHODOLOGY

The Interviews

Interviews as Social Events. The American researchers' original intention for conducting the interviews was to assess the children's levels of conceptual understanding at the midpoint in the school year. To this end, a variety of tasks were used to investigate the children's place-value conceptions and computational algorithms. At that time, it was assumed that inferences about the children's arithmetical interpretations could be transferred unproblematically to the classroom. With hindsight, it is apparent that such interviews are social events in which the researcher and child negotiate their expectations about the purposes of their interaction, their roles, their interpretations of the tasks, and their understanding of what counts as a legitimate solution and an adequate explanation (Mishler, 1986; Newman, Griffin, & Cole, 1989; Richards, 1991; Voigt, 1992). In an extreme case, the researcher and the child can have different conceptions of the nature of mathematical activity and, thus, "talk past each other" throughout the interview.

The researcher who conducted the interviews was aware that the child might construe the interview situation as one in which he or she was obliged to try to figure out what the researcher had in mind all along (Cobb, 1986). Further, because the researcher had visited the classroom regularly and had interacted with the children throughout the first half of the school year, he and they had already established bases for communication. In general, the negotiation of the purpose of the interview appeared to proceed relatively smoothly. Further, with one exception of which the researcher was aware, his and the children's understandings about the nature and purpose of mathematical activity seemed to be taken-as-shared. These observations do not, of course, invalidate the argument that the children's interpretations of and solutions to tasks were influenced by their participation in the interviews as social events. Thus, even though there was every indication that the children's participation in the interviews and the small-group sessions were commensurable (cf. Lave & Wenger, 1991), the possibility that children might have attempted to fulfill different obligations in the two situations was explored when accounting for apparent inconsistencies in their mathematical activity across situations. Consequently, interpretations and solutions that, from the cognitive perspective, appeared to be inconsistent constituted occasions to clarify the nature of the two situations as social events (cf. Yackel, chap. 4, this volume).

During the analysis of the interviews, frequent reference is made to the conceptual interpretations that the children seemed to make, and to the arithmetical objects they are inferred to have constructed. This way of talking is a shorthand used for ease of explication. The intended meaning in each case is that, in the course of his or her participation in the interview, the child acted in ways that justify making particular cognitive attributions. The situated nature of these inferences is particularly apparent in those instances in which the researcher attempted to support the child's mathematical activity. In several of these instances, children did in fact seem to make conceptual advances that they would not, in all probability, have made on their own.

One standard way of accounting for these advances is to use Vygotsky's construct of the "zone of proximal development." However, this notion elevates interpersonal social processes above intrapersonal cognitive processes (Cole, 1985). Thus, analyses that use this construct typically focus on the adult's role in scaffolding the child's activity. As a consequence, the treatment of both the child's interpretations and his or her contributions to interactions is relatively limited. The zone of proximal development was therefore replaced by a construct that is more relevant to the purposes of the investigation, that of the "realm of developmental possibilities." This construct delineates the situated conceptual advances a child makes while participating in an interaction such as that in which an adult intervenes to support his or her mathematical activity. The zone of proximal development is concerned with what the child can do with adult support, whereas the realm of developmental possibilities addresses the way in which the child's conceptions and interpretations evolve as he or she interacts with the adult. The latter construct, therefore, brings the cognitive perspective more to the fore and, thus, complements sociological analysis of the situations in which that development occurs.

The Interview Tasks. The tasks were adapted from those developed by Steffe and his colleagues (Steffe, Cobb, & von Glasersfeld, 1988; Steffe, von Glasersfeld, Richards, & Cobb, 1983). Although an interview schedule was developed in advance, the interviewer was free to either skip tasks or to pose impromptu tasks on the basis of ongoing interpretations of the child's mathematical activity. The main types of interview tasks that bear on the issues addressed in the case studies were:

1. Counting tasks: The researcher used two small felt cloths to present addition and missing addend tasks. For example, the researcher might tell the child that 9 squares are hidden beneath one cloth, that there are 13 squares hidden by both cloths, and ask the child to find how many squares are hidden under the second cloth (i.e., a task corresponding to $9 + \_ = 13$). These tasks were originally designed by Steffe et al. (1983) to investigate the qualitatively distinct types of units of one that children can create and count.

2. Thinking Strategy tasks: The researcher used plastic numerals to present the following sequence of horizontal sentences: $9 + 3 = \_$, $9 + 4 = \_$, $9 + 5 = \_$, and $6 + 6 = \_$, $7 + 5 = \_$, $8 + 4 = \_$.

3. Uncovering tasks: The researcher first established with the child that each strip contained 10 squares. After presenting further warm-up tasks, he placed a board that was completely covered by a large cloth in front of the child and gradually pulled back the cloth to reveal a collection of strips and individual squares. Each time, he asked the child how many squares there were in all. Two tasks of this type were posed. The cumulative sums after each uncovering for the first task (Fig. 3.1a) were: one strip (10), three squares (13), two strips (33), four squares (37), three squares (40), one strip (50), two squares (52), two strips (72).

The cumulative sums after each uncovering for the second task (Fig. 3.1b) were: four squares (4), one strip (14), two strips (34), one strip and two squares (46), two strips and five squares (71). These tasks were designed to investigate the quality of the units of ten that the child could create and count.

4. Strips-and-squares tasks: The researcher used the strips-and-squares materials to present a variety of addition and missing addend tasks. For example, he might first hide a collection of two strips and nine squares beneath a cloth, and make a visible collection of three strips and nine squares. The child would then be told that there are 68 squares in all and be asked to find how many are hidden (i.e., a task corresponding to $39 + \_ = 68$). These tasks were designed to investigate the quality

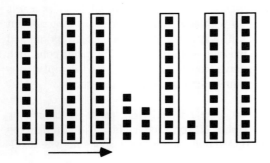

FIG. 3.1a. The first uncovering task asked during interviews.

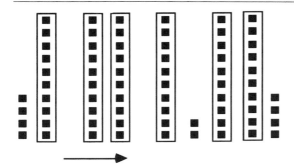

FIG. 3.1b. The second uncovering task asked during interviews.

of the units of ten the child could create. In contrast to other task types, the child's solutions could involve a reliance on situation-specific imagery in that he or she might re-present the strips and squares hidden beneath the cloth.

5. Number sentences: The researcher used plastic numerals to present the following horizontal number sentences: $16 + 9 = \_$, $28 + 13 = \_$, $37 + 24 = \_$, and $39 + 53 = \_$. In doing so, he made a routine intervention if a child consistently attempted to count on by ones. In the case of $37 + 24 = \_$, for example, he first asked the child if he or she could figure out $37 + 20 = \_$, and then asked if this result could help the child solve the original task. His goal in doing so was to investigate whether the child could curtail his or her counting activity by conceptualizing 24 as 20 and 4 more. Like the strips-and-squares tasks, these tasks were designed to investigate both the quality of the units of ten the child could create and the sophistication of his or her computational methods. However, in contrast to the previous tasks, nothing in the presentation of the sentences encouraged the creation of imagery that might support the construction of units of ten and one.

6. Worksheet tasks: The researcher asked the child to complete a facsimile of a school textbook page of two-digit addition tasks. The worksheet included four tasks that involved the same number combinations as the horizontal sentences:

$$
\begin{array}{cccc}
16 & 28 & 37 & 39 \\
+9 & +13 & +24 & +53 \\
\hline
\end{array}
$$

The rationale for these tasks is based on a previous investigation that found that children who have received traditional textbook-based instruction frequently consider that horizontal and vertical tasks involving the same numbers are separate, unrelated tasks. These children experienced no contradiction when they found that $16 + 9$ was 25 and that

$$
\begin{array}{c}
16 \\
+9 \\
\hline
\end{array}
$$

was, say, 15 or 115 (Cobb & Wheatley, 1988). Consequently, a comparison of children's solutions to the two sets of tasks indicates the extent to which they have established academic, codified arithmetic and informal, counting-based arithmetic as two separate contexts. This issue was of considerable significance given that the children in the teaching experiment class had been in traditional first-grade classes and that this was the first occasion on which they had seen the vertical format during the second-grade year. The investigation focused on whether they would attempt to recall the standard algorithm they had previously been taught, or whether they would adapt methods they had used to solve other interview tasks.

The Small-Group Sessions

Initial Analyses. As the analyses of whole-class discussions presented in other chapters illustrate, one methodological approach to the social aspect of mathematical activity is to identify interaction patterns. Krummheuer's discussion of formats of argumentation, Voigt's specification of thematic patterns, and Wood's delineation of patterns of interaction are all based on this approach. At the outset, it seemed reasonable to follow a similar approach when analyzing the video recordings and transcripts of the children's small-group activity. However, concerted attempts to identify interaction patterns of this type that held up across a number of small-group sessions proved unsuccessful. It was sometimes possible to identify patterns in two children's interactions as they solved a sequence of similar tasks in a particular session. However, the nature of the children's interactions often proved to be markedly different in the very next session.

These initial attempts to make sense of the small-group tapes and transcripts gave rise to an alternative approach. In particular, an exploratory analysis of one pair of children's activity across five consecutive sessions indicated that the expectations the children had for each other's activity and the obligations implicit in their own activity appeared to be consistent across situations, provided the relative sophistication of their individual mathematical interpretations was taken into account. In other words, although there were dramatic differences in the nature of the children's interactions both within sessions and from one session to the next, their expectations and obligations appeared to be stable once their construals of tasks and of each other's mathematical actions were considered. Further, these inferred obligations and expectations suggested a way to characterize the social relationship they established for doing mathematics in an empirically grounded way.

The viability of this approach was confirmed when conducting the four case studies in that it was possible to identify cognitively situated expectations and obligations for each of the four pairs of children that held across the 10-week period covered by the video recordings. As this approach involves both cognitive analyses of the children's individual activity and sociological analyses of their interactions, it necessarily addresses a central theoretical concern of the project: the coordination of psychological and sociological perspectives in mathematics education.

Episode-by-Episode Analyses. For each case study, the children's January interviews and their small-group sessions were analyzed in chronological order.

3. LEARNING AND SMALL-GROUP INTERACTION

Within each small-group session, children's attempts to solve individual tasks were treated as distinct but related episodes. The analysis of each of these episodes involved inferring:

1. The children's expectations for and obligations to the small-group partner.
2. The mathematical meaning the children gave to their own activity, the partner's activity, and the task at hand.
3. The learning opportunities that arose for each child.
4. The conceptual reorganizations that each child made (i.e., their mathematical learning).

This general approach can be clarified by considering the first of over 100 small-group sessions that were analyzed. Here, the two children, Ryan and Katy, solved a sequence of horizontal addition number sentences that were sequenced so they might relate the current task to a prior result and thus move beyond counting by ones. The video recording begins midway through the session as they are explaining their solutions to 48 + 18 = _ to the teacher. They have just solved 47 + 19 = 66.

1	Teacher:	48 plus 18.
2	Katy:	That's just the same.
3	Teacher:	What's just the same?
4	Katy:	See, if you take one from 19, put it with the 47, it makes this just the same (points to 48 + 18 = _ and 47 + 19 = <u>66</u>).
5	Teacher:	Do you see that, Ryan? She's saying that 47 plus 19 equals 66. [She then repeats Katy's explanation in more detail]. She said it's just the same number.
6	Katy:	Yes, cause you just take one from the 19 and add it to the 47, that makes . . .
7	Teacher:	48.
8	Katy:	48.
9	Teacher:	48, right.
10	Ryan:	I know what she's trying to say, she's trying to say take one from here (points to 19) and add it to here (points to 47).
11	Teacher:	Right.
12	Ryan:	That would be the same answer.

1. With regard to the nature of their interactions, Katy repeatedly attempted to explain how she had solved the task. However, the teacher's initial intervention, "What's the same?" might have been crucial. Thus, the most that can be inferred is that Katy was obliged to explain her thinking in the teacher's presence. Ryan, for his part, seemed obliged to try to understand Katy's explanations, as indicated by his comment, "I know what she's trying to say." However, as was the case with Katy, the teacher's interventions might have been critical. Her comment, "Do you see that, Ryan?" might have been made with the intention of indicating the obligations she expected him to fulfill.

2. With regard to the mathematical meanings the children established, Katy's use of the compensation strategy did not seem novel and was consistent with her performance on thinking-strategy tasks administered in the January interview. As Ryan used the compensation strategy to relate $8 + 8 = 16$ and $9 + 7 = \_$ in his interview, it seems reasonable to infer that he did construct a compensation relationship when he interpreted Katy's explanation.

3. Ryan had not solved the task at the beginning of the exchange and the possibility of relating successive tasks did not seem to occur to him. The manner in which he clarified what "she's trying to say" suggests that he eventually comprehended how Katy had related tasks. This, in turn, indicates that a learning opportunity might have arisen for him when he interpreted her explanations.

4. Ryan's final interpretation of the task appeared to be more sophisticated than the one he would have made if left to his own devices. However, the extent to which he reorganized his mathematical activity is unclear. His advance might have been specific to this particular exchange or, alternatively, he might have made a major conceptual reorganization. The viability of these alternative possibilities can only be determined by taking account of his solutions to subsequent tasks.

The goal of the analysis was to develop a coherent account of Ryan and Katy's small-group activity across the 10 weeks of the data corpus. The inferences made while analyzing individual episodes were, therefore, viewed as initial conjectures that could be revised in light of the children's activity in subsequent episodes. In the next episode, Katy again used the compensation strategy, relating $49 + 17 = \_$ to $48 + 18 = \underline{66}$.

1	Katy:	Easy, it's the same thing.
2	Ryan:	No it isn't.
3	Katy:	Oh yes it is.
4	Ryan:	It's one higher, it's one higher.
5	Teacher:	Let's take a look.
6	Katy:	This one's [48] one less than that number [49], so if you take one from here and add it to here that makes 49. There's 17 left so it has to be the same number.
7	Ryan:	It can't be because . . .
8	Katy:	It is, Ryan.
9	Teacher:	Why can't it be?
10	Ryan:	You're adding two.

1. As was the case in the first episode, Katy again seemed obliged to explain her reasoning to Ryan, at least in the teacher's presence. However, on this occasion, Ryan did not appear to try to make sense of her explanation. Instead, he challenged her answer and attempted to explain how he had related the two tasks. In the first episode, there was no indication that he had developed a solution when the teacher asked Katy to explain her thinking. Here, in contrast, he had solved the task in a way that he believed he could justify. It therefore seems reasonable to conjecture

that in situations in which the children's independent solutions were in conflict, Katy was obliged to explain her reasoning to Ryan with the expectation that he would attempt to make sense of what she was saying. For his part, Ryan was obliged to both challenge Katy's explanation and explain his solution. The extent to which either child attempted to understand the other's explanation is open to question.

2. Katy's interpretation of the task appears to be consistent with the inferences made when analyzing the first episode. Further, Ryan's contention that "You're adding two" clarifies the conceptual advance he made when interpreting Katy's explanation in the first episode. Perhaps he had understood how Katy had mentally transformed 47 + 19 into 48 + 18 in the first episode without reflecting on why, for Katy, it followed that the two sums would be the same. To square this conjecture with his performance in the January interview, it might be further speculated that he could only create compensating relationships of this type with smaller numbers, perhaps 10 or less (cf. Neuman, 1987).

3. As Ryan did not appear to try to make sense of Katy's explanation, there is no indication that a learning opportunity arose for him in this episode. Further, although Katy was obliged to explain her thinking in response to Ryan's challenges, she did not seem to modify her interpretation of the task when doing so. Consequently, it seems unlikely that a learning opportunity arose for her.

4. The conceptual reorganization that Ryan made in the first episode has been clarified by considering the interpretations he made in the second episode. His advance appears to be highly situation-specific. It can also be noted that in the second episode Ryan did attempt to relate successive tasks, whereas in the first episode he seemed unable to do so independently. Thus, he had come to the realization that the tasks could be solved in this way. As a consequence, the development of thinking strategy solutions was taken-as-shared by the two children.

As this brief example illustrates, the analysis involved a continual movement between particular episodes and potentially general conjectures. The general claims made about the children's small-group work over the 10-week period are empirically grounded in that they emerged while interpreting the particulars of their activity. It is interesting to note that this analytical approach bears a striking resemblance to mathematical activity as characterized by Lampert (1990). Following Lakatos (1976), Lampert argued that "Mathematics develops as a process of 'conscious guessing' about relationships among quantities and shapes, with proof following a 'zig-zag' path starting from conjectures and moving to the examination of premises through the use of counterexamples or 'refutations'" (p. 30). In the case studies, specific instances of the children's mathematical activity that conflicted with current conjectures became refutations in that they led to an examination of suppositions and assumptions that underpinned the conjectures. Lampert also observed that the zig-zag between conjectures and refutations involves a process of argumentation. Similarly, the process of developing the case studies involved both individual reflection and collective argumentation within the research group.

The analogy between the development of mathematical knowing and the analysis of the case studies is not particularly surprising once one notes, as did

Billig (1987), that argumentation almost invariably involves a movement between the general and the particular, between conjectures and refutations. The analogy holds because the case studies were developed by using an interpretivist or hermeneutic approach to understand an interpretive activity, that of learning mathematics. Thus, the approach used to develop explanations of the children's mathematical activity is consistent with the way in which children's mathematical activity is itself assumed to develop.

Analyses of Analyses. The episode-by-episode analyses of the video recordings and transcripts was in fact only the first of three phases of the analysis. In the second phase, the inferences and conjectures made while viewing the recordings themselves became "data" and were (meta-)analyzed to develop interrelated chronologies of:

1. The children's obligations and expectations (i.e., their social relationships).
2. Each individual child's mathematical activity and learning.

In the final phase of analysis, these chronologies were further synthesized to create outlines of the written case studies.

It should be stressed that this delineation of phases is made retrospectively. At the outset, the issue of how to go about analyzing the relatively large data corpus in a systematic way was itself problematic. The methodology outlined previously emerged while attempting to make sense of the video recordings, and it was only with hindsight that the consistency between this approach and Glaser and Strauss' (1967) constant comparative method for constructing ethnographies became apparent.

Viability of the Case Studies. Each case study presents a coherent account of children's individual and collective mathematical activity over a 10-week period. As is the norm in qualitative research, sample episodes are given to illustrate general claims and assertions (Atkinson, Delamont, & Hammersley, 1988; Taylor & Bogdan, 1984). Nonetheless, a difficulty arises in that the interpretations of these episodes might not seem justified if single episodes are considered in isolation. Frequently, alternative explanations can be ruled out only by referring to general features of a small-group's social relationship or to the child's realm of conceptual possibilities. However, the latter is corroborated by the very regularities across episodes that the sample episodes are meant to illustrate. Thus, the interpretation of specific episodes and the delineation of general regularities are reflexively related in that each is formulated only in relation to the other. This interdependence of the general and the particular can be viewed as a strength in that it acknowledges the reflexive relationship between the children's mathematical activity and the contexts in which they acted—their activity was both context specific and context renewing. Interpretivist approaches of this type can be contrasted with alternative methods of analysis based on the assumption of a one-to-one mapping between observed behavior and cognition.

3. LEARNING AND SMALL-GROUP INTERACTION 37

It should also be noted that the case studies do not purport to reveal the true essence of the children's mathematical activity. Instead, the viability of the case studies depends on the extent to which they are reasonable and justifiable given the research interest of coordinating cognitive and sociological perspectives (cf. Erickson, 1986). It is therefore acknowledged that other plausible interpretations of the children's mathematical activity could be made for alternative purposes.

Several considerations lend credence to the claim that the case studies are reasonable and justifiable. First, the researcher who developed the case studies was an observer in the classroom for approximately two thirds of the mathematics lessons throughout the school year. In addition, he frequently interacted with the four pairs of children and viewed all of the small-group video recordings while the classroom teaching experiment was in progress. Second, an ethnographic approach was used to conduct an episode-by-episode constant comparative analysis. Consequently, assertions can be justified by backtracking through the various phases of the analysis (i.e., the chronologies, the initial interpretations and conjectures, and, if necessary, the video recordings and transcripts). It is this record of the process of constructing the case studies that gives them their empirical grounding (Gale & Newfield, 1992). Third, the case studies were critiqued by other researchers. Two of these colleagues had participated in the classroom teaching experiment and had viewed many of the video recordings. This enabled them to respond to specific claims and to judge whether the case studies "rang true" at a more global level. Three other colleagues were familiar with neither the teaching experiment classroom nor many of the taken-for-granted practices of American elementary mathematics education. Consequently, they were able to offer a variety of alternative interpretations. Fourth, sample episodes are used to illustrate assertions. Finally, some of the small-group episodes were analyzed by other members of the research group for different purposes (e.g., Krummheuer's minicycles, Voigt's thematic patterns, Yackel's small-group and whole-class explanations). These independent analyses provided a means of triangulating the case studies.

THEORETICAL CONSTRUCTS

The following description of the constructs that emerged while developing the case studies can be used as a glossary to complement the one at the end of the book. The reader might wish to clarify the intent of the constructs that are summarized next by referring to this section of the chapter when reading the case studies.

Cognitive Constructs

The cognitive constructs address three interrelated aspects of the children's mathematical activity. These are:

1. The quality of the units of ten and one the children created.

2. The process of creating these units.
3. The possible developmental history of and metaphors implicit in particular solutions.

Each of these three aspects is discussed in turn in the following paragraphs.

Tens and Ones. The cognitive constructs developed to account for individual children's socially situated activity are based on the work of Steffe et al. (1983, 1988). However, the approach used was pragmatic in that no attempt was made to document the numerous fine-grained distinctions reported by Steffe et al. Instead, constructs were introduced and refined only when they served to clarify aspects of the children's mathematical activity that were significant given the purposes of the investigation.

The first issue addressed when conducting the cognitive analyses was that of ascertaining whether the children could routinely create and count abstract units of one when they interpreted and solved tasks. In the case of the counting tasks presented in the interviews, the inference that a child created abstract units was typically made if he or she counted to solve the missing addend task corresponding to $9 + \_ = 13$ (i.e., Nine squares visible. Thirteen squares in all. How many are hidden beneath the cloth?) Steffe et al. (1983) argued that the child's activity of putting up four fingers while counting "9, 10, 11, 12, 13" indicates that he or she was not merely counting fingers as perceptual items. Instead, the child is inferred to have solved the task by finding how many counting acts he or she performed, and raising fingers to record these acts. This ability to enumerate the units created while counting indicates that they could be objects of reflection for the child. It is in this sense that the child's units of one are said to be abstract in quality.

A distinction that proved useful when interpreting solutions that involved counting by 10 was that between 10 as a numerical composite and 10 as an abstract composite unit. This distinction can be illustrated by comparing two solutions to a strips-and-squares task in which 3 strips and 4 squares are visible and the child is told that 30 squares are hidden and asked to find how many squares there are in all. One way in which children solve this task is to move their hands with all 10 fingers extended as they count "34—44, 54, 64." Steffe et al. (1988) interpreted solutions of this sort by arguing that each of the three counting acts is a curtailment of counting 10 units of one. In other words, each counting act signifies a composite of ten units of 1. However, the way in which the children move all 10 fingers as they count indicates that these composites are not single entities or units. Instead, the children appear to count numerical composites of 10 ones. Other children solve the same task by sequentially putting up three fingers as they count "34—44, 54, 64." For Steffe et al. (1988), the children's acts of putting up an individual finger indicated that the composites of 10 ones they count are single entities or units for them. Thus, the conceptual entities the children are inferred to have created are abstract composite units of ten, each of which is itself composed of 10 ones.

A further issue addressed was that of clarifying whether the children's creation of abstract units of ten involved part–whole reasoning. Consider, for example, a strips-and-squares task corresponding to 48 + _ = 72 in which there are 4 strips and 8 squares visible, and 72 squares in all. Children often begin by focusing on the strips of 10 and reasoning that because 40 and 30 are 70, there must be 3 strips of 10 under the cloth. Some of these children then add the 8 visible squares to the 70 and realize the answer must be less than 30. They often experience a conflict in this situation in that, from their point of view, there appear to be both more than and fewer than 30 squares beneath the cloth. Children who are able to resolve this experienced contradiction appear to create an abstract composite unit by conceptually combining ones from the two parts (i.e., the visible and hidden collections). For example, a child who realizes that 30 is too much might first reason that 40 and 20 are 60, and that 8 and 2 more are 10, making 70 in all. The child might then continue by noting that as there are 72 squares in all, there must be 2 more squares hidden beneath the cloth, making 24 hidden squares in all. In the course of this solution, the child seems simultaneously to conceptualize the strips and squares in two different ways: parts comprising 6 abstract units of ten and the two collections of ones, and a whole of 7 abstract units of ten and 2 ones.

In summary, the cognitive analysis of solutions that involved the creation of tens differentiated between:

- Numerical composites of ten.
- Abstract composite units of ten.
- Part–whole relations involving abstract units of ten and one.

In view of what has been said about the socially situated nature of the children's conceptual activity, it is, in principle, impossible to give definitive behavioral indicators for the various cognitive constructs. The specific solutions that have been discussed are merely intended to be illustrative examples. Depending on the nature of the child's ongoing activity and of the interactions in which he or she engages, different interpretations might be made when analyzing solutions similar to these. For example, in the course of an interaction, a child might figure out that he or she can be effective by using a tens number–word sequence to count strips and a ones number–word sequence to count individual squares. It is then conceivable that the child could solve the strips-and-squares task corresponding to 34 + 30 = _ by counting "34—44, 54, 64" while putting up individual fingers. In such a case, each counting act signifies a strip viewed as a single thing rather than a composite of 10 squares. It was noted when discussing the methodology that single episodes are open to a variety of competing interpretations. This is particularly the case when conducting cognitive analyses. Alternative conjectures are winnowed down by attempting to develop a consistent account of a child's socially situated cognitive activity across episodes.

Image-Supported and Image-Independent Units. The cognitive constructs discussed thus far deal with the quality of the conceptual entities that children create while solving tasks. It also has proved important to infer whether a child's creation of these entities in a particular situation was image supported or image independent.[1] The first of these possibilities can be illustrated by considering a child's solution to a story problem involving money that corresponds to 38 + 24 (e.g., "Juan has 38¢ and Lucy has 24¢. How much money do they have together?"). Here, the child reasons without using materials, "38 and a dime make 48, another dime, 58—59, 60, 61, 62." Assuming that the child counted abstract composite units of ten, it can be noted that the child's situation-specific imagery of dimes might have played a crucial role in supporting the creation of these units. Consequently, in the absence of evidence to the contrary, it would be inferred that the process of creating the units was image supported. More generally, children's creation of numerical composites and abstract composite units is inferred to be image supported if, in interpreting a task or another's mathematical activity, they appear to rely on situation-specific imagery of some type. Thus, solutions to strips-and-squares tasks were typically inferred to be image supported in that the child's realization that items were grouped in strips of 10 might have been critical.

By way of contrast, consider a solution to the money story problem corresponding to 38 + 24 in which a child counts "38—48, 58, 59, 60, 61, 62" either on a hundreds board or by putting up fingers. Here, there is no indication that the child relied on situation-specific imagery. Consequently, if this appeared to be a routine counting solution for the child, it would be inferred that his or her creation of abstract composite units was image independent. A second example is provided by the following solution to the horizontal addition sentence 34 + 30 = _ : "34—44, 54, 64." Here, there is nothing particular to the task itself that might lead the child to visualize collections of 10 items when interpreting the numerals. Of course, it is possible to train children to create figural images of this sort and, indeed, to count in this way to solve a narrow range of tasks. However, in the absence of evidence indicating that this has occurred, it seems reasonable to infer that the child did not rely on situation-specific imagery when creating abstract units of ten. It would, therefore, be concluded that the process of constructing the units was image independent.

Collection-Based and Counting-Based Solutions. Whereas the distinction between image-supported and image-independent solutions deals with the process of constructing units, that between collection-based and counting-based solutions as described by Cobb and Wheatley (1988) addresses the possible developmental history of particular solutions. This distinction can be illustrated by considering two solutions to a missing addend task corresponding to 48 + _ = 72 (i.e., 4 strips and 8 squares visible; 72 squares in all; how many are hidden beneath the cloth?). In the first solution, which was discussed previously, the child begins by reasoning that 40 and 20 are 60, and that 8 and 2 are 10, making 70 in all. The child then completes the solution by noting that as there are 72 in all, 2 more squares must be hidden by the cloth, and 22 and 2 is 24. In the absence of evidence to the

---

[1]Imagery, as I use the term in this chapter, refers to figural imagery. This usage should not be confused with that of Johnson (1987), who argued that thought is inherently imaginistic. Dörfler (1991), Presmeg (1992), Sfard (1994), and Thompson (1994) discussed the implications of Johnson's work for mathematics education. In doing so, they demonstrate that highly abstract mathematical thought involves nonfigural imagery.

contrary, it seems reasonable to infer that the child relied on situation-specific imagery to create part–whole relations involving units of ten and one. In addition, it can be observed that the child's task was to find how many tens and ones would have to be added to 40 and 8 to make 70 and 2. The metaphor implicit in the solution seems to be that of manipulating collections.

This collection-based solution can be contrasted with one in which a child solves the same task by counting "48—58, 68, 69, 70, 71, 72" while putting up fingers and then saying that 24 squares are hidden. In this case, the task the child solved was that of finding the difference between 48 and 72. A crucial distinction between the two solutions is that the first was collection based whereas the second seemed to involve the curtailment of counting on. In general, a solution is said to be counting based if it appears to involve the curtailment of counting by ones. In contrast, collections-based solutions typically involve partitioning two-digit numbers into a "tens part" and a "ones part" (e.g., 48 conceptualized as 40 and 8).

As further clarification of collection-based solutions, it should be noted that children often solve addition sentences such as 39 + 53 = _ by reasoning that 30 and 50 are 80, nine and three are 12, and 80 and 12 are 92. Children who use this partitioning algorithm need not create either numerical composites of ten or abstract composite units. Instead, 30 and 50 could be unitary conceptual entities (i.e., metaphorically, 30 is an unstructured composite of ones). As a further point, it can also be observed that, in this solution, the act of partitioning a two-digit numeral such as 39 symbolizes the act of partitioning the signified number—the result is 30 and 9 rather than 3 and 9. This distinction between partitioning the numeral per se and partitioning the signified number also proved to be relevant.

Summary. The cognitive constructs that proved useful when conducting the case studies deal with three aspects of the children's socially situated mathematical activity. The first aspect concerns the quality of the conceptual entities they created when interpreting and solving tasks. Here, the central distinctions are those between ten as a numerical composite, ten as an abstract composite unit, and part–whole relations involving abstract units of ten and one. The second aspect concerns the process of creating these entities, and here the distinction was made between image-supported solutions and image-independent solutions. The third aspect concerns the possible developmental history of particular solutions, and here counting-based solutions were contrasted with collection-based solutions. The remaining distinction discussed is that between partitioning a two-digit numeral per se, and partitioning the number it signifies.

Sociological Constructs

The analysis of the children's small-group interactions focused on their obligations and expectations, and on the taken-as-shared mathematical meanings they established. Each group's social relationship was characterized in terms of regularities in children's obligations and expectations. Additional constructs that emerged in the course of the analysis deal with the types of interactions in which the children engaged, and the relations of power and authority established in the groups.

Types of Interaction. Differences in the types of interactions in which the children engaged influenced the learning opportunities that arose for them. This was the case both when the two children were still in the process of solving a task and when one or both children had arrived at a solution. In the latter case, either one child attempted to explain his or her thinking to the other, or the children attempted to resolve conflicts among their interpretations, solutions, and answers. The relevant distinction in these situations in which one or both children had already developed a solution is that between interactions that involve univocal explanation and those that involve multivocal explanation. In the first of these types of interactions, one child judges that the partner either does not understand or has made a mistake, and the partner accepts this judgment. The interaction proceeds smoothly as the first child explains his or her solution and the partner attempts to make sense of the explanation. The term *univocal* is used to emphasize that the perspective of one child dominates. It should be noted that even in these interactions, explaining is a joint activity in that the partner has to play his or her part by accepting the first child's judgment and attempting to understand the explanation.

Interactions involving multivocal explanation often occur when a conflict has become apparent and both children insist that their own reasoning is valid. One child might again assume that the partner has made a mistake. However, in contrast to univocal interactions, the partner challenges this assumption by explicitly questioning the explanation. In general, multivocal interactions are constituted when both children attempt to advance their perspectives by explicating their own thinking and challenging that of the partner. The case studies indicate that these interactions were usually productive, provided the children had established a viable basis for mathematical communication. By way of comparison, univocal interactions frequently did not give rise to learning opportunities for either child.

With regard to the interactions that occurred while both children were still in the process of developing solutions, the relevant distinction is that between direct and indirect collaboration. The children engaged in direct collaboration when they explicitly coordinated their attempts to solve a task. For example, to solve a multiplication task corresponding to $6 \times 6 = \_$, one child might count modules of six while the partner puts up fingers to record how many modules have been counted. In general, interactions involving direct collaboration only occurred when the children made taken-as-shared interpretations of a task and of each other's mathematical activity. Interactions of this type can be contrasted with those that involve indirect collaboration. In those situations, one or both children think aloud while apparently solving the task independently. Although neither child is obliged to listen to the other, the way in which they frequently capitalize on the other's comments indicates that they are monitoring what the partner is saying and doing to some extent. From the observer's perspective, the learning opportunities that arise seem fortuitous in that what one child says or does happens to be significant for the other at that particular moment within the context of his or her ongoing activity. Those occasions when indirect collaboration occurred in the case studies were frequently productive, whereas the instances of direct collaboration did not usually give rise to learning opportunities.

Mathematical Authority and Social Authority. The discussion of interactions involving univocal explanation illustrated an instance in which one child was the established mathematical authority of the group. In that situation, one child judged that the partner either did not understand or had made a mistake, and the partner accepted this judgment without question. It is important to stress that the notion of mathematical authority refers to the relationship the children have interactively constituted rather than to a single child's beliefs about his or her own role. For example, a child might believe that he or she is the mathematically more advanced and routinely attempt to help the other understand. However, regardless of what the child believes, he or she is not the mathematical authority of the group unless the partner accepts the child's judgment. Thus, in this account, both children participate in the interactive constitution of one child's role as the mathematical authority. In those instances in which such a relationship was observed, there was a clear power imbalance between the children in that one child was obliged to adapt to the other's mathematical activity in order to be effective in the group. In the case studies, such relationships did not appear to be productive for either child.

A second type of power imbalance can arise when one child regulates the way in which the children interact as they do and talk about mathematics. *Power*, as the term is used here, refers to whose interpretation of a situation wins out and becomes taken as shared. Thus, for example, it might be observed that two children engage in multivocal interactions only when a discussion of conflicting solutions fits with one child's personal agenda. The manner in which the child controls the emergence of interactions of this type indicates that he or she is the social authority of the group. In the case studies, the one instance in which a child was the established social authority of the group gave rise to an inequity in learning opportunities.

ORGANIZATION OF THE CASE STUDIES: LEARNING AND INTERACTION

As stated at the outset, the primary reason for conducting the case studies was to clarify the relationship between the children's situated conceptual capabilities and their small-group interactions. The view that emerged while conducting the case studies was that the children's mathematical activity and the social relationships they established were reflexively related. On the one hand, the children's cognitive capabilities, as inferred from both the interviews and their activity in the classroom, appeared to constrain the possible forms that their small-group relationship could take. Although it was possible to imagine that a particular pair of children might have established a variety of alternative relationships in other circumstances, a wide range of possibilities did not seem plausible. On the other hand, the relationships that the pairs actually established constrained the types of learning opportunities that arose and thus profoundly influenced the children's construction of increasingly sophisticated mathematical ways of knowing. These developing mathematical capabilities, in turn, constrained the ways in which their small-group relationships could evolve.

This reflexive relationship guided the organization of the case studies. Each case study commences with an analysis of the children's interviews. Although the interviews are treated as social situations, the primary focus is on their conceptions of 10. Next, the small-group relationships the children established are documented by identifying the types of interaction in which the children engage. In addition, attention is given to both possible power imbalances and to the extent to which the children were able to establish a viable basis for mathematical communication. As part of this social analysis, the occurrence of learning opportunities is also documented. This then provides a link to the subsequent cognitive analysis of each child's mathematical activity and learning during small-group interactions. Finally, a summary of the case study is given to both illustrate the reflexive relationship between individual and collective activity and to clarify the features of productive and unproductive relationships.

CASE STUDY—RYAN AND KATY JANUARY 14 – MARCH 23, 1987

Ryan: January Interview

During his interview, Ryan routinely counted on to solve missing addend tasks, indicating that he could create and count abstract units of one. In addition, he routinely counted by tens and ones to solve the uncovering tasks. For example, having reached 46 on the second uncovering task, he counted "56, 66, 67, 68, 69, 70, 71" when a further 2 strips and 5 squares were uncovered. This suggests that he could create image-supported abstract composite units of ten, at least when he gave linear, counting-based meanings to number words and numerals .

His solutions to the strips-and-squares tasks indicate that the creation of image-supported abstract composite units was within the realm of his developmental possibilities, at least when he established collection-based meanings. In one task, he was shown a collection of 3 strips of 10 and 9 individual squares [39], told there were 61 squares in all, and asked to find how many were covered. He explained his answer of 2 tens and a 1 as follows:

1 Ryan: Only 2 [strips of 10] under there and 1 [square], and then you add this nine [the visible 9 squares] to there and if you take that 1 [the 1 hidden square] and put it to here [with the 9 visible squares] that would make 10, and then those [the visible and hidden strips and squares] all added together would make 61.

3. LEARNING AND SMALL-GROUP INTERACTION

The crucial feature of this solution was his realization that one of the additional tens needed to make 60 could be created by composing ones from the visible and screened collections. This suggests that the tens were abstract composite units for him. The limitation of his solution was that the units of ten and one that he established while building up the hidden collection did not seem to constitute a single numerical entity for him. Instead, he attempted to keep track of the additional strips and squares needed to make 61 by relying on figural imagery and assumed that the one square he added to the 9 visible squares to make a unit of ten was also the 1 of 61. This inference is corroborated by the difficulties he experienced when solving a follow-up task in which he was shown 4 strips and 6 squares [46], told there were 71 squares in all, and asked to find how many were hidden.

1 Ryan: Three's under there.
2 Interviewer: How many blocks [i.e., individual squares]?
3 Ryan: 4—there's only 2 bars under there.
4 Interviewer: How did you get 4?
5 Ryan: I counted these [the 6 visible squares] and I counted those together and then I knew it made 10.

In this case, Ryan focused on the seven composite units of ten and seemed to lose sight of the 71. Despite these difficulties, these two solutions indicate that the construction of abstract composite units of ten was within his realm of developmental possibilities when he made collection-based interpretations of number words and numerals. There was, however, no evidence that he could establish part–whole relations that involved abstract units of ten.

Ryan initially attempted to solve the horizontal addition sentences by counting on by ones. However, the interviewer intervened before he could do so for 37 + 24 = _ and they together interactively constituted a solution which involved finding 37 + 20 and then adding on 4. Ryan solved the next task, 39 + 53 = _, by partitioning both numerals:

1 Ryan: 5 and 3's 8 and then I took this 9 and I took 1 from the 3 and add it to the 9 and 2 is left, then that's . . . I knew 92.

As he said "5" and "3" rather than "50" and "30", it is unclear whether the act of partitioning the numerals symbolized the partitioning of the numbers they signified. Nonetheless, he subsequently used the same solution method and gave similar explanations when he solved the last two tasks on the worksheet, those corresponding to 37 + 24 = _ and 39 + 53 = _. Taken together, these solutions suggest that the development of a conceptually based partitioning algorithm might have been within his realm of developmental possibilities.

Katy: January Interview

Katy's performance during the first part of the interview was very similar to that of Ryan. For example, she solved the uncovering tasks by counting by tens and ones, indicating that she could create abstract composite units of ten by relying on situation-specific imagery. Her most sophisticated solution occurred on a strips-and-squares task in which 4 strips and 8 squares were visible [48], she was told there were 72 squares in all, and asked to find out how many were hidden.

1 Katy: 3 rows of 10 . . . I think there's 30. . . . If there's 72 altogether, there's only a couple underneath here, I don't know how many.
2 Interviewer: A couple of what?
3 Katy: I think there's 22 underneath here.
4 Interviewer: How did you figure that out?
5 Katy: There's 8 right here—8 and 2 is 10, but there's 72, so there has to be 24, 24—and there's 48 right here, altogether makes 70, 71, plus another one right underneath here makes 72.

Katy, like Ryan, anticipated that she could create a unit of ten by composing ones from the visible and hidden collections. Further, as was the case with Ryan, this seemed to be an image-supported solution. The crucial difference between the two children's solutions concerned the way in which they kept track as they built up the hidden collection in re-presentation. In contrast to the difficulties that Ryan experienced, Katy reasoned "but there's 72, so there has to be 24." It would, therefore, seem that the construction of part–whole relations involving abstract units of ten and one was within the realm of her developmental possibilities.

It can be noted that Ryan constructed a partitioning algorithm with the interviewer's support when he solved the horizontal sentences. In contrast, Katy used a similar algorithm to solve these tasks spontaneously, indicating that the act of conceptually partitioning a two-digit number was relatively routine for her. With regard to the worksheet tasks, she solved each by using the standard addition algorithm and explained that she had learned this from her elder sister. However, it is not possible to infer from her explanations the extent to which the steps of this algorithm carried numerical significance for her.

Regularities in Small-Group Activity

Katy and Ryan's small-group relationship was reasonably stable and was characterized by an unresolved tension in their expectations for each other. For example, during the sessions in January and at the beginning of February, the two children typically began a new task by engaging in independent activity. Katy usually arrived at an answer first and then told it to Ryan almost immediately. As she explained to Ryan on January 27, "I'm supposed to help you with these [tasks]." This unsolicited assistance generally involved attempting to complete Ryan's solution if she judged that it was correct, and interrupting to explain her solution if

3. LEARNING AND SMALL-GROUP INTERACTION

she believed that he had made a mistake. However, Ryan rarely listened to her explanations until he had solved the task himself. Further, he voiced frustration at her interruptions, and sometimes tried to disrupt her explanations. In general, he considered that Katy's interventions infringed on his obligation to develop his own solutions and continually challenged Katy's view of herself as the mathematical authority in the group. As a consequence, there were few if any occasions when they engaged in either direct or indirect collaboration.

The tension in the children's expectations for each other frequently continued after Ryan had produced an answer. Katy often assumed that it was still her responsibility to explain her solution to him, and she also took it for granted that he had made a mistake when their answers were in conflict. For his part, Ryan frequently interrupted Katy's explanations by interjecting counterarguments. Katy, in turn, often responded by modifying her explanations to accommodate his challenges. As an example, consider an exchange that occurred on January 21 when Ryan proposed that the answer to a balance task that involved adding 8 fours was 28 and Katy thought that the answer was 32. The children had agreed that the answer to the immediately prior task of adding 6 fours was 24.

1 Ryan: 28.
2 Katy: You know that's 24 [the prior task], 25, 26, 27 . . .
3 Ryan: [Interrupts] No, all they're doing is adding 1 four [onto the previous task].
4 Katy: This one is adding 2 fours [onto the previous task]. I'll show you how I got it. I got . . .
5 Ryan: 28, see, look, that makes 16, that makes . . .
6 Katy: [Interrupts] There's 4 fours. Pretend like you don't see them (covers 4 of the 8 fours with her hand), and then you add on.
7 Ryan: These fours, and that makes 16, (counts on 4 more fours while pointing to the boxes on the activity sheet), 32.
8 Katy: 32, so that's what the number is.
9 Ryan: You didn't say it was 32.
10 Katy: No, because I was showing you how I got it.

Katy's initial comment, "You know that's 24," indicated that she intended to explain how she had related this task to the prior one. However, a different solution that seemed to be the product of their joint activity emerged in the course of the exchange as they accommodated to each others' challenges. Nonetheless, from Ryan's perspective, he had fulfilled his obligation of solving the task himself. It can also be noted in passing that the children's last two comments indicate that the tension in their social relationship had an explicitly competitive aspect.

As the sample episode illustrates, Ryan and Katy typically engaged in multivocal rather than univocal explanation when their solutions and answers were in dispute. The sample episode also illustrates that learning opportunities could arise for both children in the course of these interactions. In particular, learning opportunities could arise for Katy when Ryan was still solving a task and she monitored his ongoing activity. Learning opportunities could also arise for her when she attempted to explain her solution once Ryan had solved the task and she accommodated to his shifting challenges and counterarguments. By the same token, learning opportunities could arise for Ryan as he tried to make sense of and formulate challenges to Katy's explanations. Further, the interactions involving multivocal explanation can also be viewed as occasions for group learning in that the children frequently elaborated their taken-as-shared basis for mathematical activity while reciprocally influencing and adapting to each others' arguments.

The comments made earlier both characterize the socially constituted situations in which learning opportunities tended to arise for Ryan and Katy and hint at the generally productive nature of their social relationship with regard to mathematical learning. However, as summary statements, these comments cannot do justice to either the richness of the children's individual and joint mathematical activity, or to the diversity and unpredictability of the learning opportunities that arose for them during their interactions. An exchange that occurred on January 26 illustrates these points particularly well. The children were solving the task, "How many do you add to III::: [36] to make IIIII:. [53] ?"

1	Ryan:	(Starts to put out bars of multilinks).
2	Katy:	(Counts on from 36 to 53 on her fingers) 17.
3	Ryan:	Look, 36 (points to 3 ten-bars and 2 three-bars). And how many do we have on that? (points to the picture of 53 on the activity sheet)
4	Katy:	53. So you add 2 more tens.
5	Ryan:	2 more tens and take away one of these (points to a three-bar).

Here, as was typical, Katy solved the task first and told Ryan her answer. She then attempted to help him complete his solution, telling him to add 2 more tens to the collection of 36 multilinks. Ryan usually ignored attempts of this sort to assist him. However, on this occasion, Katy's intervention made sense to him in the context of his ongoing activity and, in the very act of interpreting Katy's suggestion, he anticipated how he would physically manipulate the multilinks to transform a collection of 36 multilinks into one of 53. Her comment, therefore, gave rise to a learning opportunity for him in that he completed his solution by re-presenting rather than by actually manipulating multilinks. Nonetheless, his solution was highly image dependent in that "ten" seemed to refer to a ten-bar as a physical object rather than to it as an instantiation of an abstract unit.

From Katy's perspective, Ryan's answer of adding 2 tens and taking away three, couched as it was in terms of physical actions, conflicted with her answer of 17. As was typical in such situations, she assumed that he was in error and attempted to explain her solution to him.

6	Katy:	Come here, come here, I think you're not getting this right. All right, you have this many numbers (points to the picture of 36) and that makes 36, and that makes 37, 38, 39 , 40 . . .

3. LEARNING AND SMALL-GROUP INTERACTION

7	Ryan:	(Interrupts) Look, look...
8	Katy:	(Ignores him and completes her count) . . . 50, 51, 52, 53.
9	Ryan:	Well this is 36 (points to the activity sheet), and we have to take away one of these things [a strip of three squares in the picture of 36].
10	Katy:	Oh no you don't.
11	Ryan:	(Ignores her) and then we add 2 of these things [two strips of 10].
12	Katy:	Here, I'll explain it to you how I got the number.
13	Ryan:	That's how I did it.

As this exchange illustrates, Ryan was not obliged to accept Katy's explanation when he believed that he could justify his own solution. Conversely, Katy believed that he had made a mistake and, as a consequence, was not obliged to listen to his explanation.

14	Katy:	Here, you have that many numbers, 36, and you add 10 more, makes 46 (holds up both hands with all 10 fingers extended), 47, 48 . . . 53 (puts up 7 fingers as she counts).
15	Ryan:	Katy, look, you have to take away a 10 [remainder of his statement is inaudible].
16	Katy:	I'll show you how I got my number. See, you have 36, and add 10 more makes 46 (holds up both hands with all 10 fingers extended), 47, 48 . . . 53 (puts up 7 fingers as she counts). Do you agree with 17?

Here, Katy curtailed counting on by ones and created a numerical composite of 10 when she explained her solution for the second time—in holding up 10 fingers, she anticipated the result of counting to 46 by ones. Although it is not possible to explain why she reorganized her counting activity on this rather than on another occasion, the fact that Ryan was manipulating bars of 10 multilinks and that she had previously directed him to add on 2 tens may have been significant. In any event, the way in which she repeatedly explained her solution while fulfilling her obligation of helping Ryan gave her the opportunity to reflect on her original solution.

An interesting aspect of the entire exchange is that both children made conceptual advances even though they did not appear to consciously listen to each other's explanations. Instead, learning opportunities arose when something that one said or did proved to be significant at that particular moment within the personal context of the other's ongoing activity. Apparently, fortuitous learning opportunities of this sort occurred repeatedly when Katy and Ryan engaged in multivocal explanation.

With the exception of the very last session in which they worked together, the regularities in the children's obligations and expectations were stable throughout February and March. Thus, Katy remained obliged to assist Ryan, and Ryan continued to try and solve tasks for himself. For example, on March 4, Katy explained to Ryan why she had asked him what his answer was by saying, "I have to make you sure." In general, she was still obligated to check that his answers were correct. The stability of their obligations and expectations was also indicated by the continual tension in their views of their respective roles the group. In this regard, open conflicts arose from time to time. For example, on both March 4 and 5, Ryan complained to the teacher when Katy solved a task quickly, presumably because he thought that it might be difficult for him to develop his own solutions.

Although it was possible to identify general regularities in the children's obligations and expectations, several novelties were observed in their interactions in March. For the most part, these seem to have stemmed from Katy's attempts to cope with the tension between her own and Ryan's obligations. For example, the following exchange occurred on March 4 when they solved the number sentence $5 \times 6 = \_$.

1	Katy:	5 sixes (draws 5 circles and then looks at them). 12 and 12's 24, 24, . . . watch, Ryan, watch, watch, watch, watch, watch.
2	Ryan:	What?
3	Katy:	Do you agree this 6 and 6 is 12, 6 and 6 is 12, 12 and 12 makes 24—24, 25, 26, . . . 30. You agree?
4	Ryan:	(No response as he continues making his own drawing.)
5	Katy:	Do you agree or not?
6	Ryan:	(Continues until he finishes his own drawing and then looks up.)

As was the case in January, Katy produced an answer first and attempted to explain her solution while Ryan continued to solve the task in his own way. However, Katy's approach of asking him whether he agreed with her solution instead of explaining it to him directly was novel. In making this request, she did not seem to act as a mathematical authority but instead appeared to implicitly indicate that his judgment was of interest to her. As previously stated, Ryan typically ignored her attempts to directly explain her solution to him before he had produced an answer. This time, however, he tried to make sense of her explanation once he had drawn six (rather than five) circles and, in the process, a learning opportunity arose for him.

7	Katy:	Do you agree?
8	Ryan:	(Looks at his own drawing of 6 circles) 6 and 6 is 12.
9	Katy:	(Interrupts) 6 and 6 is 12, and 12 and 12 is 24, and
10	Ryan:	(Interrupts) Like 12 and 12's 24, 25, 26, . . . 33, 34
11	Katy:	(Interrupts) No, see that's 6.
12	Ryan:	(Ignores her and completes his count of 2 additional sixes) 35, 36.
13	Katy:	You don't see this 6 (points to one of the circles Ryan has drawn). Watch, you're supposed to have 5 sixes, 1, 2, 3, 4, 5, 6, you don't see this one.

The exchange concluded with Ryan agreeing that there should only be five circles and that the answer was 30.

Another learning opportunity arose for Ryan on March 4 when Katy again departed from her usual routine of giving a direct explanation. On this occasion, she first solved the number sentence $\_ \times 3 = 18$ by making a tally mark each time she counted the same three fingers until she reached 18.

1	Katy:	I tried sixes. What's 3 plus 3?
2	Ryan:	6.
3	Katy:	What's 3 plus 3?
4	Ryan:	6.
5	Katy:	What's 3 plus 3?
6	Ryan:	6.
7	Katy:	What's 6 plus 6 plus 6?
8	Ryan:	OK, 6 plus 6 is twelve, 13, 14, . . . 18.
9	Katy:	See, it's 6, 3 times 6.
10	Ryan:	Yeah, but I'm supposed to do the problem [myself].

Here, Katy reconceptualized her counting solution and, instead of explaining directly, she asked Ryan a series of questions. Ryan, for his part, seemed obliged to respond and, further, appeared to pause and reflect on his sequence of responses before agreeing that the answer was six. Nonetheless, his final comment indicates that he did not consider that he had solved the task himself and he subsequently developed an alternative solution.

It should again be acknowledged that the exchanges that occurred in these last two examples were atypical. They do, however, illustrate that the children's interactional competencies began to improve as they attempted to cope with the tension in their expectations for each other. In addition, the episodes illustrate the kinds of learning opportunities that could arise when they found ways to temporarily alleviate this tension. In both examples, Ryan attempted to make sense of solutions that were more advanced than those that he typically developed when working independently. Katy, for her part, had the opportunity to reconceptualize her solutions as she attempted to explicate her thinking for Ryan. A learning opportunity of this type clearly arose for her in the second of the two examples that have been considered. It should, however, be noted that in contrast to their interactions while engaging in multivocal explanation, the children did not develop novel, joint solutions in either of these exchanges. Further, Ryan typically believed that he had solved a task himself when he modified his thinking by making counterarguments while engaging in multivocal explanation. This was not the case when he responded to a sequence of questions that Katy posed.

It was not until the last session in which they worked together, on March 23, that the children substantially modified their obligations and thus at least temporarily reorganized their small-group relationship. In this session, the children first attempted to solve the balance task 10, 10, $\_, \_, \_, \_, \_, \_ / 40$ by using a guess-and-test approach. However, Ryan then exclaimed, "It's 4, it's 4," and explained how he had related this to the previous task, 10, 10, 4, 4, 4, 4 /36.

1 Katy:	How can it be? 24, 28, (counts 3 modules of 4 by 1), 40. (Looks surprised, then turns and points at Ryan as if amazed.)
2 Ryan:	OK, if this one [the previous task] makes 36, then I knew you'd have to add 4 onto it [because 36 plus four makes 40]. That's how I knew.
3 Katy:	You're smart.

Katy then counted by fives to solve the next task, 10, 10, _, _, _, _ / 40. Ryan used this result to solve the subsequent task, 10, 10, _, _, _, _ /44.

1 Ryan:	Ah ha . . . we need 4 more . . . 44 . . . sixes. Yep, it works. Bet your lucky charm it works.
2 Katy:	26, (counts on 3 modules of 6 by ones), 44 (looks surprised). I don't get this.

The teacher then joined the group as Ryan explained:

3 Ryan:	The first one I knew [the answer to the previous problem] . . . just take off this [4 from the 44] and then all these [the other boxes] are the same so you add 1 more onto it, cause 37, 38, 39, 40. If it was fives it would be only 40, but we know we have 4 boxes and we add 1 in each box, we have 4 [more].
4 Katy:	Wait a minute, that [20 plus 12] makes 32, right, plus 12 makes 44, right.

These two solutions were the first that Ryan developed that were beyond the realm of Katy's development possibilities. Her comment, "You're smart," indicated that she was beginning to modify her view of herself as a mathematical authority who was obliged to assist Ryan by explaining her solutions to him. When Ryan produced each of these solutions, it was she who was obliged to listen and he who was obliged to explain. In the remainder of the session, Katy did not assume that Ryan was in error when their answers were in conflict, and their expectations for each other seemed to be more compatible. This change in the children's social relationship stemmed at least in part from their reassessment of their own mathematical competence relative to the other. Ryan, in fact, volunteered to the teacher "I know everything just about" right before he explained the second of his solutions. Their prior social relationship can be seen to predicate its own modification in that it both enabled and constrained each child's mathematical development.

Mathematical Learning

Katy. It will be recalled that during her January interview, Katy used the standard addition algorithm to solve the worksheet tasks presented in vertical column format, and used a partitioning algorithm when she solved horizontal number sentences. She was not observed using the standard algorithm again until February 16. Because none of the computational tasks she and Ryan attempted to solve were presented in column format, it would seem that her use of this algorithm in the interview was highly situation specific. In contrast, she used her partitioning algorithm from the outset when solving both balance tasks and number sentences.

Katy's first novel solution involving tens and ones occurred in a previously discussed episode on January 26. There, she explained she had solved a task corresponding to $36 + \_ = 53$ by saying that you have 36 "and you add 10 more makes 46 (holds up her hands with all 10 fingers extended), 47, 48, . . . 53." This solution was an advance over those she developed in her interview in that it did not appear to involve a reliance on situation-specific figural imagery of, say, a collection of hidden strips and squares. On the following day, January 27, she curtailed the solution process still further when solving a sequence of number sentences. First, she explained to Ryan that she had solved $41 + 19 = \_$ as follows: "We have 41, and then we add 10 more to make 51 (holds up both hands with all 10 fingers extended), 52, 53, 54, . . . 60." However, she then solved $31 + 19 = \_$ by reasoning, "31 plus 19 . . . is . . . 31 makes 41, 42, 43, . . . 50." This time, she did not hold up her 10 fingers, presumably because she no longer needed to create the record she would actually establish if she were to count on 10 ones. It would, therefore, seem that she created image-independent abstract composite units of ten.

As a further development, it was at this point in the session on January 27 that Katy was first observed counting by tens and ones on the hundreds board. She began to solve $39 + 19 = \_$ by saying "39 plus 10, 39 plus 10 . . . " and then "39, 49. . . ." However, on both occasions, she was interrupted by Ryan who was counting by ones on a hundreds board. Katy then took the board from him and solved the task: "39, 49, that's 10 (points on the hundreds board), 49—50, 51, 52, . . . 58" (counts on the board). In doing so, she seemed to interpret Ryan's counting solution in terms of her prior conceptualization of 19 as 10 and 9 more, and to express this conceptualization by counting by tens and ones on the hundreds board. Solutions of this type subsequently became routine for her when she solved missing addend and missing minuend tasks as well as addition and subtraction tasks.

Another solution that Katy produced on January 27 further substantiates the inference that she could now create image-independent composite units of ten when she gave number words and numerals counting-based interpretations. First, she explained to Ryan that she had related $60 - 31 = \_$ to the previous task, $60 - 21 = 39$, by saying, "See, you're taking away another ten." When Ryan disputed this, she attempted to convince him by counting by ones from 39 to 29 on a hundreds board. When Ryan again challenged her answer, she further explicated her reasoning by counting on a hundreds board as follows:

1 Katy: (Counts from 60 to 50 by ones) 10, that's 10, now we have 29 more to go, (counts from 50 to 40 by ones) 10, that's 2 tens, now we need 1 more ten, (counts from 40 to 30 by ones) 10, and then 1 more makes 29.

This explanation demonstrates that Katy could monitor her activity of creating and counting abstract composite units of ten, at least when she used the hundreds board.

An exchange that occurred with the researcher 1 week later on February 3 both indicates a limitation of the conceptual reorganization that Katy had made and clarifies the role that the hundreds board played when she used it to count by tens and ones. In this episode, she first solved the task, "How many do you have to take away from 74 to leave III::. [35]" by routinely counting from 74 to 35 on the hundreds board, "10, 20, 30, 1, 2, 3, ... 9—39." At this point, a researcher intervened.

1 Researcher: Katy, could you do that problem without using the hundreds board?
2 Katy: We know we have that number [74] and we have that many left [35], so if you have that many [35], you just go up.
3 Researcher: How would you go up?
4 Katy: Like, 35—45, 55, 65, 75, but that's too much. So you take away a one and then you'll have . . . 74.
5 Researcher: So how many do you have to add to 35 to make 74?
6 Katy: You have 35—45, 55, 65, 75 (slaps her open hands on the desk each time she performs a counting act), then you take one away, then it's too much, then you take one away and you will have . . . 74.

These two solution attempts suggest that, in the absence of the hundreds board, the abstract composite units of ten that she created were not objects of reflection for her. In other words, although both the counting sequences she completed during the exchange with the researcher carried the significance of incrementing by tens, she did not seem able to step back and monitor herself as she incremented. By way of contrast, her final explanation of $60 - 31 = \_$ on January 27 indicated that she was aware of what she was doing when she counted by tens and ones on the hundreds board. It therefore seems reasonable to infer that her use of the hundreds board was crucial in making this reflection possible. In particular, it appears that the hundreds board came to symbolize a number sequence for her (not merely number–word sequence) once she could create image-independent composite units of ten. Thus, for example, on February 2 she immediately gave 57 as her answer to $37 + 20 = \_$ and, to explain, simply moved a finger down two rows from 37 on a hundreds board while saying "I know 37 plus 20 make 57." Here, the act of moving down two rows and reading the resulting numeral symbolized adding 20 for her. More generally, she seemed to give certain regularities in the numerals on the hundreds board numerical significance (e.g., 7, 17, 27, 37. . . signified incrementing by 10). The ability to see the hundreds board as being arithmetically structured in this way and to consider this structuring as self-evident was a further aspect of the conceptual reorganization she had made. As a consequence, the abstract composite units she created when counting by tens and ones were simply there in the hundreds board as objects of reflection for her, and she could therefore monitor her counting activity.

The most sophisticated counting solution that Katy produced without using the hundreds board occurred on March 2, when she solved a money story problem. She first found that the value of a collection of coins shown diagramatically was 58¢, and then interpreted the task as a missing addend corresponding to 58 + _ = 99.

1 Katy: (Counts on her fingers) 58—59, 60, 61, 62, . . . 86, 87, 88—30 more makes 88—89, 90, 91, . . . 98, and that makes 40,—99, so you have 41.

In contrast to the explanations she gave on February 3 when questioned by the researcher, Katy monitored her counting activity while solving this task. However, at this time she did not create composite units of ten, but instead counted by ones and established numerical composites of 30 and then 40 as she went along.

It has already been noted that Katy did not use the standard addition algorithm until February 16, indicating that it was specific to situations in which computational tasks were presented in vertical column format. Her subsequent use of this algorithm in an increasingly wide range of situations might also be a consequence of the conceptual reorganization that she made at the end of January and the beginning of February. Then, she developed the ability to construct abstract composite units of ten in an image-independent manner when she made counting-based interpretations of number words and numerals. Her increased use of the standard algorithm might indicate that she also became able to construct image-independent units of this type when she gave numerals collection-based meanings. Consider, for example, the first occasion on which she was observed using the standard algorithm after the January interview. She explained that she had found the sum of 5 twelves while solving a balance task by adding 36 and 24 as follows:

1 Katy: See, you've got these 3 [twelves] to make 36, and this makes 24, and that together [the 4 and the 6], that makes 10, so add the 1 [from the 10] up there, 4, 5, 6—see, I carried.

If she could construct abstract composite units of ten in the absence of situation-specific figural imagery, then the sum of 6 and 4 could simultaneously be 10 units of one and 1 unit of ten. As a consequence, the numeral-manipulation acts of partitioning the 10 and carrying the 1 could symbolize conceptual actions on arithmetical objects for her. The standard algorithm, which presumably had been taught to her as a sequence of instructions for manipulating numerals, could then become part of her computational repertoire; its use would be consistent with the relational beliefs that seemed to inform her mathematical activity in the classroom.

In summarizing Katy's mathematical learning, it should be recalled that she and Ryan frequently engaged in multivocal explanation. Learning opportunities arose for her in these situations as she monitored and assessed Ryan's attempts to solve tasks, and as she adapted her explanations to accommodate his frequent challenges and counterarguments. In addition, apparently fortuitous learning opportunities arose when something that Ryan said or did happened to be pertinent at that moment within the personal context of her ongoing activity. The observed changes in Katy's computational methods indicated that she made a major conceptual reorganization during the time that she worked with Ryan. For example, the way in which she curtailed counting by ones in a variety of situations indicated that she became able to create abstract composite units of ten in an image-independent manner when she gave number words and numerals counting-based meanings. As a consequence of this development, the hundreds board became numerically structured in terms of units of ten and one for her. Its use could then facilitate her attempts to monitor relatively sophisticated counting-based solutions. Finally, her increasing use of the standard algorithm indicated that she might also have become able to create abstract composite units of ten without relying on situation-specific imagery when she gave number words and numerals collection-based meanings.

Ryan. The analysis of Ryan's interview in January indicates that the construction of an addition algorithm that involved partitioning two-digit numbers was within his realm of development possibilities. He did, in fact, construct such an algorithm as he attempted to make sense of Katy's partitioning solutions. For example, on January 28, he initiated a joint solution that involved partitioning numbers. The task was presented in a balance format and involved finding the sum of 3 twenty-ones.

1	Ryan:	20 plus 20.
2	Katy:	Is 40.
3	Ryan:	And then two more.
4	Katy:	Makes 42. (Starts counting from 42 on the hundreds board.)
5	Ryan:	And then you have...look you have 3 [21s].
6	Katy:	Three 21s.
7	Ryan:	40 plus 20, that would be 60, and then you have 61 . . .
8	Katy:	(Interrupts) 61, 62, 63 . . . extra ones.

Ryan's comments as he and Katy solved the next task, that of adding 4 twenty-ones, confirm that the 2 of 21 signified 20 for him.

1	Katy:	21 four times . . . 4 21s . . . 2 and 2 is 4, and 2 and 2 is 4, and 4 ones makes 44.
2	Ryan:	The 2s are 20s, don't you know that!
3	Katy:	21.
4	Ryan:	If you take away the ones, those are 20s.

Although Ryan's partitioning solutions were initially restricted to balance tasks, he did partition numbers with increasing frequency when solving tasks of this type in the following weeks. For example, on March 23 he explained that he had solved a balance task that involved adding 3 fifteens as follows:

1	Ryan:	You take two [of the 15s] and that makes 30, and you have 15 more and you add 10, that'll make 40, and then 45.

3. LEARNING AND SMALL-GROUP INTERACTION

He then explained how he had solved the next task of adding four 15s.

1	Ryan:	30, take 5 away, you have 10, and that would make 50, and then . . .
2	Katy:	(Interrupts) Add the 2 fives.
3	Ryan:	The 2 fives together will make a ten.
4	Katy:	And 50 plus 10 will make.
5	Ryan:	60.

Ryan's solutions to the uncovering tasks in the January interview indicated that the construction of ten as an abstract composite unit might have been within his realm of developmental possibilities when he gave number words and numerals counting-based meanings. There was little indication that he constructed image-independent units of this type, even though he repeatedly observed and, on occasion, attempted to contribute to solutions in which Katy counted by tens and ones. Consider, for example, the following exchange which occurred on January 27 when they solved $39 + 19 = \_$.

1	Katy:	39 plus 10, 39 plus 10 . . .
2	Ryan:	(Interrupts) 39 plus 19.
3	Katy:	39, 49 . . .
4	Ryan:	(Interrupts) No. (Counts by ones on the hundreds board) 59.

Ryan seemed to make these interruptions with the intention of correcting Katy, presumably because he thought that she had read the task statement as $39 + 10 = \_$. Both in this instance and on other occasions, the possibility that she had created a unit of ten did not occur to him.

Ryan also continued to count by ones when he used the hundreds board. On March 16, the researcher in fact asked him if a solution in which Katy counted by tens and ones on a hundreds board to solve $16 + \_ = 72$ was acceptable. He said, "Yes, it's all right," and actually enacted her solution successfully at the second attempt, but then proceeded to count by ones to solve subsequent tasks. In general, he seemed to know what Katy did in the sense that he could imitate the way she counted and, further, he realized that these counts yielded correct answers. However, because he could not create image-independent composite units, he did not understand what counting by tens and ones had to do with solving the task. As a consequence, Katy's solutions did not make sense to him.

A second exchange between Ryan and the researcher on March 16 suggested that he was also unable to construct image-independent composite units of ten when he established collection-based meanings. In this episode, Ryan challenged a solution in which Katy used the standard algorithm to add 23 and 51.

1	Katy:	That [two and five] makes seven, that [three and one] makes four, 74.
2	Ryan:	But you don't always know that, you can't always do that.

3	Researcher:	Why can't you always do that?
4	Ryan:	Well, that's an easier way, but sometimes you might get mixed up and get the wrong answer.
5	Researcher:	Do you know why you get mixed up?
6	Ryan:	Because you don't know which number you are going to use. She just picked the numbers.

Ryan's comments indicate that the standard algorithm had no rhyme or reason to it for him. It was previously suggested that Katy's increasing use of this algorithm occurred when she could construct abstract composite units of ten in the absence of situation-specific imagery. The difficulty Ryan had in making sense of her explanations indicates that he could only construct units of this type when, as in the interview, he could rely on situation-specific imagery.

There is one further aspect of Ryan's mathematical activity that seems worth mentioning. Occasionally, he unexpectedly produced surprisingly sophisticated solutions. One of these cases, which occurred on March 23 when he related successive multiplication balance tasks, was already discussed. In one of these solutions, he used 10, 10, 4, 4, 4, 4 / 36 to solve 10, 10, _, _, _, _, _ / 40, explaining that he "knew you'd have to add an extra 4 onto it" because 40 was 4 more than 36, and because the 2 tens were the same in both tasks. He developed an equally sophisticated solution 3 weeks earlier on March 2, when he and Katy solved a money story problem corresponding to 58 + _ = 99. Then, he gave 41 as his answer and explained his solution as follows:

1	Ryan:	50 and 50 make 100, and ... 4 quarters make a dollar, but you already have 8, so you have to take away 9 because it's 50 but you only have 8.

Here, Ryan appeared to rely on figural imagery of coins as he related the task to $50 + 50 = 100$. In particular, he seemed to first take 8 away from 50 while using the compensation strategy [i.e., $58 + (50 - 8) = 100$], and then take away an additional one because the total was 99 rather than 100.

He developed a third surprisingly sophisticated solution 3 days later on March 5 when he and Katy solved the task, "54 are taken away and 36 are left. How many were there to begin with?" They had just agreed that 54 was the answer to the immediately preceding task, "How many do you add to 37 to make 91?" Katy counted to 90 on the hundreds board, and Ryan then exclaimed:

1	Ryan:	54, 54 [in both task statements], this [37] is 1 higher, this [36] is 1 lower, we know this [the answer] has to be 1 lower, and we have 90, and the answer's going to be 1 lower, and this [90] is 1 lower than that [91].
2	Katy:	It has to be 90!
3	Ryan:	Yes.

These solutions to a variety of different types of tasks seem disparate at first glance. However, in each case Ryan seemed to explicitly coordinate numerical changes in one or more of the parts with those in the whole. This suggests that during January and February, Ryan became able to take numerical part–whole structures as givens and act on them conceptually. This and the other constructions he made while working with Katy had important consequences for the nature of their social relationship.

Summary

The analysis of the children's January interviews indicated that Katy was the more conceptually advanced of the two children. For example, she was able to routinely partition two-digit numbers but Ryan was not. Differences in the children's conceptual capabilities also became apparent when they worked together in the classroom in January. Thus, it was noted that Katy typically proposed answers first, either because a task was routine for her but not for Ryan, or because her relatively sophisticated solutions were more efficient than Ryan's. In addition, she usually prevailed when her own and Ryan's solutions were in conflict, in part because she was able to adapt her explanations in response to his challenges. These differences in the children's cognitive capabilities gave rise to a tension in their roles; Katy felt obliged to explain her solutions to Ryan, whereas he felt obliged to solve tasks himself in personally meaningful ways. A tension of this sort can clearly be coped with in a variety of alternative ways, indicating that the relationship between the children's cognitive capabilities and their social relationship was not deterministic. Rather, their individual cognitive capabilities constrained the possible forms that this relationship could take, and thus the learning opportunities that could arise for them as they attempted to fulfill their obligations and be effective in the classroom.

The analysis of the social relationship they established indicates that they did little to modify their obligations and thus accommodate to each other's expectations. Instead, the tension in their obligations manifested itself in frequent contentious exchanges. Despite, or perhaps because of, this discord, their ongoing interactions were reasonably productive in terms of the mathematical learning opportunities that arose. For example, Ryan's implicit refusal to concur with Katy's view of herself as the mathematical authority in the group contributed to the emergence of interactions involving multivocal explanation in which they challenged each other's thinking and justified their own solutions by making arguments and counterarguments. Further, in these interactions Katy's assumption that she was the mathematical authority gave rise to learning opportunities for her as she attempted to explicate her thinking, and for Ryan as he interpreted and challenged her more sophisticated solutions.

The analysis of each child's learning indicated that they both made significant progress during the time that they worked together. For example, Katy became able to construct ten as a composite unit without relying on situation-specific imagery when she established both collection-based and counting-based meanings. For his part, Ryan constructed a partitioning algorithm. On the one hand, these cognitive developments occurred within the constraints of an evolving social relationship that was itself constrained by the children's developing capabilities. On the other hand, these developments led to modifications in that relationship. For example, once Ryan constructed a partitioning algorithm Katy was no longer the one who almost invariably produced an answer first. Further, Ryan's development of solutions that were difficult for Katy to understand led her to revise her view of herself as the mathematical authority in the group.

CASE STUDY—HOLLY AND MICHAEL JANUARY 13–MARCH 23, 1987

Michael: January Interview

Michael's solutions to the counting tasks indicated that he could create and count abstract units of one. When considered in isolation, several of his solutions to the uncovering and the strips-and-squares tasks seem to indicate that he could also create abstract composite units of ten by relying on situation-specific imagery. However, if the interview is viewed as a social event, it seems more plausible to suggest that he learned to be effective in the interview by differentiating between the number–word sequences he should use to count a collection of strips, and those he should use to count individual squares.

One of Michael's first attempts to count by tens and ones occurred on the second uncovering task when he had reached 46 and two additional strips and five squares were uncovered. He initially counted by ones but the interviewer interrupted him when he had one strip left to count by pointing to the strip and saying, "61 and 10 more." Michael immediately replied, "71." In this situation, the interviewer intervened to investigate whether Michael could create ten as an abstract composite unit and thus curtail counting by ones. However, a holistic analysis of the entire interview suggests that Michael interpreted interventions of this type as assessments that his activity was inappropriate. In this particular case, he was able to be effective when he counted the remaining strip by giving the successor of 61 in the number–word sequence 51, 61, 71, 81.

Taken in isolation, several of Michael's solutions to the strips-and-squares tasks also appear to be quite sophisticated. For example, in one task in which 4 strips and 5 squares were visible [45], he was told that there were 65 squares in all, and asked to find how many squares were covered. He solved the task by putting up 2 fingers as he counted "55, 65—2 strips, and 10, 20 squares." One possible explanation is that his counting acts signified abstract composite units. Alternatively, it could be that in the course of his interactions with the interviewer he had learned to solve tasks of this type by producing a number–word sequence that went from the number of visible squares, 45, to the number of squares in all, 65. In this account of his solution, each counting act by 10 signified a strip viewed as a single entity rather than as a composite of 10 units of one.

Michael's attempts to solve a previous task are consistent with this latter explanation. Then, he had answered by stating the number of hidden strips and, during the ensuing exchange with the interviewer, could have learned to give an answer that was acceptable to the interviewer by counting the imagined strips, "10, 20 . . . " Further, his solution to a subsequent task fits with this explanation. Here, 3 strips and 5 squares [35] were visible, there were 61 squares in all, and Michael was asked to find how many were hidden. First, he produced a number–word sequence that went from 35 to 61: "36—46, 56, 66, 65, 64, 63, 62, 61." The interviewer then asked him how many squares were hidden, and he counted the strips and squares signified by his counting acts as follows: "There's 2 strips of 10, that makes 20 squares and . . . 65, that's 21, 64, that's 22, and 63 that's 23 squares, and 62 that's 24 squares, and 61 that's 25 squares." For him, each act of counting by one signified a square regardless of whether it was performed in the forward or backward direction. It seems reasonable to conclude that, despite his intentions, the interviewer's frequent interventions on the uncovering and strips-and-squares tasks served to support Michael's development of solutions that were based on number–word regularities rather than the creation of numerical meaning.

The conclusion that the construction of ten as an abstract composite unit was not within the realm of Michael's developmental possibilities is also consistent with his solutions to the horizontal sentence and the worksheet tasks. In those, he either counted by ones or attempted to use his version of the standard algorithm. For example, he counted by ones to solve the first two horizontal sentences, $16 + 9 =$ _ and $28 + 13 =$ _, but said that the next task, $37 + 24 =$ _ was "Too big." The interviewer then asked him "What's 37 plus 20?" and Michael added by columns to get 57. Thus, in this situation, the possibility of conceptualizing 20 as 2 tens did not occur to him. The interviewer then asked if he could use this result to help him solve the original task, $37 + 24 =$ _, but Michael again attempted to add by columns and thought the answer would be either 53 or 54. It would therefore seem that, in the absence of supportive imagery, Michael's most sophisticated counting solutions involved counting by ones.

Holly: January Interview

During the first part of the interview, Holly created abstract units of one as she counted on to solve missing addend tasks. However, in contrast to Michael, she seemed to routinely create and count abstract composite units of ten when solving the uncovering and strips-and-squares tasks. For example, on the second uncovering task, the interviewer uncovered two strips and five squares after Holly had reached 46. She immediately counted by tens and ones to find how many there were now in all, "56, 66, 67, 68, 69, 70, 71." Her solutions to other tasks in the interview indicate that situation-specific imagery was in fact crucial to her construction of abstract composite units of ten.

Initially, she attempted to solve the horizontal sentence and worksheet tasks by either counting by ones or by using her version of the standard algorithm. However, the solutions that she developed with the interviewer's support indicated that the act of conceptually partitioning two-digit numbers was within her realm of developmental possibilities. For example, the interviewer intervened as she began to solve the worksheet task corresponding to 37 + 24 = _ by asking her to first find 37 + 20 = _. She then used this result to solve the original task. Further, she solved the next task, 39 + 53 = _, by adding 50 to 39 to make 89, and then adding on 3 to make 92. The way in which she spontaneously modified this solution method while solving the remaining two tasks adds credibility to the view that the act of partitioning a numeral symbolized the conceptual act of partitioning the number for her. For example, she explained that the answer to 59 + 32 = _ was 91 because "First I took the 50 and the 30 and made it an 80, and then I took the 9 and made it 89, and then I add the 2 and had 91." She then solved 22 + 18 = _ in the same way. By way of contrast, there was no indication that numerical partitioning was within the realm of Michael's developmental possibilities.

Regularities in Small-Group Activity

The small-group relationship that Holly and Michael established was stable for the entire time that they worked together after the January interviews. Further, in contrast to Katy and Ryan, the expectations that Michael and Holly had for each other were generally compatible. As a consequence, their interactions were relatively harmonious, and conflicts about their respective roles were rare.

The analysis of the January interviews indicated that Holly was the more conceptually advanced of the two. In line with this conclusion, Holly was almost invariably able to develop a way of solving tasks when she and Michael worked together in the classroom, whereas Michael sometimes did not have a way to proceed. Their interactions in these situations indicated that they both took it for granted that Holly was the mathematical authority in the group. This can be illustrated by examining an exchange that occurred on January 13 as they solved a sequence of balance tasks. Just prior to the episode of interest, Holly had explained to Michael that she had solved the task 35 + 45 = _ by first adding 30 and 40 to make 70, and then adding on the additional 10. Against this background, the two children attempted to solve a task that they both interpreted as 36 + 46 + 10 = _.

1 Michael: 36, 46 . . . that would be 90, plus 10 more equals 100.
2 Holly: 36, 46 . . . 82 . . . add 10 more . . . it's 92.
3 Michael: Plus 10 more.
4 Holly: I know, I added the 10.

Although it is not clear how Michael arrived at his intermediate answer of 90, this seems to have been, at best, an educated guess. In the absence of multilinks or a

3. LEARNING AND SMALL-GROUP INTERACTION

hundreds board, he was unable to calculate the sum of such numbers throughout the time that he worked with Holly.

As Holly's subsequent explanation will make clear, she solved this task as she did the previous one, by using her partitioning algorithm. The analysis of her January interview indicated that the conceptual act of partitioning a two-digit number was within her realm of developmental possibilities. Here, it seemed to be a routine interpretive act for her, at least when she solved balance tasks. Michael, however, did not have a way to proceed and attempted to fulfill his obligation of developing a solution by accommodating to her mathematical activity. His interjection, "Plus 10 more" indicates that he was in fact attempting to make sense of and contribute to Holly's solution. However, this proposal conflicted with Holly's solution in that she had already added 10. A power imbalance was realized in interaction as they dealt with this conflict. Holly's response, "I know, I added the 10," indicates that she did not interpret Michael's interjection as a challenge that required her to give a justification. Both here and in similar situations, she assumed that Michael did not understand when his interpretation was incompatible with her own. Michael, for his part, accepted these judgments without question, thus participating in the interactive constitution of Holly as the mathematical authority in the group.

In general, Holly was obliged to explain her thinking and thus help Michael understand in situations in which incompatibilities in their interpretations became apparent. For example, the sample episode proceeded as follows:

4 Holly: I'll explain. That's 70 (points to the activity sheet).
5 Michael: Plus 6 more and 6 more equals 12, plus 12 more.
6 Holly: No, no, no, you don't understand. See, there's 70 to start with because there's 30 plus 40 equals 70. Add 10 more equals 80, 86 . . .
7 Michael: 87, 88, 89, 90 (counts on his fingers).
8 Holly: That's 92.

Michael's attempt to contribute to Holly's explanation again indicates that he was obliged to try and understand how she had solved the task. In attempting to accommodate to her mathematical activity, he proposed that 2 sixes be added to 70. Although this suggestion was consistent with the way in which she had solved the previous task, it did not fit with her final explanation to this task (i.e., she first added 10 to 70). As before when their interpretations were in conflict, Holly assumed that Michael did not understand her solution. Michael, for his part, did not question this judgment, but instead continued to try and accommodate to Holly's mathematical activity. In doing so, he further contributed to the establishment of Holly as the mathematical authority in the group.

The sample episode concluded when Michael indicated that he agreed with Holly's answer.

9	Holly:	So it equals 92.
10	Michael:	OK.

Holly paused before repeating her answer of 92, indicating that she might have been waiting for Michael to indicate agreement. Her activity in other episodes in which Michael did not indicate agreement supports this inference. The following episode both exemplifies this point and illustrates additional regularities in their interactions.

The exchange occurred on March 2, when they solved a money story problem. The task statement showed a picture of coins with a total value of 37¢ and asked, "Krista has this much money. She has 25¢ less than Larry. How much money does Larry have?" Michael interjected as Holly counted the pictured coins, again illustrating that he was obliged to develop a way of solving the task.

1	Holly:	10, 20, 25, 26 . . .
2	Michael:	(Interrupts) 30.
3	Holly:	30, 35, 36, 37.
4	Michael:	37.

It soon became apparent that Michael thought that 37 was the answer to the task. Holly, the mathematical authority in the group, was obliged to explain her thinking and Michael was obliged to listen.

5	Holly:	No it isn't, this is Krista's money, then add 25 more for Larry.
6	Michael:	37.
7	Holly:	37—47, 57, 58, 59, 60, 61, 62—62.
8	Michael:	This is weird.

In contrast to the previous episode, Michael did not indicate that he accepted her solution, and, in such situations, Holly was obliged to explain her thinking again.

9	Holly:	Add 10 cents, that makes 47, then add another 10, 57, then 58, 59, 60, 61, 62.
10	Michael:	Okay, okay.
11	Holly:	Do you understand it?
12	Michael:	(nods) [Yes].

As this exchange illustrates, Holly was obliged not only to give an explanation, but to ensure that Michael understood how she had solved the task (or at least indicate that he did). Further, although she had the right to decide when he did not understand and needed assistance, he had the right to decide whether or not an explanation she gave was adequate.

The interactional sequence just illustrated occurred repeatedly as Michael and Holly worked together. In its three phases, (a) either Michael could not solve a task or Holly judged that he did not understand, (b) Holly gave an explanation, and (c)

Michael indicated acceptance or nonacceptance of her explanation. This sequence can be contrasted with certain regularities that were identified in Ryan and Katy's interactions. Recall that Ryan typically challenged Katy's explanations when she intervened after inferring that he had made a mistake. In doing so, Ryan was questioning Katy's view of herself as the mathematical authority in the group. Katy, for her part, often adapted her explanations to accommodate to Ryan's challenges, thereby contributing to the emergence of interactions involving multivocal explanation. These often contentious exchanges were potentially productive for both children in terms of the learning opportunities that arose. In contrast to these interactions, Holly and Michael had institutionalized their respective roles as the more and the less mathematically competent members of the group. For example, Michael did not question Holly's assumptions that he did not understand, but instead attempted to accommodate to her mathematical activity. In general, whereas Ryan and Katy frequently engaged in multivocal explanation, Holly and Michael's interactions were typically univocal. As a consequence, they did not experience the learning opportunities that arise in interactions that involve argument and counterargument.

In considering the learning opportunities that did arise for them, it can be noted that Holly might, on occasion, have become increasingly aware of aspects of her mathematical activity when she explained her solutions to Michael. However, she did not seem to consciously decenter and attempt to infer the possible intent of Michael's mathematical activity before judging that he did not understand. Instead, her own interpretation seemed to be the sole measure against which she made these judgments. This limited her opportunities to learn by comparing and contrasting their mathematical activity.

Learning opportunities could arise for Michael as he participated in these interactions when he attempted to fulfill his obligation of understanding her explanations. For example, in the first of the two sample episodes that have been discussed, he began to anticipate certain steps in Holly's partitioning algorithm. The crucial issue concerns the nature of that learning; did his attempts to participate in Holly's solutions involve the partitioning of numbers experienced as mathematical objects, or was he merely partitioning two-digit numerals? This question is addressed later, when his mathematical learning is analyzed in more detail.

It is important to stress that although the interactional sequence previously outlined occurred with some frequency, it was typically limited to those occasions in which Michael either did not have a way to proceed, or his interpretation conflicted with Holly's and she assumed that he did not understand. Typically, Michael was not obliged to listen to Holly's explanations in other situations. As an example, consider an exchange that occurred on January 14 when they solved the horizontal sentence $27 + 9 = \_$. Holly solved the task on her own while Michael went to get a bag of multilinks. She began to explain her solution to him when he returned.

    1 Holly: Do you know why it's 36? Listen, 28, 29 . . . 36 (counts on her fingers). I counted on my fingers, 36.

2 Michael: (Begins to put out collections of multilinks.)
3 Holly: 36, OK. Watch, 28, 29 . . . 36 (counts on her fingers).

Holly expected Michael to accept her solution, and she explained it for a second time when he failed to do so. Again, Michael did not respond but instead continued to try and solve the task by using the multilinks. At this point, Holly began to observe and eventually contributed to his solution, thus accommodating his mathematical activity.

This episode is representative in that Michael was generally not obliged to abandon a solution attempt and listen to Holly's explanations unless she judged that he did not understand. Further, the manner in which Holly gave up her attempt to explain her solution illustrates that, in general, she accepted that it was his right to solve tasks himself when, in her judgment, he was capable of doing so. By way of contrast, the analysis of Ryan and Katy's small-group interactions indicated that there was a continual tension between Katy's attempts to explain her solutions and thus assist Ryan, and Ryan's attempts to solve tasks for himself. The social relationship that Holly and Michael had established was such that Holly, the mathematical authority in the group, could only intervene if she inferred that Michael was not able to solve a task successfully on his own. They therefore dealt with the tension between the obligation for each to develop their own solutions and that of explaining their thinking by giving priority to the first of these two obligations. This, it should be noted, was a reciprocal arrangement in that Michael typically waited for Holly to complete her solution on the infrequent occasions that he arrived at an answer first.

The analysis of Ryan and Katy's small-group interactions also indicated that apparently fortuitous learning opportunities frequently arose for both children. These were occasions when something that one child said or did happened to be significant at that particular moment when interpreted by the other within the personal context of his or her ongoing activity. Learning opportunities of this kind appeared to be far less common in Michael and Holly's interactions. Michael repeatedly observed Holly solve tasks, but her relatively sophisticated solutions often seemed to be beyond his realm of developmental possibilities. Recall, for example, that there was no indication from his January interview that he might construct either a partitioning algorithm or ten as an abstract composite unit. On occasion, he actually attempted to solve a task in the way that she might, but gave up when things did not make sense in terms of actions on mathematical objects. Conversely, Michael's activity was usually very routine from Holly's point of view and, as a consequence, she rarely reorganized her own activity when making sense of what he was doing.

As noted at the outset, Holly and Michael's social relationship was stable and relatively harmonious when compared with that of Ryan and Katy. Thus, at first glance, their interactions seemed to better exemplify what is typically meant by cooperating or collaborating to learn. However, the analysis of the learning opportunities that arose suggests that this was not the case. The univocal interactions in which they participated seemed far less productive for their mathematical development than exchanges characterized by argument and counterargument.

Mathematical Learning

Holly. Holly's solutions to number–sentence and worksheet tasks during the January interview indicated that the construction of a partitioning algorithm was within her realm of developmental possibilities. In addition, she created abstract composite units of ten by relying on situation-specific figural imagery when she solved the strips-and-squares tasks. It is against this background that her learning as she worked with Michael is considered.

As noted in passing when discussing the children's social interactions, Holly used the partitioning algorithm routinely when she solved balance tasks on January 13. For example, she explained to Michael that she had solved a task corresponding to $35 + 45 = \_$ as follows:

1 Holly: I'll explain. See, you take the 5 off the 35 and you make it 30. If you take the 5 off the 45 you make a 40, and that equals 70. And you do it like this. You take the 40 and the 30 and you put them together and you get 70, like 3 plus 4 equals 7 so 40 plus 30 equals 70, and add 5 more, that equals 75, add 5 more, that equals 80.
2 Michael: I agree.

In giving this explanation, she invoked the metaphor of acting in physical reality (e.g., "take the 5 off," "you put them together"). This suggests that the act of partitioning a two-digit numeral symbolized the partitioning of a number experienced as an arithmetical object for her.

Holly used this algorithm in all the sessions involving balance tasks, but did not use it in any other type of task. For example, on January 14, she solved a sequence of addition number sentences by either counting by one or using multilinks. It will be recalled that Ryan's partitioning solutions were also limited to balance tasks. This suggests that both children gave different meanings to two-digit numerals when they interpreted balance tasks and, say, number sentences. Thus, it would seem that they established numbers as arithmetical objects that could be conceptually manipulated only when they interpreted the notation used to present balance tasks.

In this regard, one metaphor that was generally taken as shared by the classroom community when discussing balance tasks was that of a number in a box. For example, the teacher and students often spoke of the goal of a balance task to be that of finding the number that went in an empty box. Holly, in fact, used this metaphor on January 13 when she explained her interpretation of a task to Michael: "This is a take-away box, this is a take-away." One characteristic of the box metaphor is that its use implies the bounding of whatever is placed inside it (Johnson, 1987). Consequently, Holly's and Ryan's interpretation of tasks in terms of this taken-as-shared metaphor may have supported their construction of numbers as discrete, bounded entities composed of abstract units of one. To the extent that this was the case, their creation of relatively sophisticated arithmetical entities was supported by a task-specific convention of interpretation. This conclusion does not, of course, imply that a taken-as-shared metaphor carries with it a mathematical meaning that is self-evident to the adult; as will be seen when discussing Michael's mathematical learning, numerals in boxes did not signify discrete arithmetical objects for him. Instead, it seems that a metaphor such as that of a number in a box can support some students' active, image-supported construction of arithmetical objects, but that the mathematical significance of a metaphor is relative to a student's conceptual possibilities.

With the exception of her partitioning solutions to balancing tasks, Holly's solutions when she established collection-based numerical meanings tended to be relatively unsophisticated and to involve the use of multilinks. In contrast, she created abstract composite units of ten on several occasions when she established counting-based meanings. The first such solution occurred on January 26 and seemed to involve situation-specific imagery. She gave the following explanation to a researcher after solving the strips-and-squares task, "You start with 73 and take away II:: [24]. How many are left?"

1 Holly: 73 take away two 10s, you get 53.
2 Researcher: Uh huh [Yes].
3 Holly: And take away 1 more, and take away 4 more, and you get . . . 49.

In this solution, which involved the curtailment of counting, Holly seemed to organize the 24 she would count back into abstract composite units of ten. Significantly, she solved the very next task in which the minuend was given as a picture and the subtrahend as a numeral by counting by ones on a hundreds board. The task in question was, "You start with IIIII:: [54] and take away 35. How many are left?" The use of a picture to show the 24 she would count back in the first task, therefore, seemed to play a crucial role in supporting her construction of abstract composite units. In this regard, her solution was similar to those she produced in the January interview.

One week later, on February 2, Holly again created abstract composite units of ten when she and Michael worked with Katy (Ryan was absent from school). During the session, Katy routinely counted by tens and ones on the hundreds board. For example, she explained her solution to a balance task corresponding to 37 + 25 = _ as follows:

1 Katy: (Points on the hundreds board) 25 plus 10 makes 35, plus 10 makes 45, plus 10 makes 55. 1, 2, 3, 4, 5, 6, 7 . . . 62.

As it so happened, Holly had made three collections of 10 tally marks and 7 individual marks when she interpreted Katy's solution. This enabled her to make sense of Katy's activity of counting by tens and ones, and she began to solve tasks by creating and counting abstract composite units of ten. For example, she subse-

3. LEARNING AND SMALL-GROUP INTERACTION 69

quently explained to Michael that she had solved a balance task corresponding to 37 + 57 = _ as follows:

1 Holly: (Points on the hundreds board) I started at 57 and I counted down 10, 20, 30, and then I counted these 1, 2, 3 . . . 7 and I came up with 94.

Although solutions of this type were clearly within the realm of her developmental possibilities, she counted exclusively by ones when she used the hundreds board to solve strips-and-squares tasks the following day (February 3). It would, therefore, seem that the advances she made on February 2 were specific to her interactions with Katy and that, further, her grouping of tally marks into collections of 10 might have been crucial. In particular, these collections gave her a basis in imagery that she could use to make sense of Katy's solutions.

Imagery also seemed to play a role in two of the remaining three solutions in which Holly created abstract composite units while establishing counting-based numerical meanings. One of these solutions was described when discussing regularities in the children's social interactions. Then, it was noted that, on March 2 Holly counted by tens and ones to solve the money story problem, "Krista has this much money. [Picture shows coins whose total value is 37¢.] She has 25¢ less than Larry. How much money does Larry have?"

1 Holly: 37—47, 57, 58, 59, 60, 61, 62—62.
2 Michael: This is weird.
3 Holly: Add 10 cents, that makes 47, then add another 10, 57, then 58, 59, 60, 61, 61.
4 Michael: Okay, okay.
5 Holly: Do you understand?
6 Michael: (Nods) [Yes].

Holly's explicit reference to coin values in her second explanation ("add 10 cents") indicates that situation-specific imagery supported her structuring of 25 into units of ten and of one. Later in the same session, she used a hundreds board to solve the task, "Scott has this much money. [Picture shows coins with total value of 42¢.] He has 30¢ more than Susan. How much money does Susan have?" To do so, she first counted the values of the pictured coins and then reasoned:

1 Holly: Then take away 42, 41, 40, 39, 38, 37, 36, 35, 34 . . . I can't do this. We need the hundreds board (picks one up). 42 (points on the hundreds board)—12 (seems to move her finger up a column).

The final occasion on which Holly created counting-based composite units of ten again involved the use of the hundreds board. On March 23, she explained that she had solved a balance task corresponding to 16 + 16 + 16 = _ as follows:

2 Holly: Listen, you start on 16 [On the hundreds board], and you go down 1 10, go to 26, then you go down to 36, then you count 1, 2, 3 ... 12, you count 12, count 6 more, 1, 2, 3, 4, 5, 6, and you have 54. I don't get this.

Presumably, Holly did not "get it" because 54 was not the answer that she had produced when she previously counted by ones to solve the task. The important feature of this solution is that she attempted to count by tens and ones. As has already been observed, Holly had used the partitioning algorithm to solve balance tasks since the beginning of January. Consequently, her interpretation of 16 as 10 and 6 was relatively routine. Thus, as was the case on the other occasions when she counted on the hundreds board by tens and ones, she established abstract composite units of ten when she interpreted the task and then expressed them by counting.

In summary, the manner in which Holly created abstract composite units when she gave linear, counting-based meanings to number words and numerals was consistent with her performance in the January interview. More generally, she did not appear to make any major conceptual advances while working with Michael despite repeated indications that certain constructions were within her realm of developmental possibilities. In this regard, the analysis of the children's social interactions indicated that the learning opportunities that arose for Holly when she and Michael engaged in univocal explanation were relatively limited. Significantly, Michael rarely challenged her explanations or requested further clarification by asking questions that would oblige her to elaborate her explanations. In contrast, Katy frequently adjusted her explanations to accommodate Ryan's challenges. It is possible that many of the solutions that Holly attempted to explain might have been beyond Michael's realm of developmental possibilities and that he was therefore unable to participate by challenging or by asking clarifying questions. In any event, a comparison of the learning opportunities that arose for Holly and for Katy indicates that the social context within which children give explanations can profoundly influence the contribution that those experiences make to their conceptual development.

Michael. There was no indication that Michael either partitioned numbers or created abstract units of ten in even a highly image-dependent manner during the time that he worked with Holly. For example, his observations of Holly and, during one session, Katy counting by tens and ones on the hundreds board did not lead to a significant conceptual advance. He did, however, consciously attempt to solve tasks in this way on two occasions. The first of these attempts occurred on March 5 when he solved the task, "How many do you add to 37 to get 91?" He first began to count by ones on a hundreds board but suddenly stopped, saying "I've got a faster way." He then counted "1, 2, 3 ... 9" as he pointed on the hundreds board to the squares 38, 39, 40, 50, 60, 70, 80, 90, and 91. Next, he counted (and pointed to squares): "1 (38), 2 (39), 10 (40), 20 (50), 30 (60), 40 (70), 50 (80), 60 (90)." Finally, he decided to check his answer by counting by ones and arrived at an answer of 54.

3. LEARNING AND SMALL-GROUP INTERACTION 71

Michael's goal in these solution attempts was to trace a path from 37 to 91 on the hundreds board. However, only the column "10, 20, 30 . . ." seemed to be structured in terms of tens for him. In order to learn to count by tens and ones on the hundreds board while watching others, he would have to come to the realization that moving down a column from, say, 37 to 47 was a curtailment of a count of the squares 38, 39, 40 . . . 47. Only then could it signify an increment of 10. A pause or break in the rhythm of counting by ones does occur after 40 when the child moves from the end of one row to the beginning of the next. A difficulty arises because there are no similar pauses either before 37 or after 47. Consequently, in contrast to the count of an intact row (e.g., 51, 52 . . . 60), there are no breaks in sensory-motor activity that would facilitate the isolation of a count of the squares 38, 39, 40 . . . 47 as a discrete, conceptual entity. It was this construction that appeared to be beyond the realm of Michael's developmental possibilities.

Michael's attempts to make sense of the sophisticated collections-based solutions that Holly produced seemed equally unsuccessful. It has already been noted that he attempted to participate when Holly used her partitioning algorithm to solve balance tasks on January 13. Then, he made several interjections that indicated that partitioning a two-digit numeral might signify the partitioning of a number as an arithmetical object for him. For example, the following exchange occurred as they solved a task that they interpreted as $36 + 46 + 10 = \_$.

1 Holly: I'll explain. That's 70.
2 Michael: Plus 6 more and 6 more equals 12, plus 12 more.

Here, the partitioning of 36 and 46 into 30 and 6, and 40 and 6 respectively, was taken as shared. Michael also anticipated a step in Holly's solution when she began to explain it for a second time.

3 Holly: No, no, no, you don't understand. See, there's 70 to start with because there's 30 plus 40 equals 70. Add 10 more equals 80, 86 . . .
4 Michael: 87, 88, 89, 90 . . .
5 Holly: (Interrupts) 92.

The possibility that Michael made a major conceptual reorganization while participating in these exchanges seems unlikely when his solutions in subsequent episodes are taken into account. For example, 2 months later, on March 23, Holly attempted to solve a balance task corresponding to $16 + 16 + 16 = \_$ by counting on a hundreds board by tens and ones. Michael then counted on the hundreds board by ones, but his answer conflicted with Holly's.

1 Michael: I think that I can settle this. 6 and 6 is 12 (attempts to count 6 more on the hundreds board)—19. 19 plus 3 ones, 20, 21, 22, so the answer must be 22.

When Michael interpreted Holly's count by tens and ones, he partitioned the numeral 16 per se, producing a 1 and a 6 as the result. This suggests that the accommodations he made during the exchange on January 13 might have involved noticing that she called the results of partitioning the numeral 36 as 30 and 6 rather than 3 and 6. Although there was no reason why she might do this from his perspective, he went along with it so that he could participate in the development of a joint solution.

Thus far, Michael's conceptual capabilities have been characterized almost exclusively in negative terms. An indication of his developmental possibilities can, however, be inferred from an exchange that occurred on January 14. Here, Michael attempted to solve $47 + 19 = \_$ by using multilinks while Holly counted on and arrived at the answer of 66. He first made two collections, one of 4 ten-bars and the other of a ten-bar and a nine-bar, and then counted them to get 59. At this point, Holly, in her role as the mathematical authority, intervened.

1 Holly: You forgot the extra 9.
2 Michael: (Makes a nine-bar.)
3 Holly: Not the 9, not the 9, not the 9. I mean 7. That's 47 (pats the collection of 4 ten-bars).
4 Michael: (Breaks 2 multilinks off the nine-bar) 47.

Finally, they counted the multilinks and agreed that the answer was 66. They then solved several other tasks together and, in the process, Holly initiated Michael into a mathematical practice that was generally taken as shared by the classroom community—that of establishing collections of multilinks arranged in bars of 10 as a way of giving meaning to two-digit number words and numerals. This practice seemed to make sense to Michael, and he routinely used the multilinks in this way for the remainder of the school year. This, of course, does not imply that the bars of 10 multilinks signified abstract composite units for him. The difficulties he experienced in other situations in which he established collection-based numerical meanings suggests that the composite units he established when he manipulated bars of the multilinks were perceptually based; counting bars 10, 20, 30 . . . was a curtailment of the count of the individual multilinks by one. Michael's initiation into this classroom mathematical practice was the most significant advance he made while working with Holly after the January interview.

Summary

The analysis of Michael's and Holly's interviews indicated that there were significant differences in their conceptual capabilities. These differences were clearly reflected in their social relationship in that Holly was the mathematical authority. In this regard, their relationship differed from that of Ryan and Katy. There, Ryan, the less conceptually advanced of the two children in January, contested Katy's view of herself as the authority in the group. One of the ways in which he did this was by challenging Katy's thinking when she attempted to explain her solutions to him. Although Michael's failure to make such challenges cannot be accounted for exclusively in cognitive terms, it was noticeable that the solutions Ryan challenged were frequently within his realm of developmental possibilities. In contrast, those Michael accepted without challenge often proved to be beyond his realm of developmental possibilities. Obviously this observation is something of a truism—it is easier to challenge ideas that make some sense. This, of course, is not to say that the relatively small differences in Ryan and Katy's conceptual possibilities when compared with those of Michael and Holly meant that Ryan was destined to challenge Katy's thinking and Michael to view Holly as a mathematical authority. Instead, the point is that, other things being equal, it was more difficult for Michael than for Ryan to challenge his partner's mathematical activity and thus contribute to the establishment of situations for multivocal explanation.

The analysis of Ryan and Katy's social interactions indicates that multivocal explanation can be relatively productive in terms of the learning opportunities that arise. In comparison, the univocal explanation that Michael and Holly engaged in seemed far less productive. Further, the analysis of Michael and Holly's social relationship indicates that learning opportunities do not necessarily arise for children in small-group work merely because they explain their solutions to others; it seems crucial to consider the social situations within which they develop their explanations. Situations in which it is taken as shared that the explainer is an authority appear to be much less productive than those in which the explainer attempts to accommodate to anticipated or actual challenges and criticisms. By the same token, the analysis of Michael's mathematical activity indicates that learning opportunities do not necessarily arise for students in small-group work when they listen to another's explanations. This can be the case even when the explainer takes seriously his or her obligation to help the listener understand, as did Holly.

This case study of Michael and Holly's mathematical activity indicates that the differences in their developmental possibilities constrained the nature of their social relationship. The analysis also indicates that the learning opportunities that arose as they worked together were relatively limited. As a consequence, neither the differences in their conceptual capabilities nor the general nature of their social relationship changed. Instead, their conceptual possibilities and the social relationship they established tended to stabilize each other.

CASE STUDY—ANDREA AND ANDY JANUARY 13–MARCH 23

Andy: January Interview

Andy counted on to solve missing addend tasks during the first part of the interview, indicating that he could create and count abstract units of one. However, he did not appear to create abstract composite units of ten when he solved the uncovering and strips-and-squares tasks. Nonetheless, learning opportunities arose for him as he interacted with the interviewer. For example, on the first uncovering task, two strips were uncovered when he had reached 52. He counted the squares on the strips by ones and gave 71 as his answer. At this point the interviewer intervened:

1 Interviewer: (Points to the first strip uncovered) What's 52 and 10 more?
2 Andy: (Counts the individual squares) 53, 54, 55 . . . 62.
3 Interviewer: And another 10 (Points to the second strip.)
4 Andy: 72.

Andy also curtailed counting by ones with the interviewer's support when he solved the second uncovering task. Then, he first counted two strips by ones after he had reached 14, giving 36 as his answer. The interviewer next uncovered a strip and two squares, asking "What's 36 and 10 more?" Andy immediately replied "46," and then counted the two squares. Finally, two strips and five squares were uncovered and, without further prompting, Andy counted "49, 50, 51, 52, 53, and then another, 63, and then 73."

It will be recalled that Michael made similar advances while solving the uncovering tasks. There, it is argued that he counted strips as single entities rather than composites of 10 units of one. Certain of his solutions appeared to be relatively sophisticated when considered in isolation, because he learned to differentiate between the number–word sequences he should use to count strips, and those he should use to count single squares. In contrast, the advance Andy made appeared to involve the construction of numerical composites of ten. For example, he was asked to solve a strips-and-squares task in which 3 strips and 5 squares [35] were visible, there were 75 squares in all, and he was to find how many were hidden. He counted "35—45, 55, 65, 75," each time raising and lowering both hands with all 10 fingers extended, and then moving a plastic numeral to record his counting act. He then completed his solution as follows:

1 Andy: 4 more, 40. I counted 4 times 10, 10, 20, 30, 40; I counted 4 times.

Andy's 10 extended fingers seemed to signify a composite of 10 units of one rather than 1 unit of ten. As a consequence, he had to devise a relatively cumbersome way to keep track of his counting, whereas children who create abstract composite units of ten typically put up 1 finger to record each act of counting by ten.

Andy solved a subsequent task in a similar way, but experienced difficulties when the last strips-and-squares task was presented. In this task, 3 strips and 6 squares [36] were visible, there were 61 squares in all, and he was asked to find how many were covered. As before, he moved his open hands and used plastic numerals to make records. He counted "36—46, 56, 66" and said that "3 strips and 1 extra" were covered. The interviewer then removed the cover to reveal 2 strips and 5 squares, and Andy said that the task did not make sense. Thus, the possibility of creating a numerical composite of ten from the individual squares of the visible and screened collections did not arise for him. Andy's failure to create abstract

3. LEARNING AND SMALL-GROUP INTERACTION

composite units of ten while solving the strips-and-squares tasks is consistent with the way in which he attempted to solve all the horizontal sentence and worksheet tasks presented by counting by ones. The interviewer did intervene to help him to develop a more sophisticated solution to $28 + 13 = \_$, but without success. Andy said that 28 plus 10 was 38, but then counted on by ones from 28 when the original task was posed, giving 40 as his answer. Thus, it would seem that neither the partitioning algorithm nor the construction of abstract composite units of ten were within his realm of developmental possibilities. However, the advances he made earlier in the interview indicate the construction of image-supported numerical composites of ten was a developmental possibility for him.

Andrea: January Interview

Andrea's solutions to the counting tasks were very similar to those that Andy developed and, as in his case, it is inferred that she could create and count abstract units of one. Her solutions to the uncovering and strips-and-squares tasks indicate that she, like Michael, learned to count strips as single entities rather than as composites of 10 ones. For example, the first strips-and-squares task presented asked her to find how many squares were hidden given that 3 strips and 3 squares [33] were visible, and that there were 63 squares in all. When she said that this was "too tough," the interviewer asked her to solve a similar task with 3 strips [30] visible and 70 in all.

1 Andrea: [No response].
2 Interviewer: Okay, how many do we have here? (Points to the visible strips.) Just count them.
3 Andrea: 10, 20, 30.
4 Interviewer: And there's 70 all together.
5 Andrea: 40, 50, 60, 70 (puts up a finger each time she counts)—4 of tens.
6 Interviewer: How many ones?
7 Andrea: I'm not sure . . .
8 Interviewer: (Removes the cover.)
9 Andrea: None.

In comparison with the other interviews, the extent to which the interviewer had to intervene before she could begin to solve this relatively elementary task was quite striking. The possibility of counting imagined strips did not seem to arise for her until she had counted the visible strips at the interviewer's direction. Interpreted within the context of her other solutions, her answer of "4 of tens" seems to mean that she had counted 4 single entities, each of which is called a ten. The inference that she counted the squares as singletons is also consistent with her response that there were no ones when the cover was removed.

The interviewer next re-posed the task he had asked at the outset, that in which 3 strips and 3 squares [33] were visible and there were 63 squares in all. This time,

Andrea put up fingers as she counted, "33—43, 53, 63—3 of tens." She was then asked to solve a similar task in which 4 strips and 3 squares [43] were visible and there were 65 squares in all. Here, in contrast to the previous tasks, 43 and 65 were not in the same counting-by-tens number–word sequence. Andrea dealt with this novelty by ignoring the individual squares and counting, "40—50, 60—so there's 2 tens in there and none of ones." In this case, she attempted to be effective in the interview by using a tens number–word sequence to count strips as single entities. In general, there was no indication that the construction of either numerical composites of ten or abstract composite units of ten were within her realm of developmental possibilities.

Andrea solved the first two horizontal sentences presented, $16 + 9 = \_$ and $28 + 13 = \_$, by counting on by ones, but attempted to solve the sentences involving larger numbers by mentally placing the numerals in columns. When she attempted to use this approach to solve $39 + 53 = \_$, she added the four individual digits to arrive at an answer of 20. On the worksheet tasks, she used the standard algorithm which, she explained, had been taught to her by a sixth-grade friend. She made just one error while using this algorithm, putting the "1" in the ones place and carrying "2" when she solved the task corresponding to $39 + 53 = \_$. This and her solution to the horizontal sentence $39 + 53 = \_$ suggest that the activity of doing arithmetic did not necessarily have to involve acting on actual or symbolized arithmetic objects for her; instead, it could at times be a matter of following procedural instructions. By January of the school year, she was the only child in the class who seemed to hold instrumental beliefs of this type.

Regularities in Small-Group Activity

In contrast to the prior two case studies, neither Andrea nor Andy attempted to act as a mathematical authority. Further, they both considered that it was their right to develop their own solutions. This is illustrated by their indifference to Holly's escalating sequence of attempts to assist them during the one session in which they worked with her on March 4. Here, the task was $\_ \times 3 = 18$.

1	Holly:	6. It takes 6. . . . It makes . . . it takes 3 6 times to make it, make it 3.
16	Holly:	By sixes, by sixes. . . . By sixes guys. . . . Try sixes. Sixes worked for me.
25	Holly:	It's sixes! It's by sixes. . . . By sixes guys! It's by sixes. . . . You don't believe me. It's by sixes.

Holly then solved the task again, writing 6 threes on a piece of scrap paper as she did so.

| 37 | Holly: | (Standing up) By sixes. (Holds her scrap paper out for Andy to see.) Count them. (Andy is not looking.) Okay. (Sits down.) |

3. LEARNING AND SMALL-GROUP INTERACTION

38 Andy: Yeah, but you're probably not doing it right.

As Andy's final comment indicates, neither he nor Andrea accepted that Holly had the right to interrupt their solution attempts. More generally, the analysis of Andrea and Andy's interactions indicated that there was an implicit agreement that neither child could begin to solve the next task until the other had indicated acceptance of his or her answer to the current task. As a consequence, they rarely established situations for either direct or indirect collaboration when they solved tasks.

In contrast to the other seven children whose mathematical activity is analyzed in these case studies, Andy did not always attempt to solve the assigned mathematical tasks. Instead, he appeared to have a variety of alternative agendas, some of which were mathematical in nature. For example, for several weeks he made drawings to record his and Andrea's solutions. In doing so, he typically drew a large rectangle to symbolize a ten and a small square to symbolize a one. As he explained to the researcher on February 3:

1 Andy: I only do pictures just to remind us, because we might need one to figure out the other one, that's why I make pictures of them.

While he was working on these drawings or arranging them in a file he had constructed, Andrea would typically solve tasks on her own with his tacit approval. On other occasions, he simply decided to take a break from the solving tasks. For example, in one instance he announced to Andrea, "I'm going to give my brain a rest," after he had spent a considerable amount of time constructing what was for him a novel solution.

Although there did not seem to be a strict pattern to the occurrence of those occasions when Andy engaged in what is traditionally called *off-task behavior*, certain general tendencies that clarify his motivations could be observed. For example, it was noticeable that he rarely solved several consecutive tasks that were routine for him. Thus, he never solved a sequence of addition tasks by counting on a hundreds board or by making and then counting collections of multilinks. He did, however, frequently persist in his attempts to solve personally challenging tasks. He also tended to switch his attention from one of his other interests to that of solving an assigned mathematical task when he noticed that Andrea was having difficulty. In addition, he seemed to engage in an activity that he found intellectually stimulating when he constructed his drawings. Thus, although it would be an exaggeration to characterize Andy as a student who constantly sought intellectual challenges, it is reasonable to say that he shied away from the routine.

Andrea, for her part, seemed to assume that it was Andy's right to do something other than solve the assigned tasks if he so chose. They did, however, go through a ritualistic exchange once she had arrived at an answer in which she solicited Andy's agreement, and Andy, who typically did not listen to her explanation, indicated that he accepted her solution. This formality completed, Andrea could then begin to solve the next task. Obviously, no learning opportunities could occur in exchanges of this sort.

The potentially most productive occasions for learning occurred when their independent solutions were in conflict. Here, they challenged each other's explanations, thus engaging in multivocal explanation. However, two aspects of their mathematical activity made it difficult for them to communicate effectively in these situations. The first of these aspects concerns their differing beliefs about mathematical activity. As was noted when analyzing Andrea's January interview, conventional symbols did not necessarily have to signify anything beyond themselves for her. There were strong indications that she continued to hold this belief for the entire time she worked with Andy. In contrast, there was every indication that acts of manipulating symbols had to carry the significance of acting on mathematical objects for Andy. The difficulties that arose as a consequence of these differences in their beliefs about mathematical activity can be illustrated by considering an exchange that occurred on February 16. Andrea had solved a task that involved adding 5 twelves by using the standard addition algorithm. She explained her solution to Andy as follows:

1 Andrea: 2, 4, 6, 8, 10 (adds the 5 twos in the ones column). You put the "oh" there (writes "0" in the ones place), the 1 there (writes "1" at the top of the tens column). 1, 2, 3, 4, 5, 6 (counts the 6 ones in the tens column). Do you agree that's 60?

From Andy's point of view, "That's impossible." As the exchange continued, he repeatedly challenged Andrea's justifications, thus contributing to the establishment of a situation for multivocal explanation. Her solution was unacceptable to him because he did not understand how 10, the sum of the ones column, could become an "oh" and a 1.

17 Andrea: 2, 4, 6, 8, 10. Now listen. Put that "oh" there.
18 Andy: Yeah.
19 Andrea: The one up there.
20 Andy: That isn't "oh."
21 Andrea: 1, 2, 3, 4, 5, 6.
22 Andy: That's not "oh." That's 10.
23 Andrea: No.
24 Andy: That isn't "oh" though. That's 10. 1, 2, 3 . . . 9, 10.
25 Andrea: Now listen.

Andrea repeatedly described the steps of the standard algorithm in a similar way in response to Andy's challenges. For her, a procedural description of this kind constituted a mathematical justification. She therefore concluded from Andy's failure to accept her solution that he was not listening. On the other hand, from Andy's point of view, Andrea had not adequately justified why she wrote "0" and "1."

At this point, the children had reached an impasse. Andrea therefore sought the teacher's assistance and they both explained their interpretations of the conflict to her.

3. LEARNING AND SMALL-GROUP INTERACTION 79

27 Andrea: He doesn't—he doesn't get. He doesn't get that right.
28 Andy: (Simultaneously) I don't understand. All the twos add up to 10. She takes 1 away from the 10.

Andy's last statement indicates that he interpreted Andrea's action of writing a 1 at the top of the tens column as taking away a 1 from the 10. As a consequence, her solution seemed irrational to him. Andrea responded to his challenge by again repeating her procedural description.

30 Andrea: 2 plus 2. 2, 4, 6, 8, 10. I know that's 10.
31 Andy: It is 10.
32 Andrea: You put the darn "oh" there. 1, 2, 3, 4, 5, 6.
33 Andy: (Simultaneously) Show me how you are doing it. I don't understand.

Andy's comment "show me how you are doing it" was an explicit request for an explanation. Consistent with his January interview, it seemed beyond his realm of developmental possibilities to interpret Andrea's procedural descriptions in terms of actions on arithmetical objects. Conversely, Andrea was unable to do more than specify how she manipulated numerals. From her point of view, this should have been sufficient to legitimize her solution. As a consequence, the two children arrived at an impasse that they were unable to resolve. Their differing beliefs about what counted as a mathematical explanation made it impossible for them to communicate effectively and negotiate a taken-as-shared interpretation of Andrea's solution in this situation for multivocal explanation. Consequently, a conflict in their interpretations that might have given rise to learning opportunities instead proved unproductive and, at times, resulted in feelings of mutual frustration.

The second aspect of their mathematical activity that limited learning opportunities concerns their differing interpretations of particular tasks. As an illustration, consider the following exchange that occurred on February 3. Andy had solved the task, "How many do you take away from 74 to leave III::. [35]?" by first drawing a picture for 74. He had then reasoned that he would have to take away 4 tens and add a one to transform this drawing into one for 35.

1 Andy: Listen, listen. You leave 30 right (points to his drawing of 74). You take away 4 [tens] from 70, and you add a one.
2 Andrea: That would be 75, baby [i.e., 74 plus one]!
3 Andy: No 3, 4 (points to the ones in his drawing). It's 4, there is already 4 [ones] there! There is already a 4 there, see? And then you would add 1. 74! So you take the 4 [tens] from 74.
4 Andrea: Take away 1?
5 Andy: Listen, you don't take away 1, that will leave you 3 [ones] and then you would have 33.

Clearly, the two children seemed to be talking past each other in this exchange. The fact that they did not differentiate linguistically between the rectangles that signified 10 and the squares that signified 1 certainly did not improve their chances of communicating successfully. However, as the episode continued, it became apparent that there was an additional difficulty.

15	Andy:	Okay, take away 4 [tens], take away these 4 (points to his drawing). You take away 4 [tens] and you add 1 [one]. That will be 5 [tens] just like over here (points to the diagram on the activity sheet), just like 5.
16	Andrea:	I don't get it.

The teacher then joined them and Andy subsequently tried for a third time.

32	Andy:	What I did was, like I drew 70 and 4 more.
33	Teacher:	Right.
34	Andy:	You add a one and you take away 4 tens.
35	Andrea:	He put a "w" on the ones [referring to the fact that Andy had spelled one as wone].
36	Teacher:	[Referring to his spelling] That's okay, I understand what he's saying, and you do too, don't you?
37	Andrea:	(Makes a gesture with her hand indicating that she is uncertain about Andy's explanation.)

As Andrea's final gesture indicates, she had no way of knowing whether his solution was reasonable. One of Andrea's routines in such situations was to try and change the topic of conversation, as she did in this case by mentioning Andy's spelling. The analysis of her mathematical learning indicates that she was often unable to construct task interpretations compatible with those that were taken-as-shared by the classroom community. This was particularly evident for certain balance and strips-and-squares tasks. In the sample episode, for example, it is doubtful that she ever conceptualized the task as that of finding out how many she would have to take away from 74 to leave 35. In the absence of an adequate basis for mathematical communication, their participation in multivocal explanation did not give rise to learning opportunities for either child.

In summarizing the observations made about Andrea and Andy's social relationship, it can first be noted that neither acted as or viewed the other as a mathematical authority. On the one hand, they rarely established situations for either direct or indirect collaboration, but instead waited for the other to complete his or her independent solution attempt. On the other hand, they typically challenged each other's explanations once they had both arrived at answers, thus contributing to the emergence of interactions involving multivocal explanation. Recall that Ryan and Katy also tended to engage in interactions of this type and that these were generally productive for their mathematical learning, whereas those in which Andrea and Andy participated were not. This difference can be attributed to the difficulties that

Andrea and Andy had in establishing an adequate basis for mathematical communication. For Andrea, arithmetic could involve manipulating symbols that did not necessarily signify anything beyond themselves. In contrast, for Andy, arithmetic was an activity that involved acting on experientially real numerical objects. These differing beliefs about mathematical activity greatly hampered their attempts to negotiate taken-as-shared understandings. In addition, attempts to make sense of each other's arguments were often unsuccessful for conceptual reasons. For example, Andy was frequently unable to interpret Andrea's procedural solutions in terms of actions on arithmetical objects. Conversely, Andrea often seemed unable to conceptualize tasks in ways compatible with the taken-as-shared interpretations of the classroom community. As a consequence, she and Andy often seemed to talk past each other.

As a final observation, it can be noted that when compared to Ryan and Katy, few fortuitous learning opportunities arose for Andy and Andrea. The same two reasons advanced to account for the relatively unproductive nature of interactions involving multivocal explanation seem relevant here. In other words, the occurrence of fortuitous learning opportunities is limited by incompatible beliefs about the general nature of mathematical activity and by discrepancies in developmental possibilities.

Mathematical Learning

Andy. Andy's solutions to the uncovering and strips-and-squares tasks in the interview indicated that the construction of ten as a numerical composite was within his realm of developmental possibilities when he could rely on situation-specific imagery. He did, in fact, seem to create composites of this type on several occasions when he worked with Andrea. Typically, these solutions involved either drawings or the use of paper strips and squares to support his image-based constructions. The first such solution occurred on January 27, when he and Andrea solved a sequence of horizontal sentences. Andy initially intended to solve $50 - 9 = \_$ by using his paper strips and squares but the researcher intervened to ask him if he could solve it another way. He then extended all 10 fingers and moved both hands slightly five times as he counted, "10, 20, 30, 40, 50." At this point, he paused and appeared to reflect before saying, "I almost got it." Eventually, he abandoned this solution and instead accepted Andrea's counting-down-from solution.

This solution attempt seemed to involve an advance for Andy, even though he failed to arrive at an answer. In the interview, he counted by ones to solve horizontal sentences, whereas here he created and counted numerical composites of ten. However, his initial intention to use paper strips seemed essential. His 10 extended fingers could then have signified a collection of the 10 ones on a strip, and the movement of his hands the act of counting them. Recall that he made similar movements when he solved covered collections tasks during the interview. The difficulties he experienced once he had counted to 50 also indicate that he created numerical composites of ten. Composites of this type cannot be conceptually reorganized, because the image of a strip or of two open hands is integral to them. However, Andy would have to make precisely this kind of reorganization and break up one of the numerical composites he counted if he were to take away nine.

The other novel solutions that Andy developed also indicated that solutions that involved reorganizing numerical composites were currently beyond his realm of developmental possibilities. For example, his solution to a strips-and-squares task corresponding to 74 − _ = 35 on February 3 has already been described. Then, he first made a drawing for 74 and next argued that 4 tens should be taken away and a one added. However, he did not reconceptualize these 4 tens and the unit of one he added as 39, even after he had explained his solution three times. The difficulties he had in making such a reconceptualization were even more apparent on February 12, when he solved a strips-and-squares task corresponding to 44 + _ = 72. He first made a drawing for 44, then drew 3 more tens, and finally circled 2 of the ones. When questioned by a researcher, he said that he had added 3 tens and taken away 2 ones.

1 Researcher: Okay. And what number would that be?
2 Andy: That would be 72.
3 Researcher: Okay. If you add 3 tens and take away 2 ones, so how many do you have to add then?
4 Andy: You have 3 tens and take away 2 ones.
5 Researcher: And what number is that?
6 Andy: That would be 72 altogether.
7 Researcher: Yes, so what number is that if you add 3 tens and take away 2 ones?
8 Andy: 72, makes 72.

As previously noted, Ryan was able to reconceptualize a solution of this kind when the researcher intervened. Significantly, the construction of abstract composite units of ten was within Ryan's but not Andy's realm of developmental possibilities.

Thus far, the discussion has focused on some of the difficulties that Andy encountered. It is, therefore, essential to stress that he did develop novel solutions that were at least as sophisticated as those observed in the interview. Two solutions in particular illustrate both his creativity and his current conceptual limitations. The first of these solutions occurred on March 5, when he and Andrea attempted to solve the task, "How many do you add to 27 to make 82?" In this case, drawings of strips and squares were not used to present the task, and, initially, Andy did not seem to have a way to proceed. However, he had an insight when the teacher pointed to 27 and then to 82 on a hundreds board as she read the task statement. He immediately put his finger on 27 and moved down the board one square at a time as he counted "1, 2, 3, 4, 5, 6." He then began to make collections of multilinks and eventually announced to Andrea, "Yes! I found out!"

1 Andy: Oh boy . . . 27. Okay, here's 27 (puts 2 ten-bars and a seven-bar on his desk). There's 27. (Looks at the remaining multilinks in

		his lap.) There's another 10. There's another 10. There's another 10. And there's another 10, and a 10 (puts 5 ten-bars on his desk). Okay, and those (places 5 individual cubes on his desk).
2	Andrea:	(Counts the individual cubes) 1, 2, 3, 4, 5.
3	Andy:	Count all this up and it'll make 82.

The insight that Andy had when the teacher pointed to the hundreds board seemed to involve the realization that he could start with a collection of 27 and add tens to make one of 82. He appeared to view each complete row on the hundreds board as a numerical composite of ten and counted down a column to see how many such composites he should make. He then added 5 bars of 10 to a collection of 27 multilinks and, finally, added individual cubes until he reached 82. However, despite the creativity of his solution, there was no indication that he conceptualized the multilinks he added as a single numerical entity. Instead, they seemed to be a succession of tens for him. Thus, he did not count the additional bars and individual cubes to complete his solution, but instead described each bar as another 10.

The apparent difficulty he had in monitoring his activity in this episode can be contrasted with a solution he had produced 2 days earlier on March 3. At that time, he solved a balance corresponding to $25 + 25 + 10 + \_ = 81$ by counting on by ones from 60, writing down "10," "10," and "1" as he did so. He then explained his answer of 21 as follows:

1	Andy:	25 and 25 is 50, then I counted up to that [81] and I got . . . well, I used my fingers, 51, 52, 53 . . . 59, 60, and then I counted another 10 up, and another 1, and I put 1 on it, and 10 and 10 is 20 and I put 1 on it.

As was the case with other solutions in which he created numerical composites of ten, this solution was situation specific. He was able to structure counting by ones into modules of ten because he had already counted from 50 to 60. As before, he spoke of a succession of tens: "I counted another 10 up, and another 1." However, in this case, he made a record of his counting activity, and this enabled him to reconstruct how he had counted. In a previous case study, it was noted that Katy's use of the hundreds board supported her reflection on her counting activity. Both there and in the case of Andy's solution, the use of written notation played an important role in helping the children reflect on their mathematical activity.

As will be seen when Andrea's mathematical learning is discussed, she used the standard addition algorithm increasingly frequently when she worked with Andy. There was no indication that her repeated explanations constituted learning opportunities for him. In a previously discussed exchange, Andy said that he did not understand how Andrea had added 5 twelves and asked her to tell him what she was doing after she had described the steps of her procedural solution several times. He subsequently expressed his views about Andrea's use of the algorithm on several other occasions, including the last session in which they worked together, on March 23. In this instance, Andrea added 30 and 18.

1	Andrea:	"Oh" plus 8 is 8.
2	Andy:	48.
3	Andrea:	So you put the 8 there and 3 plus 1 is 4, so you put the 4 there and 48.
4	Researcher:	Okay and so you . . .
5	Andy:	I don't understand her doing that.
6	Researcher:	Doing that, uh huh.
7	Andrea:	Okay.
8	Researcher:	Okay.
9	Andy:	She's weird.
10	Andrea:	Thank you, Andy.

The difficulty that Andy had in making sense of Andrea's explanations provides a further indication that the construction of ten as an abstract composite unit was not within his realm of developmental possibilities. This inference is also consistent with the fact that he was never observed attempting to count by tens and ones on a hundreds board. The only instance when he used a hundreds board to do anything other than count routinely by ones occurred in the previously described solution to a missing addend task corresponding to 27 + _ = 82. Then, he counted down a column by ones before adding bars of 10 multilinks to a collection of 27.

In summary, there was no indication that Andy significantly reorganized his arithmetical thinking during the time he worked with Andrea. Nonetheless, he did create numerical composites of ten in an increasingly wide range of situations while attempting to solve personally challenging problems. His interactions with Andrea rarely seemed to play an important role in either his formulation of these problems or in his development of novel, situation-specific solutions.

Andrea. Several solutions that Andrea produced during her interview indicated that her beliefs about the nature of mathematical activity were instrumental to a considerable extent. Numerous incidents that occurred as she interacted with Andy corroborated this inference. One such episode has already been discussed. In that, she explained how she had used the standard algorithm to add 5 twelves. A second example occurred on February 12, when she solved the task, "How many do you add to III:::: [38] to make 66?" She counted the array of strips and squares and then wrote 38 + 66 in vertical column format before using the standard algorithm. However, she stopped when she reached the tens column and erased "66."

1	Researcher:	And why did you rub that out?
2	Andrea:	Because it [the answer] was too much.
3	Researcher:	Oh? What, the answer was too big?
4	Andrea:	Yeah.
5	Researcher:	How do you know that?
6	Andrea:	Well, it was a hundred and something.
7	Researcher:	And why would a hundred and something be too big?
8	Andrea:	I don't know (shrugs her shoulders).

She next counted the three strips in the diagram on the activity sheet and used the standard algorithm to add 30 and 38. At this point, she asked the teacher for assistance, saying "We're having a little trouble." However, when she explained why she had added 38 and 30, she did not make any reference to the task statement but instead described what she had done procedurally.

33	Andrea:	See, uh, I put 30 plus 38. 30 plus 38 (points to the numerals written in column format).
34	Teacher:	Why did you put these numbers first of all. See, I just got here, I don't know what you have been up to.
35	Andrea:	Well, here is 30, 1, 2, 3 (counts the three strips in the diagram).
36	Teacher:	Okay, I see, I'm sorry. I was on the top one. All right, go ahead.
37	Andrea:	1, 2 . . . 8 (counts the individual squares). There's 8 ones.
38	Teacher:	(Nods).
39	Andrea:	So I put down 38. 30 and 38.
40	Teacher:	Um, um.
41	Andrea:	And it goes, um 8 and 0 makes 8, and 3 and 6, and 3 and 3 makes 6. That makes 68.

As the episode continued, the teacher asked Andrea what she was trying to find out and eventually attempted to establish a basis for communication with her by using multilinks. At the teacher's direction, she and Andy made a collection of 66 multilinks, removed 38 of them, and found that there were 28 left. The teacher then suggested that they check whether this was the answer to the original task, saying "Try it and see if that will work." In response, Andrea used the standard algorithm to add 38 and 66, concluding "That'll be 94." At this point, the teacher seemed to give up in her attempt to establish a basis for mathematical communication with Andrea, saying simply "Give her a hand there Andy, see if you can figure that out." Andy then counted the collection of 66 multilinks, and Andrea commented "66. I know, but what do we add?" Thus, she still believed that she could solve the task using the standard addition algorithm; it was merely a matter of identifying the two numbers she was supposed to add. Andy ignored her question, separated 38 multilinks from the collection of 66, and picked up the remainder. Andrea concluded from this that 38 was the answer to the task and attempted to write this on the activity sheet.

126	Andy:	No, no, no.
127	Andrea:	Stop it, I'm trying to write something (she writes "38").
128	Andy:	Not 38. (Counts the collection of 28). 28, you add 28 to it.
129	Andrea:	Okay.
130	Andy:	Add 28.
131	Andrea:	(Adds 38 and 28 using the standard algorithm and gets 66.) Yeah, you're right, I messed up on that one.

Incompatibilities of this kind between Andrea's task interpretations and those of both Andy and the teacher occurred repeatedly. As the episode also illustrates,

Andrea's decisions of which computations to perform often seemed almost arbitrary. There was, however, a pattern to those occasions when she computed in this way. A comparison of Andrea's solutions to different types of missing addend and missing subtrahend tasks indicates that she had, in fact, constructed a procedure that was effective when the whole was shown as a picture of strips and squares. For example, during the session on February 3, she solved the first task, "36 are added to make IIIIIII:. [83]. How many did we begin with?" by shading in three strips and six squares and then counting the remainder. Similarly, during the same session, she solved the task "How many do you have to add to 57 to make IIIIIII:. [83]?" by first shading five strips and seven squares. However, tasks in which the known part rather than the whole was shown as a picture were problematic for her. Thus, in this session on February 3, she explained to the teacher that she was having difficulty in solving "How many do you have to take away from 74 to leave III::. [35]?" because there was "only 35." Similarly, she performed apparently arbitrary computations in the sample episode discussed earlier in which she attempted to solve "How many do you add to III::::: [38] to make 66?" In situations such as these, when she could neither interpret the task in terms of numerical relationships nor rely on a perceptually based procedure, her only recourse appeared to be to add or subtract almost at random.

The way in which Andrea turned to the teacher for assistance in the sample episode was also quite typical. In contrast to the other children, she seemed to view the teacher as the mathematical authority in the classroom. Her expectation that the teacher would decide what did and did not count as an acceptable solution is compatible with her instrumental beliefs and her attempts to be effective by constructing procedural instructions. In particular, because the steps of a procedure of this kind do not signify the mental manipulation of arithmetical objects, there is nothing beyond the procedure to refer to when giving a justification. Consequently, she and other children who construct such procedures must necessarily appeal to an authority to know whether their mathematical activity is acceptable.

It should again be stressed that Andrea was the only child in the class who held instrumental beliefs and viewed the teacher as a mathematical authority in the traditional sense at the time of the January interviews. It seems essential to take her current conceptual possibilities into account when explaining why her beliefs were outside the mainstream of the classroom microculture. In this regard, recall that her interpretations of strips-and-squares tasks that involved missing addends and missing minuends were often incompatible with the taken-as-shared interpretations of the classroom community.

A similar difficulty also occurred on balance tasks and money story problems that involved missing addends and minuends. For example, on March 2, she attempted to solve a story problem that showed collections of coins with total values of 35¢ and 52¢. The question asked was, "How much more money does Jay [52¢] have than Barb [35¢]?" Andrea first made collections of 35 and 52 multilinks before telling Andy, "I don't know how to do this." She then sought the teacher's assistance but, as in previous episodes, they seemed to talk past each other. The teacher first asked Andrea if she could use Jay's money to "match Barb's money." Her expectation seemed to be that Andrea would remove 35 multilinks from the collection of 52 and find how many were left. However, Andrea simply matched each collection of multilinks with the corresponding collection of coins in the problem statement. Their task interpretations appeared to be incompatible in that the teacher included the 35 in the 52, whereas for Andrea they were two unrelated numbers. In an attempt to resolve this misunderstanding, the teacher reread the problem statement, emphasizing that it said "How much *more* money does Jay have than Barb?" Andrea then proposed:

1 Andrea: So what you do. You count this [Barb's money], I mean you count this [Jay's money] after that [Barb's money].

Thus, she and the teacher continued to talk past each other.

In contrast to the difficulties she experienced while solving strips-and-squares, balance, and money story problems, Andrea was able to solve other types of missing addend and missing minuend tasks routinely. For example, on January 27, she solved missing-addend number sentences by counting by ones on a hundreds board. The crucial difference between the two sets of tasks was that she established counting-based meanings when interpreting those that she solved routinely, and collection-based meanings when interpreting those with which she had difficulty. This observation is consistent with her performance in her January interview. There, it was inferred that the construction of ten as a numerical composite was beyond her realm of developmental possibilities even when she could rely on situation-specific imagery.

This analysis of Andrea's task interpretations clarifies why she maintained her instrumental beliefs. She was frequently unable to enter into the mainstream discourse of the class and, further, could not make sense of Andy's collection-based explanations. In such situations, it was reasonable for her to assume that others were following unknown procedural instructions. This would, in turn, imply that she should construct such procedures if she was to fulfill the obligation of developing her own solutions. An episode that occurred on March 2 illustrates how she attempted to be effective in this way. In this situation, she first attempted to solve two balance tasks, 50, __ / 25, 25, 10, 1, 1 and 50, __ / 25, 25, 10, 10, 1, 1, by using the standard algorithm to add the numbers on the right hand side of each, thus getting answers of 62 and 72, respectively. At this point, the teacher joined the group and, in the process of explaining Andy's solution to the second of these tasks, covered the 50 and the two 25s with her fingers because they balanced each other. While the teacher and Andy conceptualized the situation in terms of numbers balancing each other, Andrea constructed a situation-specific procedural instruction when she interpreted their explanations. First, she solved 50, __ / 25, 25, 10, 10, 1, 1 by covering the 50 and the two 25s exactly as the teacher had done, thus getting an answer of 22. She then followed this procedure of ignoring the 25s when she solved the next two tasks, 25, 25, 25, 10, 10, 1, 1/ 75, __ and 50, __ / 25, 25, 25, 10, 10, 1, 1, and arrived at answers of 22 in each case rather than of 22 and 47, respectively. Given that she could not participate in the mathematical reality that

Andy and the teacher took as shared, it was entirely reasonable for her to construct an instructional procedure of this sort as she attempted to be effective in the classroom.

In general, there was no evidence that Andrea made any conceptual advances of note during the time that she was observed working with Andy. Further, the only occasion when a procedural instruction that she constructed was not merely local and ad hoc occurred on January 28, when she worked with Linda and Kathy in Andy's absence. It will be recalled that, during her interview, Andrea solved the worksheet tasks by using the standard algorithm but had difficulty in using it to solve other types of tasks, including horizontal sentences. In this episode, she broadened the applicability of this algorithm as she solved a sequence of balance tasks. One of these tasks involved adding 3 nineteens.

1 Linda: What is 9 and 9?
2 Andrea: 9 and 9 is . . . 18 . . . and 9 more . . .
3 Kathy: And . . .
4 Andrea: (Counts on her fingers) 19, 20, 21 . . . 27. (Points to and counts the 3 "1"s of the 3 nineteens) 27, 26 . . . 27, 28, 29, 30.

In this case, Andrea partitioned the numeral 19 per se to produce 1 and 9 rather than 10 and 9.

As the episode continued, Andrea rejected Linda's proposal that they should add the 3 nines to 30. Instead, she attempted to justify her own answer of 30 by using the standard algorithm. In doing so, she first added the 3 nines but then wrote the "2" of 27 in the ones place and the "7" above the 3 ones. She then commented "I can't get that one" and seemed genuinely surprised that she had not obtained an answer of 30. She had presumably inferred from Linda's initial question, "What's 9 and 9?" that Linda was using the standard algorithm and had attempted to use this procedure mentally when she partitioned the nineteens. From her point of view, her justification was merely a matter of spelling out the steps of her algorithmic procedure in more detail, and she was therefore surprised by the unexpected outcome.

Later in the episode, Andrea related Linda and Kathy's partitioning solutions to her own use of the standard algorithm. In this case, Kathy first solved a task that involved adding 4 twenty-ones by counting 20, 40, 60, 80 and then adding 4 more to make 84. She then redescribed her solution, and it was at this point that Andrea had an insight.

1 Kathy: 2, 4, 6, 8 . . . 1, 1, 1, 1.
2 Andrea: I know something, I know something. Okay, 2 and 2, and 2 and 2. 4, 6, so . . . 60, hey look . . . 2 and 4. 2 plus 2 plus 2 more is 6, and 8, and it's 84.

There was no indication that Andrea partitioned the twenty-ones as experientially real arithmetical objects; the result of partitioning 21 was still 2 and 1 for her.

Instead, her insight seemed to involve the construction of relationships between two number–word sequences (e.g. 6—60, 8—80). By constructing a procedure in which she renamed the results of partitioning numerals 20, 40, 60, 80, she was able to be effective when she interacted with Linda, Kathy, and other children who used the partitioning algorithm. Significantly, it was after this episode that Andrea used the standard algorithm to solve an increasingly wide range of tasks.

As was the case with the partitioning algorithm, counting by tens and ones on the hundreds board was beyond the realm of her developmental possibilities. The only occasion when she did anything other than count by ones on a hundreds board occurred on March 5. At that time, she first solved the task "How many do you have to add to 27 to make 82?" by counting by ones. Later, while waiting for Andy to complete his solution, she traced a path from 27 to 82 on the hundreds board by touching the squares 37, 47, 57, 67, 77, 87, 86, 85, 84, 83, 82. She counted as she did so and concluded, "Okay, I think it's 11." Interestingly, during the same session, the researcher asked her if she knew what 62 and 10 more would make without counting. She very deliberately looked at the hundreds board before she replied, "72." In this case, she presumably knew that she would move down one row on the hundreds board if she counted 10 units of one. This, of course, is less conceptually sophisticated than creating and counting numerical composites of ten.

Summary

Andrea's and Andy's performances in their January interviews appeared to be similar in many respects. With the exception of the finding that the construction of numerical composites of ten was within Andy's but not Andrea's realm of developmental possibilities, there was no indication that one child could reason at a significantly more sophisticated level than the other. Significantly, neither seemed to regard either the self or the other as a mathematical authority in the group. As a consequence, they routinely engaged in multivocal explanation when their solutions or answers were in conflict.

The analysis of their interviews also indicated that Andrea's but not Andy's beliefs appeared to be instrumental to a considerable extent. This inference was corroborated by observations made when the children worked together in the classroom; they seemed to have different understandings of what counted as an explanation and a justification. As a consequence, they experienced great difficulty in establishing a taken-as-shared basis for mathematical communication. Their participation in interactions involving multivocal explanation was, therefore, not particularly productive in terms of the learning opportunities that arose for them. In this regard, their interactions contrasted sharply with those of Ryan and Katy.

The analysis of Andrea's mathematical activity in the classroom indicated that she was unable to interpret certain tasks in ways compatible with the taken-as-shared interpretations of the classroom community. This further contributed to the difficulties she and Andy had in communicating effectively. For example, Andy's novel collections-based solutions seemed to be beyond her realm of developmental possibilities. As a consequence, learning opportunities did not arise for her when she listened to his explanations and, further, the questions and challenges she raised did not give rise to learning opportunities for him. Thus, in one instance, Andy attempted to solve a missing addend task corresponding to $27 + \_ = 82$ by adding bars of 10 multilinks to a collection of 27 until he made a collection of 82. This was clearly a novel solution for him and he failed to find out how many multilinks he had added to reach 82. Because Andrea did not interpret this as a missing addend task, she could not ask him how many he had added or intervene in some other way that might help him reflect on his mathematical activity.

The analysis of the children's interactions also indicated that learning opportunities rarely arose for either child when Andrea explained her solutions. One reason for this was that Andy could not interpret her use of the standard algorithm in terms of actions on arithmetical objects. In addition, she often carried out apparently arbitrary computations when she did not have a procedural instruction she could follow. Finally, it should be noted that the learning opportunities were further limited because Andy did not necessarily feel obliged to develop his own solutions.

Not surprisingly, given the relatively few learning opportunities that arose, both children made very limited progress in the time that they worked together. Andrea used the standard algorithm in an instrumental way to solve a wider range of tasks, and Andy created numerical composites of ten in a wider variety of situations. A vicious circle seems to have arisen in that the children's initial beliefs and conceptual capabilities resulted in the development of a social relationship that limited their opportunities for learning. Their conceptual capabilities, therefore, remained relatively stable and, as a consequence, their social relationship also changed little.

CASE STUDY—JACK AND JAMIE JANUARY 13–MARCH 23, 1987

Jamie: January Interview

In the first part of the interview, Jamie solved a missing addend task by counting on, thus indicating that he could create and count abstract units of one. His solutions to the uncovering tasks demonstrated that incrementing by ten was relatively routine for him. For example, he reasoned as follows when 2 strips were uncovered after he had reached 52: "52 and 10 more would be 62, and add 10 more would be 72." For the most part, the strips-and-squares tasks were also unproblematic for him. In the most challenging task that he was asked to solve, 3 strips and 6 squares [36] were visible, he was told that there were 61 squares in all, and asked to find how many were hidden. He first gave 24 and then 25 as his answer, explaining:

1 Jamie: Okay, like there is 30 right here (points to the visible strips), and there's 20 here (points to the cloth). And there's 5 [squares under

3. LEARNING AND SMALL-GROUP INTERACTION 91

the cloth], and put the 5 onto all of these (points to the visible strips and squares), and that makes 61 altogether.

Taken together, Jamie's solutions to the uncovering and strips-and-squares tasks indicate that he could create both counting-based and collection-based abstract composite units of ten, at least when he could rely on situation-specific imagery. Further, his solution to the final strips-and-squares task indicates that the construction of part–whole relations involving tens and ones was within his realm of developmental possibilities.

In contrast to the majority of the other children, Jamie used the same solution method to solve both the horizontal sentences and the worksheet tasks—the partitioning algorithm. For example, he gave the following explanation for the worksheet task corresponding to 39 + 53 = _:

1 Jamie: 3 . . . 30 plus 50 is 80, and 9 plus 3 is 12. Put all those together and I come up with 92.

In summary, it can be noted that although Jamie could construct counting-based abstract composite units of ten, he tended to establish collection-based rather than counting-based numerical meanings. This was the case for the strips-and-squares, number sentence, and worksheet tasks.

Jack: January Interview

Jack's solutions to the counting tasks indicated that counting abstract units of one was routine for him. Further, like Jamie, he solved the uncovering tasks by counting tens and ones without prompting. For example, 2 strips and 5 squares [25] were uncovered when he had reached 46, and he found how many squares there were now in all by counting, "46—56, 66, 67, 68, 69, 70, 71." Taking account of his solutions to other interview tasks, it appears that he counted the strips as abstract composite units of ten.

The analysis of Jamie's interview indicates that he, like Katy, created a part–whole relationship involving units of ten and one when he solved the strips-and-squares tasks. By coincidence, Jack's solutions were similar to those of Katy's partner, Ryan. For example, he reasoned as follows when solving a task in which 3 strips and 5 squares [35] were visible, there were 71 squares in all, and he was asked to find how many were hidden.

1 Jack: Four of these [strips] under there . . . no.
2 Interviewer: Why not?
3 Jack: There would be too many of those (points to the visible squares).
4 Interviewer: So how many strips do you think are under there?
5 Jack: 3.
6 Interviewer: Is there anything else under there?

7 Jack: And 5 of these (points to the visible squares).
8 Interviewer: How many squares does that make in all?
9 Jack: 35.

Like Ryan, Jack anticipated that he could create one of the tens needed to make 70 by composing units of one from both the visible and screened collections. Also like Ryan, he had difficulty in keeping track of the whole as he built up the hidden collection, even when the interviewer supported his solution attempt. It would therefore seem that the construction of a part–whole relationship involving units of ten and one was beyond his realm of developmental possibilities.

In the absence of any assistance, Jack's only reliable way of solving the horizontal sentences was to count by ones. The interviewer, therefore, intervened when he solved 28 + 13 = __.

1 Interviewer: What's 28 and 10 more?
2 Jack: It's 41.
3 Interviewer: How did you figure that out?
4 Jack: 'Cause 28 plus 10 plus 3 more would be 41.

He then developed this solution method further as he solved the remaining horizontal sentence and the worksheet tasks. For example, his reasoning to solve 39 + 53 = __ was:

1 Jack: You have 53, 10 more makes 63, plus 10 more 73, plus 10 more 83, plus 9 . . . 92.

These solutions suggest that the image-independent construction of abstract composite units of ten was within Jack's realm of developmental possibilities when he established counting-based meanings. There was, however, no indication that he could use a partitioning algorithm. As a general point of comparison, it is also interesting to note that Jamie established only collection-based meanings when he solved the horizontal sentence and worksheet tasks, whereas Jack established only counting-based meanings.

Regularities in Small-Group Activity

As was the case with the other three groups, the social relationship that Jack and Jamie established remained relatively stable during the 10 weeks that they were observed working together. Throughout this period, their solution attempts generally followed one of three interactional scenarios. If a task involved a relatively routine computation for Jamie, then he proceeded without necessarily explaining his solution, and Jack had to adapt to Jamie's mathematical activity as best he could. In contrast, Jamie took note of Jack's mathematical activity when Jack proposed a thinking strategy solution, and in these situations their interactions typically

3. LEARNING AND SMALL-GROUP INTERACTION

involved either indirect collaboration or multivocal explanation. Finally, Jamie typically invited Jack's participation when a task was problematic for him, and in these cases again they engaged in either indirect collaboration or multivocal explanation. Each of these three scenarios is discussed in turn.

It is apparent from the previous three case studies that the children attempted to achieve a variety of goals in the classroom. Jamie, however, was the only one of the eight children who pursued a personal agenda of attempting to complete as many tasks as possible. This was indicated by frequent comments in which he told Jack how many activity sheets they had completed or indicated that they needed to hurry up. In addition, his computation methods were typically more efficient than Jack's, and consequently he was usually the first to arrive at an answer. This was particularly the case for tasks that involved direct addition in that he, but not Jack, had constructed a partitioning algorithm. Consequently, on those occasions when he could solve a task by performing what was for him a routine computation, he typically neither explained his solution to Jack nor waited for Jack to complete his own solution. Instead, he usually began to solve the next task without checking whether Jack agreed with his answer. Jack's role was then limited to that of contributing to and making sense of Jamie's solutions as best he could. For example, on February 2, Jamie solved a sequence of balance tasks corresponding to $24 + 39 = \_$, $46 + 30 = \_$, $25 + 36 = \_$, and $36 + 25 = \_$. Jack observed and interjected comments, but did not initiate a single solution step. A researcher then intervened to ask Jamie how he had solved the last of the tasks, $55 + 34 = \_$.

1 Researcher: (To Jamie) How did you do this one?
2 Jamie: Well, 50 and 30 make . . .
3 Jack: 5 and 3 make 70, 7.
4 Jamie: 80, and . . . and 6 more makes 86 and 5 more makes 91.
5 Researcher: Did you agree, Jack?

As the exchange continued, Jack attempted to explain how the task could be solved by using the partitioning algorithm. The meaning that partitioning a numeral might have had for him is discussed later in the chapter, when the analysis of his mathematical learning is presented. For the present, it suffices to note that he continued to try and understand Jamie's efficient computational methods.

The following episode provides a further illustration of the way in which Jack attempted to adapt to Jamie's mathematical activity, and also exemplifies the general asymmetry in their respective roles. The episode occurred on February 16, after Jamie had used his partitioning algorithm to solve a sequence of balance tasks corresponding to $3 \times 12 = \_$, $4 \times 12 = \_$, and $5 \times 12 = \_$. Jack challenged his answer of 36 to the first task as Jamie began to solve the next task, $5 \times 11 = \_$.

1 Jamie: 11 plus 11 is 22.
2 Jack: 32, that's 32 (points to Jamie's answer of 36).
3 Jamie: 22 plus 22 makes 46, no 44.
4 Jack: 32. Jamie, that's 32. 32.

5 Jamie: What? (writes the answer of 55 on the activity sheet).

The way in which Jamie repeatedly ignored Jack's challenges was relatively typical. Jack persisted in his challenge and attempted to explain how he had solved the task.

6 Jack: 32. 'Cause 10 and 10 make 20, plus, and that would make, 2 twos right there would make it . . .
7 Jamie: (Interrupts) Wait, 10 and 10 and 10.
8 Jack: . . . Would make 24.
9 Jamie: 10; 10 and 10 make 20, right?
10 Jack: Yeah.
11 Jamie: And 2 more make 24.
12 Jack: (Interrupts) 2—3, 4—5, 6.

At this point in the exchange, Jamie challenged Jack's right to interject in this way.

14 Jamie: I didn't even get to explain it to you. Don't you want me to explain it to you? [Begins to explain his solution to the balance task $4 \times 12 = \_$] 12 and 12, . . . that [4 tens] makes 40, right?
15 Jack: I know, I know.
16 Jamie: 42, 44, 46, 48.

Jamie then went on to explain his solutions to the other two tasks he had solved while Jack listened relatively passively.

In the course of this exchange, a power imbalance between Jack and Jamie was realized in two ways. On the one hand, the conflict between their answers was resolved without discussion with Jack accepting that Jamie was the mathematical authority in the group. On the other hand, there was also a tension in their views about how the conflict in their answers should be handled. Jack attempted to initiate an interaction involving multivocal explanation, whereas Jamie attempted to initiate an interaction involving univocal explanation. This tension was also resolved without discussion, with Jack accepting Jamie's interpretation of the situation. This incident was paradigmatic in that it was Jamie who usually initiated the transition from one type of social situation to another. He was, therefore, the social authority in the group in that he controlled the way in which he and Jack interacted as they worked together.

One immediate consequence of this power imbalance was that Jack's repeated attempts to initiate the renegotiation of their obligations and expectations were unsuccessful. Jamie continued to solve tasks that were routine for him as rapidly as possible. As a consequence, learning opportunities were limited for both children in these situations. This was obviously the case for Jamie, given that he simply performed a routine computation. Jack, on the other hand, seemed to be caught in something of a vicious circle. He found it difficult to participate because his computational methods were less sophisticated than those used by Jamie. Conversely, opportunities for him to construct increasingly sophisticated computational algorithms were relatively limited, precisely because he was often unable to contribute in these situations.

The only viable way that Jack could make a contribution when a task was routine for Jamie was to propose an efficient solution that involved relating the current task to one that had been previously solved. As the following episode, which occurred on February 18, illustrates, the nature of their interactions often changed dramatically when Jack was able to make a proposal of this type. In this case, he suggested using $2 \times 4 = 8$ to solve $4 \times 4 = \_$.

1	Jack:	4 times 4. Four, let's see.
2	Jamie:	Twenty.
3	Jack:	What's 8 plus 8?
4	Jamie:	16. It's 4 sets of 4, 8 . . . 16.
5	Researcher:	Why did you say "What's 8 and 8?" Jack?
6	Jamie:	'Cause 4 sets, um 4, 2 sets make 8.
7	Researcher:	Yes.
8	Jack:	You have 2 more sets. Like it's 2 and 2 make 4.

In this instance, it was Jamie who adapted to Jack's mathematical activity rather than vice versa. It can also be observed that, in comparison with the other three groups, they did not first solve the task independently and then explain their solutions, but instead engaged in indirect collaboration.

This episode is representative with regard to the learning opportunities that arose for Jamie in such situations; he was able to contribute to the development of a thinking strategy solution that was within his realm of developmental possibilities. As a consequence, he increasingly began to develop his own thinking strategy solutions. In this case, in contrast to tasks that he could solve by performing a routine computation, he typically accepted Jack's challenges and thus contributed to the emergence of interactions involving multivocal explanation. For example, on January 13, Jamie had solved a balance task corresponding to $49 + \_ = 62$ by using multilinks. He then reasoned that the answer to a task corresponding to $46 + \_ = 62$ would be 3 less than 13 because 46 was 3 less than 49. Jack challenged this solution by pointing out that 46 plus 10 only made 56, and Jamie revised his solution. At times, it was almost as though Jamie had his own private coach to support his construction of thinking strategies.

If attention now focuses on Jack rather than Jamie, it is apparent that learning opportunities could also arise for him as he monitored and challenged Jamie's solutions and participated in multivocal explanation. In addition, learning opportunities sometimes arose for him as well as for Jamie when they engaged in indirect collaboration. For example, on February 18, they solved $10 \times 4 = \_$ immediately after they had found $5 \times 4 = 20$.

1	Jack:	No, look, it's 5 more sets [of 4]. Look.
2	Jamie:	Yeah.
3	Jack:	5 more sets than 20.

4   Jamie: Oh! Oh, 20 plus 20 is 40. So it's gotta be 40.

Jamie then began to count by fours to verify the answer, and, while observing him, Jack had an insight that he subsequently explained to a researcher.

18 Researcher: Why did you say 40, Jack?
19 Jack: 'Cause, 'cause 10 fours make 40, 'cause, like um, 'cause . . .
20 Jamie: 5 fours make 20.
21 Jack: Just turn it backwards.
22 Jamie: 5, 5 fours make, I mean . . .
23 Jack: 4 sets of 10 make 40. Just turn it around.

As can be seen, Jack's insight was the realization that the result of adding 10 fours was the same as that of adding 4 tens. Jamie's calculation of the answer seemed crucial in that it was only after Jack heard "20 plus 20 is 40" that he reconceptualized the tasks as 4 tens rather than as 20 and 5 more fours.

Jack generated thinking strategy solutions far more frequently than did the other seven children discussed in these case studies. It seems reasonable to speculate that his apparent predilection for thinking strategy solutions was a consequence of his attempt to be effective while working with Jamie. If he could participate when he proposed thinking strategies, then perhaps he, more than any of the other children, actively looked for opportunities to relate the current task to one that had previously been solved. This being the case, his development of these solutions was socially situated in that it might well not have occurred had he worked with someone other than Jamie.

Thus far, interactions in which Jamie performed routine computations have been contrasted with those in which Jack proposed a thinking strategy solution. The remaining scenario to be discussed is that in which a task was problematic for Jamie. Often, he first attempted to solve the task on his own and ignored Jack's attempts to make a contribution. However, when he reached an impasse, he typically invited Jack to participate. These invitations were reasonably easy to identify in that Jamie moved the activity sheet that was generally in front of himself toward Jack. For his part, Jack typically accepted these invitations, thus implicitly acknowledging that Jamie was the social authority in the group. The interactions that occurred in these situations can be illustrated by considering their attempts to solve a balance task corresponding to $36 + 46 = 10 + \_$ on January 13.

Jamie began by calculating the sum of 36 and 46, but then invited Jack to participate when he noticed that there was a 10 on the right-hand side of the balance, saying "We're in trouble now." He then reasoned, "70—60, um . . . 70 (points twice to the activity sheet) . . . 72." This, as it transpired, was a highly situation-specific solution in that Jamie had just added 30 and 40 to make 70. He could, therefore, take account of the ten on the other side of the balance before he added the 2 sixes. At this point, Jack argued that the ten should be added to 82, the sum of 36 and 46.

1   Jack: 82 [The sum of 36 and 46].
2   Jamie: 72 [repeats his answer].

3. LEARNING AND SMALL-GROUP INTERACTION

3	Jack:	92 [82 plus 10].
4	Jamie:	I mean 62 [72 subtract 10].
5	Jack:	No, look those 2, 3 and 4, are 70. So 6 and 6, 12.
6	Jamie:	70 . . . that's easy.
7	Jack:	What?
8	Jamie:	82 [36 and 46].
9	Jack:	92 (points to the "10").
10	Jamie:	82 goes right there (points to the empty box).
11	Jack:	No, 92. Look, 10.

In this exchange, the children engaged in indirect collaboration. Jamie's last response of 82 suggests that, at this point in the episode, he interpreted the task as simply that of adding 36 and 46. However, a learning opportunity arose for him when he attempted to make sense of Jack's answer of 92.

12	Jamie:	92. I don't get it, I don't get it.
13	Jack:	(Moves the activity sheet in front of himself).
14	Jamie:	Where are you getting the 9? We've got to subtract to do this problem.
15	Jack:	To do what?
16	Jamie:	Subtract to do the problem.

Presumably, Jamie inferred that Jack had arrived at his answer of 92 by adding 10 to 82. In the process, he seemed to reconceptualize the task as that of subtracting 10 from 82. Jack continued to argue that the 10 should be added, thus contributing to the emergence of an interaction involving multivocal explanation. The ensuing dispute was not settled until the teacher intervened.

Jack and Jamie were the only pair who engaged in indirect collaboration with any regularity. This was also the only group in which one child was a social authority. From the observer's perspective, Jamie seemed to express this authority in action by using Jack as a resource he could call on to help him complete as many tasks as possible. Thus, he ignored Jack when a task was routine, but invited him to participate when a task was problematic, and also attempted to capitalize on Jack's efficient thinking strategy solutions. One immediate consequence of this power imbalance was the inequality in the learning opportunities that arose for the two children. Jack generally had to cope as best he could unless he could propose a thinking strategy solution or a task happened to be problematic for Jamie. In contrast, Jamie expressed his social authority by initiating interactions that gave rise to learning opportunities for himself.

Mathematical Learning

Jack. Jack's solutions to the horizontal sentence and worksheet tasks in the January interview indicated that the creation of abstract composite units of ten was in his realm of developmental possibilities when he established counting-based meanings. However, there was no evidence that he constructed units of this type with any regularity during the small-group sessions. In this regard, the analysis of their social relationship indicated that his learning opportunities were relatively limited when compared with those that arose for Jamie, and that he frequently had to adapt to Jamie's collections-based solutions.

It was unclear from Jack's interview whether the construction of a partitioning algorithm was within the realm of his developmental possibilities. Several of the episodes that have been presented demonstrate that he attempted to use this algorithm on several occasions. Perhaps the most sophisticated of these efforts occurred on February 19, when he and Jamie solved the task "35 are taken away and 26 are left. How many were there to begin with?" Jamie found the task problematic and eventually accepted Jack's proposal to add 35 and 26.

1	Jamie:	You know that 30. . . . You know that 30 and 20 are more than 40.
2	Jack:	I know, it's 50.
3	Jamie:	No, [the answer to the task is] not 50.
4	Jack:	20 and 20 is 40, 20 and 30 are 50.
5	Jamie:	61 is the answer.
6	Jack:	50, what's 6 plus 5? Yes, 61.

The way in which Jack inferred how Jamie had arrived at the answer of 61 suggests that the partitioning algorithm was within his realm of developmental possibilities. Nonetheless, Jack did not once use this algorithm independently in the remaining weeks that he worked with Jamie. Instead, as in the sample episode, it seemed essential that Jamie first state part of the result of the partitioning (e.g. " . . . 30 and 20 . . . "). This suggests that although the act of partitioning a two-digit numeral such as 35 did not symbolize the partitioning of the signified number for him, he could conceptualize 35 as 30 plus 5 after the event. Although it is not entirely clear why he seemed unable to make this final conceptual step, it can be noted that Jamie rarely explained his partitioning solutions. Consequently, Jack could only infer what Jamie might have done from the relatively cryptic comments he made while solving a task.

As a further point, the sample episode was the first occasion on which Jack attempted to use the algorithm to solve anything other than a balance task. He had, for example, contributed to Jamie's partitioning solution to a balance task as early as January 13. In this regard, recall that both Ryan's and Holly's use of this algorithm was initially restricted solely to balance tasks. In those cases, it was suggested that imagery associated with the taken-as-shared metaphor of a number in a box may have supported their conceptual constructions. This same metaphor may have also facilitated Jack's attempts to make sense of Jamie's partitioning solutions.

Three of Jack's thinking strategy solutions indicate that he could construct abstract composite units of ten without relying on situation-specific imagery. On

3. LEARNING AND SMALL-GROUP INTERACTION 99

January 14, he related the number sentence $60 - 21 = \_$ to $70 - 21 = 49$, arguing that it was "10 lower than 49." In addition, he used the sentence $31 + 29 = 60$ to solve $31 + 19 = \_$ on January 27, and he related a strips-and-squares task corresponding to $35 + \_ = 87$ to another corresponding to $35 + 42 = 77$ on February 12. These solutions are consistent with his construction of a relatively efficient counting-based algorithm during his January interview. In that, he solved $39 + 53 = \_$ by incrementing 3 units of ten and then adding 9 onto 53. Similarly, in using $35 + 42 = 77$ to solve $35 + \_ = 87$, he seemed to conceptualize 87 as the result of incrementing 10 onto 77. These thinking strategy solutions add credibility to the speculation that the construction of ten as an abstract composite unit was within his realm of developmental possibilities when he established counting-based meanings, but not when he established collection-based meanings.

Jack also developed a variety of collection-based thinking strategy solutions. The most sophisticated of these strategies seemed to involve taking a numerical part–whole structure as a given. For example, in a previously documented solution, he related $5 \times 4 = 20$ to $10 \times 4 = \_$ on February 18 by arguing that it was "5 more sets than 20." In this case, 20 as a numerical whole composed of fours seemed to be an object of reflection for him. As a further example, on March 5, he related a strips-and-squares task corresponding to $\_ - 55 = 27$ ("55 are taken away and 27 are left. How many were there to begin with?") to a previously solved task corresponding to $27 + 55 = 82$ ("How many do you have to add to 27 to make 82?") In this case, 82 as a whole composed of 27 and 55 seemed to be an object of reflection for him. Interestingly, the first of these sophisticated collection-based solutions was not observed until February 16. Their frequent occurrence from that point on suggests that Jack might have made a conceptual advance and could objectify numerical part–whole relationships. This appeared to be the only major advance that he made while working with Jamie.

Jamie. During his interview, Jamie used the partitioning algorithm to solve both horizontal sentence and worksheet tasks. He subsequently used this algorithm routinely to solve a wide range of tasks in the classroom. His solutions to the uncovering tasks presented in the interview indicated that he could also create abstract composite units of ten by relying on imagery when he established counting-based numerical meanings. However, there was no indication that he became able to create counting-based units of ten in an image-independent manner. His most sophisticated counting-based solution occurred on March 16, when he and Jack worked with Michael in Holly's absence. Michael initially began to count by ones on a hundreds board to solve the task, "How many do you have to add to 38 to make 86?"

1	Jamie:	No. Here, wait. Let me see it. Let me see this (takes the hundreds board from Michael).
2	Michael:	I always count by ones.
3	Jamie:	I don't. Okay, 38 . . . 38 (points to the hundreds board). Okay . . . 1, 2, 22, 32 (counts the squares 39, 40, 50, 60).
4	Michael:	I can't use it that way.

Jamie's comments to Michael notwithstanding, he rarely used a hundreds board and this was the first occasion on which he was observed attempting to count by tens and ones. He eventually completed his solution as follows:

5 Jamie: One, 2 (counts 39 and 40). . . . That would be 12 (points to 50). 12, 22, 32 (counts down the last column) . . .
6 Michael: 42.
7 Jamie: 42 (points to 80).
8 Michael: 52.
9 Jamie: 42.
10 Michael: 52.
11 Jamie: Wait. No, wait. 42, 43, 44 . . . 48 (counts from 81 to 86). 48.

Jamie's count by ten down the last column indicates that each complete row on the hundreds board was at least a numerical composite of ten for him. As was previously noted when discussing Katy's and Michael's use of the hundreds board, the rhythm of counting by ones is necessarily interrupted at the beginning and end of each row. These natural breaks or pauses in activity facilitate the isolation of counting an intact row as a discrete entity. In contrast, there are no pauses that can support the isolation of counting by ones from, say, 38 to 48 in the same way. The fact that Jamie did not count by tens starting from 38 suggests that he was yet to structure the hundreds board in this way. Significantly, he counted by ones on all but 1 of the 11 occasions that he subsequently used the hundreds board in this session. Ironically, it might have been Jamie himself, as the social authority, who limited the opportunities he had to construct increasingly sophisticated counting-based units. He almost invariably gave number words and numerals collections-based meanings and typically thwarted any discussion of alternative interpretations with Jack.

Although Jamie did not elaborate his routine computational methods while working with Jack, he did learn to construct increasingly sophisticated thinking strategies. For example, on January 21, it was only at Jack's initiative that he used a balance task that involved adding 3 sevens to solve one that involved adding 4 sevens. He then solved the next task, which involved adding 5 sevens, independently of the previous two solutions even though Jack proposed a thinking strategy solution.

1 Jack: 7 more [than the previous task].
2 Jamie: That's 14, plus 14, equals . . .
3 Jack: What's 14 plus 14?
4 Jamie: (Simultaneously) 28.
5 Jack: 28—29, 30 . . . 35 (counts on his fingers). That's what's equal.
              . . .
6 Jamie: (Writes the answer.)

In contrast, Jamie used a variety of sophisticated strategies when he solved balance tasks approximately 6 weeks later, on February 26. For example, he and Jack both used the result of adding 4 twenty-fives to find the sum of 4 twenty-fours.

S. Williams

3. LEARNING AND SMALL-GROUP INTERACTION

1	Jamie:	That makes, that makes 90 . . . 6.
2	Jack:	Just take away 1, 2 . . .
3	Jamie:	4, 4.
4	Jack:	4 from that.
5	Jamie:	It's 96.
6	Jack:	Yeah.

This relatively cryptic exchange incidentally illustrates the extent to which their use of thinking strategies had become taken as shared. In the course of his interactions with Jack, Jamie had developed what might be called a thinking strategy attitude and actively looked for opportunities to create relationships. In the following example, which also occurred on February 26, he explained how he had found the sum of 4 nines.

1	Jamie:	34. I mean, 36.
2	Jack:	Yeah. No. I don't know how you did it.
3	Jamie:	Well, if you uh, if, you just pretend these were tens, that would make 40, take away 4 more makes 36.
4	Jack:	Yeah.
5	Researcher:	Do you agree with that?
6	Jamie:	That's true. Yeah.

Later in the same session, Jamie found the sum of 6 sixes by pretending they were fives and then adding 6 to 30. As these last two solutions illustrate, Jamie's use of thinking strategies was not limited to relating successive tasks. In this regard, his ability to develop thinking strategy solutions began to surpass that of Jack.

Summary

There is every indication that Jamie was the established social authority in the group. This power imbalance between himself and Jack manifested itself in two ways. On the one hand, it was Jamie who typically initiated the transition from one type of interactional situation to another. On the other hand, he defused Jack's concerted attempts to renegotiate obligations and expectations within the group. It also appears that Jamie's personal agenda was to complete as many tasks as possible. If the pursuit of this goal came into conflict with the obligation of, say, helping Jack understand how a task had been solved, the former goal typically prevailed. As a consequence of this power imbalance, the relationship between Jack's and Jamie's conceptual possibilities and the nature of their social relationship was less direct than was the case for the other three groups.

Three distinct interactional scenarios were identified in the course of the analysis. First, Jamie typically solved tasks that were routine for him without either explaining his solutions to Jack or waiting for Jack to complete his own solutions. Jack's role then became that of attempting to make sense of Jamie's collection-based solutions. Jamie also seemed to regard himself as the mathematical authority when tasks were routine for him and, on those rare occasions when he acknowledged Jack's questions and challenges, initiated interactions involving univocal explanation. Second, the only occasions when Jamie took note of Jack's mathematical activity were when he proposed thinking strategy solutions. In these cases, they often engaged in either indirect collaboration or multivocal explanation. It was noted in passing that Jack, more than any of the other seven children, consistently attempted to develop thinking strategy solutions. Given the efficacy of these solutions with respect to his interactions with Jamie, it seems reasonable to speculate that he might have developed this "relating attitude" in an attempt to cope with the social situation in which he found himself. Third, when a task was problematic for Jamie, he typically invited Jack to participate, and they again frequently engaged in either indirect collaboration or multivocal explanation.

The learning opportunities that arose for Jack in these various types of interaction were relatively limited. The only major conceptual advance that he might have made was to objectify numerical part–whole relationships. Jamie, for his part, acted as though Jack was a resource for his learning and constructed increasingly sophisticated thinking strategy solutions. As a consequence of this advance, he became the mathematical authority in an increasingly wide range of situations. Jack then found it even more difficult to participate, and seemed to experience greater frustration as he failed to fulfill his obligation to develop his own solutions. Thus, although the regularities in their interactions remained unchanged, their social relationship appeared to be degenerating.

REFLECTIONS

The first issue discussed in this overview of the case studies concerns the relationship between the individual children's mathematical activity and their small-group interactions. Attention, then, focuses on two pragmatic issues outlined when introducing the case studies. The first concerns the extent to which the children engaged in inquiry mathematics, and the second concerns the nature of the learning opportunities that arose in the course of their small-group interactions. Next, Piagetian and Vygotskian perspectives on social interaction and learning are considered in light of the case studies. Finally, the pedagogical implications of the case studies are discussed.

Cognitive and Social Aspects of Small-Group Activity

The case studies indicate that the four groups of children established markedly different social relationships. In each case, the individual children's situated cognitive capabilities appeared to constrain the type of relationship they developed. For example, Katy but not Ryan was able to routinely partition two-digit numbers. Further, she but not Ryan became able to create abstract composite units of ten in the absence of situation-specific imagery. This, in turn, enabled her to count by tens and ones on the hundreds board. These differences in the two children's cognitive capabilities were apparent in the tension that characterized their small-group relationship in that Katy felt obliged to explain her solutions and help Ryan understand, whereas he felt obliged to solve tasks for himself. The differences in Michael's and Holly's cognitive capabilities were the most pronounced of any of the groups, and this was reflected in the social relationship they established, with Holly as the mathematical authority. Andrea and Andy appeared to be the most cognitively compatible of the four groups at the outset, in that there was no indication that one could reason in significantly more advanced ways than the other. Significantly, neither child seemed to regard either the self or the other as a mathematical authority in the group. The relationship between Jack's and Jamie's cognitive capabilities and the small-group relationship they established is confounded by Jamie's role as the social authority in the group. As a consequence, the nature of their interactions tended to reflect the difficulty of a task for Jamie rather than the differences in their cognitive capabilities.

The case studies indicate that the relationships the children established both facilitated and constrained the occurrence of learning opportunities, thereby influencing each child's mathematical development. The tension in Katy's and Ryan's views of their own and the other's role was realized in frequent argumentative exchanges that facilitated both children's mathematical development. In contrast, Holly, as the mathematical authority, was obliged to explain her thinking to Michael, who was obliged to listen and to adapt to her activity. Exchanges of this sort rarely appeared to give rise to learning opportunities for either child. The reciprocity in Andrea's and Andy's views of their mathematical competence was indicated by the way in which they frequently challenged each other's explanations. However, as a consequence of the difficulty they had in establishing a taken-as-shared basis for mathematical communication, these exchanges did not appear to be particularly productive for either child. Jamie, in his role as the social authority of the group, typically regulated the way in which he and Jack interacted. In doing so, he acted as though Jack was a resource for his own learning, with the consequence that their small-group interactions were more beneficial for him than for Jack.

The central notion that characterizes the relationship between the cognitive and social aspects of small-group activity is that of constraints (Cobb, Wood, & Yackel, 1992). The very fact that it is possible to imagine a variety of alternative social relationships that children with particular cognitive capabilities might establish indicates that the link between the cognitive and social aspects is not deterministic. This is, perhaps, most apparent in Jack and Jamie's case, in that Jamie's role as the social authority did not appear to have a cognitive basis. No doubt, the nature of their interactions would have been very different had Jamie routinely given Jack the opportunity to develop his own personally meaningful solutions. It can be noted in passing that, with the possible exception of this group, issues of status within the peer-group culture that seem endemic at higher grade levels (cf. Linn, 1992) appeared to have little influence on the children's interactions.

Although the children's interactions changed in the teacher's presence (Yackel, chap. 4, this volume), the social relationships they had established were relatively stable for the 10-week period covered by the video recordings. Further, within each group, the children's cognitive capabilities relative to those of the partner were also stable. The only exception to this observation occurred during the last session in which Katy and Ryan worked together. Then, for the first time, Ryan produced several solutions that were beyond the realm of Katy's developmental possibilities, and she appeared to revise her view of herself as the mathematical authority in the group. With this exception noted, it seems reasonable to conclude that the children's cognitive capabilities and the social relationships they established tended to stabilize each other. Further analyses of small-group activity are needed to clarify whether this was an incidental characteristic of these four groups, or whether it is a more general phenomenon.

Inquiry Mathematics

The extent to which the children engaged in inquiry mathematics is addressed by first considering the quality of their small-group activity and then comparing this to their activity in the individual interviews.

Small-Group Norms. As part of her attempt to foster the development of an inquiry mathematics tradition in her classroom, the teacher guided the development of certain small-group norms. These included explaining one's mathematical thinking to the partner, listening to and attempting to make sense of the partner's explanations, challenging explanations that do not seem reasonable, justifying interpretations and solutions in response to challenges, and agreeing on an answer and, ideally, a solution method. The case studies indicate that the teacher was reasonably successful in this regard (cf. Wood, chap. 6, this volume, for a further discussion of the teacher's role in the classroom). The most obvious exception was Jamie's frequent failure to either explain his own thinking or give Jack the opportunity to develop his own personally meaningful solutions. This was one feature of his role as the social authority of the group. In addition, it was observed that Michael rarely challenged Holly's explanations. This deference to Holly, the mathematical authority in the group, might have reflected the relatively large difference in the two children's cognitive capabilities.

Mathematical Objects and Procedural Instructions. It will be recalled that, in cognitive terms, inquiry mathematics involves creating and acting on experientially real mathematical objects. In sociological terms, it involves participating in the development of a taken-as-shared mathematical reality. The children's small-group activity provides numerous indications that the teacher was generally successful in guiding the development of an inquiry mathematics tradition. For example, although Katy demonstrated that she could perform the standard addition algorithm in her interview, she did not do so with any regularity in the small-group sessions until she had made a major conceptual reorganization and could give the steps of the algorithm numerical significance. This indicates that she considered this solution method to be inappropriate until she could explain it in terms of actions on mathematical objects. In addition, both Ryan and Andy, who had not made this conceptual reorganization, rejected their partners' use of the standard algorithm. On one occasion, Andy in fact referred to it somewhat disparagingly as "mixing up a bunch of numbers."

The case studies also indicate that counting by tens and ones on the hundreds board generally seemed to involve the creation of counting-based abstract composite units of ten. Several instances were documented in which children who could not create units of this type in the absence of situation-specific imagery attempted to imitate the partner who counted by tens and ones. In each case, the children gave up the attempt, presumably because their counting acts did not have numerical significance for them. Overall, seven of the eight children were rarely, if ever, observed solving tasks by following procedural instructions. Instead, within the constraints of their small-group relationships, they attempted to develop personally meaningful solutions that carried the significance of acting on arithmetical objects.

Andrea was the one exception to this conclusion, in that she produced instrumental solutions that involved following procedural instructions both in her interview and when working with Andy. During the teaching experiment, it was apparent that her beliefs about the nature of mathematical activity were at odds with those of the other members of the class. The case studies clarify why she did not reorganize her beliefs: It appears that she was frequently unable to make mathematical interpretations that were compatible with those of the teacher and the other children. This was particularly the case when she established collection-based numerical meanings while attempting to solve missing addend and missing subtrahend tasks. In these instances, it might well have appeared to her that the other children were solving tasks by following some unknown procedural instruction. If this was the case, her experiences in the classroom would have tended to corroborate rather than to conflict with her instrumental beliefs. She would then have no reason to reorganize her beliefs (cf. Krummheuer's discussion of different framings, chap. 7, this volume).

Interviews and Small-Group Sessions. Arguments concerning the socially situated nature of cognition raise the possibility that the children might have experienced the interviews and small-group sessions as different social situations. As it transpired, the interviews were reasonably good predictors of their mathematical activity in the classroom. This does not, of course, mean that the interviews somehow tapped the children's pure cognitions, undistorted by social influence. Instead, it indicates that the obligations they attempted to fulfill to be effective in the interviews were similar to those that they attempted to fulfill in the classroom (cf. Lave & Wenger, 1991). In Andrea's case, this could involve following procedural instructions, whereas for the other children it involved engaging in inquiry mathematics activity.

Michael was the most obvious exception to this general conclusion. In the classroom, he typically attempted to develop personally meaningful solutions that had numerical significance for him. However, in his interview, his solutions to the strips-and-squares tasks appeared to be somewhat instrumental in nature. It appears that while interacting with the interviewer, he learned to be effective by differentiating between the number–word sequences used to count a collection of strips and those used to count a collection of individual squares. In doing so, he seemed to view the strips as singletons rather than as composites of 10 units of one. The interviewer, for his part, was not aware that Michael might have been solving the tasks in this way. At the time, the interviewer in fact thought that his interventions were generally successful and that Michael had reorganized his conceptual activity and was counting abstract composite units of ten. However, an analysis of their interactions suggests that Michael interpreted these interventions as indications that his mathematical activity was inappropriate. For him, the interview then became a situation in which he was obliged to figure out how the interviewer wanted him to solve the tasks. Thus, the interviewer unknowingly supported Michael's development of solutions that were based on number–word regularities rather than numerical meanings.

The miscommunication between Michael and the interviewer emphasizes the importance of viewing interviews as social events. It is essential to coordinate an analysis of the child's cognitive activity with an analysis of the obligations and expectations that are negotiated. Only then is it possible to infer what the socially constituted tasks might be for the child and thus what his or her intentions and purposes might be. As a further point, the analysis of Michael's interview calls into question claims made in discussions of so-called scaffolding instruction that focus almost exclusively on what the adult does to support the child's activity, and on changes in the child's observed solutions to specific types of tasks.

Learning Opportunities

Univocal Explanation. Michael and Holly routinely engaged in interactions that involved univocal explanation both when their solutions were in conflict and when Holly judged that Michael did not understand. In these interactions, Holly, as the mathematical authority, was obliged to explain her solutions. Michael, for his part, was obliged to try to understand her mathematical activity. Even though Holly attempted to fulfill her obligation of helping Michael understand, there was no indication that these exchanges were productive for either child.

This conclusion indicates that the mere act of explaining does not necessarily give rise to learning opportunities. In this regard, Webb (1989) contended that giving help facilitates a student's learning only if the helper clarifies and organizes his or her thinking while explaining a solution in new and different ways. Further, Pirie (personal communication, August 1991) reported that interactions involving univocal explanations can be productive when one student is an active, participatory listener. However, Michael rarely asked clarifying questions or indicated which aspects of Holly's solutions he did not understand. Consequently, Holly merely described how she had solved a task when she gave an explanation. Although it is not possible to unequivocally explain why Michael did not play a more active role, it can be noted that the differences in his and Holly's conceptual possibilities may have been such that it was difficult for them to establish a viable basis for mathematical communication. In any event, it seems premature to conclude from Michael and Holly's case study that small-group relationships in which one child is the established mathematical authority must inevitably be unproductive.

Multivocal Explanation. Both groups of Ryan and Katy and Andrea and Andy engaged in multivocal explanations with some regularity. In Ryan and Katy's case, exchanges characterized by argument and counterargument appeared to give rise to learning opportunities for both children. In addition, they sometimes developed novel joint solutions. Further, they both seemed to believe that they had solved a task for themselves when they participated in these exchanges. This was not the case when Ryan responded to a sequence of questions that Katy posed in an attempt to lead him through her solutions. In general, the comparison of Katy and Ryan's and Michael and Holly's case studies indicates the need to take account of the social situations that children establish when assessing the role that particular activities such as explaining can play in their mathematical development (for a further description of this dyad, see Yackel, chap. 4, this volume).

In contrast to Ryan and Katy, interactions involving multivocal explanation did not appear to be particularly productive for either Andrea or Andy. They had difficulty establishing a taken-as-shared basis for communication and, as a consequence, frequently talked past each other. The difficulties they experienced appeared to reflect differences in both their beliefs about mathematical activity and their interpretations of particular tasks. It was noted, for example, that an explanation for Andrea could involve stating the steps of a procedural instruction, whereas for Andy it had to carry numerical significance. As a consequence, there were occasions when Andrea felt that she had adequately explained a solution and Andy persisted in asking her to explain what she was doing. Further, as was observed when accounting for the persistence of Andrea's instrumental beliefs, her interpretations of tasks were sometimes at odds with those of Andy, the teacher, and the other member of the class. As a consequence, exchanges that might have given rise to learning opportunities for both children instead often resulted in feelings of frustration.

Direct Collaboration. Interactions that involved direct collaboration did not appear to be particularly productive for any of the children. Consider, for example, a collaborative solution in which one child reads the problem statement, the other counts to solve it on the hundreds board, and the first child records the answer. This example is paradigmatic in that the children typically engaged in direct collaboration when a task was routine for both of them and the intended solution was taken as shared. As a consequence of this compatibility in their interpretations, the possibility that learning opportunities might arise was relatively remote (for a further description of this dyad, see Yackel, chap. 4, this volume).

Indirect Collaboration. Jack and Jamie were the only group who engaged in indirect collaboration with any regularity. In these situations, they thought aloud while apparently solving tasks independently. However, the way in which they frequently capitalized on each other's comments indicates that they were, in fact, monitoring each other's activity to some extent. Learning opportunities arose when what one child said and did happened to be significant for the other at that particular moment within the context of his ongoing activity. The occurrence of such opportunities, of course, requires a reasonably well-established basis for communication. In its absence, Jack and Jamie would merely have engaged in independent activity (for a further description of the argumentation pattern between Jack and Jamie, see Krummheuer, chap. 7, this volume).

Summary. A comparison of the learning opportunities that arose for the four pairs of children indicates that interactions involving both indirect collaboration and multivocal explanation can be productive. The former type of interaction occurs when both children are still attempting to arrive at a solution, and the latter type occurs when they attempt to resolve conflicting interpretations, solutions, and answers. In both cases, the establishment of a taken-as-shared basis for communication seems essential if learning opportunities are to occur. Significantly, both types of interaction involve a reciprocity in the children's roles, with neither being the mathematical authority. In contrast, learning opportunities were relatively infrequent for the one pair of children who routinely engaged in univocal explanation. As a caveat, it should be noted that situations of this type might be productive if one child is an active participatory listener.

Piagetian Perspectives

Piaget's (1967) discussion of the role of intellectual interchange in cognitive development focused on qualitative changes in children's logicomathematical conceptions. In the course of his analysis, Piaget delineated three requirements that interchanges between peers need to satisfy in order to contribute to cognitive reorganization (Rogoff, 1990). The first is that there be a reciprocity between the partners such that the children have an awareness of and interest in exploring alternatives to their own perspective. In the teaching experiment classroom, the teacher's attempts to initiate and guide the renegotiation of classroom social norms served to support the development of this awareness and interest. For example, children who acted in accord with the small-group norms of explaining their thinking and attempting to understand the partner's thinking were necessarily exploring alternatives to their own perspective. Although the teacher had a fair measure of success in guiding the establishment of these norms, it was observed that Jamie, the social authority in his group, had no interest in Jack's perspective when a task was routine for him. Further, Michael viewed Holly as a mathematical authority and rarely questioned or challenged her explanations. In line with Piaget's analysis, neither of these relationships was particularly productive, whereas exchanges in which Katy and Ryan responded to each other's explanations, justifications, and challenges frequently gave rise to learning opportunities.

The frequent references made earlier to the teacher's role in guiding the development of social norms indicate the need to transcend the restrictions of a purely cognitive analysis. In particular, it seems essential to locate the children's developing awareness of and interest in each other's perspectives within the broader classroom social context that they and the teacher established. Only then can their awareness and interest be viewed as both a cognitive achievement and as an aspect of the classroom microculture.

The second requirement for productive exchanges identified by Piaget is that the partners must have attained sufficient intersubjectivity to allow them to explore the existence and value of alternative perspectives. In other words, the children must have developed a taken-as-shared basis for mathematical communication (cf. Voigt, chap. 5, this volume). The comparison of Ryan and Katy's activity with that of Andrea and Andy when they engaged in multivocal explanation illustrates this point. These exchanges were productive for Ryan and Katy, who had established a taken-as-shared basis for mathematical communication. However, Andrea and Andy frequently developed feelings of frustration as they talked past each other while attempting to resolve conflicts.

The case studies indicate that the differences in the children's cognitive capabilities influenced the extent to which each pair was able to establish a taken-as-shared basis for mathematical communication. However, it should also be noted that, with the exception of Andrea, the teacher had influenced their beliefs as she guided the renegotiation of classroom norms. As a consequence, all but Andrea had a taken-as-shared understanding of what counted as a problem, a solution, an explanation, and a justification. Further, the children's mathematical development was both facilitated and constrained by their participation in the mathematical practices of the classroom community. Thus, as was the case when discussing the reciprocity of perspectives, it is essential to go beyond a purely cognitive analysis. Only then can the children's attempts to communicate effectively be related to their participation in the classroom microculture.

The third requirement proposed by Piaget is that children conserve their claims and conjectures and thus avoid contradicting themselves. As the case studies indicate, the second graders did, for the most part, avoid contradicting themselves when they engaged in mathematical argumentation. It is tempting to account for this observation by referring to the genesis of the children's development of logical thought. However, the findings of an investigation reported by Balacheff (1991) call into question a purely cognitive explanation of this type. In Balacheff's study, secondary school students first worked in groups and then discussed their solutions in a whole-class setting. Balacheff found that these students often challenged others' solutions on the basis of apparently superficial features such as their simplicity or complexity. Further, he observed that "some students can move in the same argumentation from one position to another which is completely contradictory" (p. 186). As a consequence, solutions were often accepted by the class, "provided some modification had been made, but their validity had not really been discussed" (p. 186).

Billig's (1987) arguments concerning the importance of criticism of inconsistency help clarify why the second graders generally conserved their claims and conjectures, whereas some of Balacheff's secondary school students did not do so. The students in Balacheff's investigation were not challenged when, from the adult perspective, they contradicted themselves. Consequently, they continued to be effective in the classroom when their arguments were inconsistent. In contrast, the second graders in the project classroom had learned to challenge each other when inconsistent arguments were proposed (Yackel & Cobb, 1993). The teacher had played an active role in bringing this about, by initiating and guiding the renegotiation of what counts as an acceptable explanation and justification in her classroom. Thus, in place of a purely cognitive explanation, it is again necessary to consider the differing norms or standards of argumentation established by the two classroom communities when accounting for the apparent conflict between Balacheff's findings and the small-group case studies.

In general, the case studies are consistent with Piaget's analysis of intellectual interchange. However, for each of the three requirements for productive interchanges that he identified, it seems necessary to coordinate psychological and sociological perspectives by considering aspects of the classroom microculture. In this regard, the case studies substantiate Bauersfeld's (1992) claim that an analysis of small-group interactions must take account of the classroom social context jointly established by the teacher and students.

The social aspect of productive exchanges is further highlighted by the consistency between the case studies and Halliday and Hasan's (1985) characterization of discourse in terms of its field, tenor, and mode. The *field of discourse* refers to the general nature of people's social actions and involves the specification of what is happening that requires them to use language at a given time and place. These issues were addressed in the case studies by discussing both the classroom mathematics tradition or microculture and the whole-class and small-group norms. The *tenor of discourse* refers to who is taking part, the nature of the participants, and their permanent or temporary status and roles. This aspect of discourse corresponds to the small-group relationships that were identified by analyzing the children's obligations and expectations. Finally, *the mode of discourse* refers to what part the language is playing, that is, what it is that the participants are expecting the language to do for them in a particular situation (e.g., persuade, teach, clarify, declare, etc.). Here, Halliday and Hasan seemed to be concerned with the taken-as-shared nature of a linguistic act. The case studies deal with this aspect of discourse by specifying the social situations the children established in the course of their interactions. Thus, the function of an explanation in an interaction involving univocal explanation differs from its function in multivocal explanation. In the former case it might be to impart information, whereas in the latter case it might be to convince or persuade.

Vygotskian Perspectives

Piaget's discussion of the relation between social interaction and learning was structured by his focus on qualitative shifts in children's logicomathematical conceptions. Vygotsky's (1960) treatment of these same issues was premised on a different set of assumptions, the most well-known of these is captured in the following frequently cited passage:

> Any higher mental function was external and social before it was internal. It was once a social relationship between two people. . . . We can formulate the general genetic law of cultural development in the following way. Any function appears twice or on two planes. . . . It appears first between people as an intermental category, and then within the child as an intramental category. (pp. 197–198)

Thus, for Vygotsky, individual psychological processes are derived from and generated by interpersonal relationships. Further, as this passage indicates, Vygotsky assumed that an asymmetry between the child and his or her social environment was the normative case. In his view, it is the adult's responsibility to help the child perform actions that are beyond his or her individual competence. As a consequence, Vygotsky was concerned about situations in which a peer group is left to cooperate without adult supervision (van der Veer & Valsiner, 1991). The investigation of small-group collaborative activity has, therefore, not been a primary interest of researchers working in the Vygotskian tradition.

An example of an analysis conducted in this tradition is provided by Newman, Griffin, and Cole's (1989) attempt to account for small-group learning in science in terms of the internalization of intermental actions that are socially distributed across the group. Forman's (1992) analysis of discourse and intersubjectivity in peer collaboration constitutes a second example. She argued that students have to coordinate two forms of discourse when they are asked to collaborate to learn in school. One of these is the cooperative attitude of everyday discourse, and the other is the academic discourse of the mathematics register. These and other analyses conducted in the Vygotskian tradition typically subordinate individual thought to social and cultural processes. In contrast, the view developed in the case studies is that of a reflexive relation between psychological and sociocultural processes, with neither being subordinated to the other.

The importance that Vygotsky attributed to social interaction and interpersonal relations was complemented by his emphasis on the use of cultural tools such as mathematical symbols. Thus, he argued "that children's participation in cultural activities with the guidance of more skilled partners allows children to internalize the tools for thinking" (Rogoff, 1990, p. 13). This implies that "children's cognitive development must be understood not only as taking place with social support in interaction with others, but also as involving the development of skill with sociohistorically developed tools that mediate intellectual activity" (Rogoff, 1990, p. 35).

Rogoff quoted Leont'ev (1981) to further clarify this sociocultural perspective on the relationship between cultural tools and the development of individual thought.

[For Vygotsky] the tool mediates activity and thus connects humans not only with the world of objects but also with other people. Because of this, humans' activity *assimilates the experience of mankind*. This means that humans' mental processes (their "higher psychological functions") acquire a structure necessarily tied to the sociohistorically formed means and methods [i.e., cultural tools] transmitted to them by others in the process of cooperative labor and social interaction. (Leont'ev, 1981, p. 55; quoted in Rogoff, 1990, p. 13)

It is in this sense that sociocultural theorists working in the Vygotskian tradition view cultural tools as carriers of meaning from one generation to the next. Bauersfeld (1990) noted that Vygotsky called them objective tools to emphasize that they serve as mediator objects that, by their use, relate the individual to his or her intellectual inheritance, the mathematical knowledge of the culture.

Rogoff (1990) attempted to resolve the apparent conflicts between Piagetian and Vygotskian accounts of development by suggesting that qualitative shifts of the type investigated by Piaget might require shared communication and an awareness of others' perspectives, whereas a simple explanation or demonstration might be sufficient when children learn to use culturally developed tools for thinking. The case studies allow us to address Rogoff's speculation directly because, in sociocultural terms, they document the children's appropriation or mastery of a particular cultural tool, that of our conventional positional numeral system. The children's counting on the hundreds board is particularly significant in this regard because the columns were composed of numeral sequences such as 7, 17, 27. . . . It will be recalled that two of the children, Katy and Holly, developed solutions that involved counting by tens and ones on the hundreds board during the 10-week period covered by the case studies. The discussion details both the advances they made and the changing meanings that they gave to numeral sequences on the hundreds board. In addition, the interpretations that the two girls' small-group partners, Ryan and Michael, gave to their solutions are also considered. This analysis of the ways in which the children interpreted and used the hundreds board then provides a context in which to compare Vygotskian treatments of cultural tool use with what emerges from the case studies.

Orientation: The Social Setting. As the focus is on individual children's mathematical constructions, it is important to note that Katy's and Holly's learning did not occur in a social vacuum. A small number of children in the class could already count by tens and ones on the hundreds board in January. The teacher frequently legitimized these children's explanations by redescribing them in terms that she hoped might be more comprehensible to other children, by asking these children to demonstrate their solutions to the class, and by emphasizing that this was quicker than counting by ones. Thus, the children who could only count by ones had every opportunity to realize that there was a more efficient way of using the hundreds board. Further, although the teacher continued to stress that the children should solve tasks in ways that were meaningful to them and tried to ensure that they did not feel obliged to blindly imitate this way of counting, she nonetheless gave every indication that she valued these solutions. From the observer's perspective, the teacher was attempting to establish counting by tens and ones on the hundreds board as a taken-as-shared mathematical practice. For their part, the two children who made this advance contributed to the establishment of this practice by giving explanations that the teacher could capitalize on for the benefit of other children.

Katy. Katy's solutions during her January interview indicated that she could create abstract composite units of ten by relying on situation-specific imagery. Nonetheless, she counted exclusively by ones when she used the hundreds board during the five small-group sessions that followed the interview. She was first observed counting by tens and ones on the hundreds board in the sixth session, which occurred on January 27. This advance, in turn, appeared to reflect a reorganization she made in her counting activity in the proceeding session (January 26). Then, she first solved the task "How many do you have to add to III::: [36] to make IIIII:. [53]?" by counting on by ones from 36 to 53. However, she subsequently curtailed her counting activity and created ten as a numerical composite when she explained her solution to Ryan.

1 Katy: Here, you have that many numbers, 36, and you add 10 more makes 46 (holds up both hands with all 10 fingers extended), 47, 48 . . . 53 (puts up 7 fingers as she counts).
2 Ryan: Katy, look, you have to take away a 10 [the remainder of this statement is inaudible].
3 Katy: I'll show you how I got my number. See, you have 36, and add 10 more makes 46 (holds up both hands with all 10 fingers extended), 47, 48 . . . 53 (puts up 7 fingers as she counts). Do you agree with 17?

Here, Katy's acts of holding up all 10 fingers anticipated the visual result she would have produced had she counted by ones to 46. This solution was novel when compared with those she produced during the individual interview in that it did not appear to involve a reliance on situation-specific imagery.

On the following day, January 27, Katy curtailed her counting activity still further when she solved a sequence of number sentences. Instead of holding up all 10 fingers, she solved the sentence $31 + 19 = \_$ by reasoning, "31 plus 19 . . . is . . . 31, makes 41, 42, 43 . . . 50." This suggests that she may have created an abstract composite unit of ten in an image-independent manner. Later in this same session, she was observed counting by tens and ones on the hundreds board for the first time. In this situation, she began to solve the sentence $39 + 19 = \_$ by saying "39 plus 10, 39 plus 10 . . . " and then "39, 49. . . . " However, she was interrupted by Ryan and took the hundreds board he was using to solve the task as follows: "39, 49, that's 10" (points to 39 and 49 on the hundreds board), 49—50, 51, 52 . . . 58" (counts the corresponding squares on the hundreds board). Crucially, in developing this solution, Katy appeared to conceptualize 19 as a unit of 10 and 9 more before she began to count on the hundreds board.

Counting by tens and ones on the hundreds board subsequently became one of the routine ways in which Katy solved a wide range of tasks. There was no indication that her use of the hundreds board played any appreciable role in supporting the advance she made. Instead, it seems that counting by tens and ones on the hundreds board was a consequence of her newly developed ability to create abstract composite units of ten in the absence of situation-specific imagery. However, several solutions that she subsequently developed suggest that using the hundreds board might facilitate children's conceptual constructions once they can count by tens and ones. For example, in the very first session in which she was observed counting in this way, January 27, Katy explained to Ryan that she had solved the sentence 60 − 31 = _ as follows: (Counts on the hundreds board from 60 to 50 by ones) "Ten, that's 10, now we have 21 more to go (counts from 50 to 40 by ones), 10, that's 2 tens, now we need 1 more ten, (counts from 40 to 30 by ones), 10, and then 1 more makes 29." In giving this explanation, Katy demonstrated that she could reflect on and monitor her activity of creating composite units of ten when she used the hundreds board. An exchange that occurred a week later (February 3) with one of the researchers indicates that her use of the hundreds board supported her reflection on her ongoing activity.

In this episode, Katy first solved the task, "How many do you have to take away from 74 to leave III::. [35]?" by routinely counting from 74 to 35 on the hundreds board, "10, 20, 30, 1, 2, 3 . . . 9—39." At this point, the researcher intervened.

1 Researcher: Katy, could you do that problem without using the hundreds board?
2 Katy: We know we have that number [74] and we have that many left [35], so if you have that many [35], you just go up.
3 Researcher: How would you go up?
4 Katy: Like, 35—45, 55, 65, 75, but that's too much. So you take away a 1 and then you have . . . 74.
5 Researcher: So how many do you have to add to 35 to make 74?
6 Katy: You have 35—45, 55, 65, 75 (slaps her open hands on the desk each time she performs a counting act), then you take 1 away, then it's too much, then you take 1 away and you will have . . . 74.

In both of these attempts to solve the task without using the hundreds board, the composite units of ten she created while counting did not seem to be objects of reflection for her. In other words, although both the counting sequences she completed during the exchange with the researcher seemed to carry the significance of incrementing by tens, she seemed unable to step back from her ongoing activity and monitor what she was doing. By way of contrast, her explanation of 60 − 31 = _ indicates that she was aware of what she was doing as she created abstract composite units of ten. It seems reasonable to speculate that her use of the hundreds board was crucial in making this reflection possible.

Holly. Holly's solutions to interview tasks indicated that, like Katy, she could create abstract composite units of ten by relying on situation-specific imagery. In addition, she used a partitioning algorithm to solve balance tasks throughout the time that she was observed working with Michael. In this case, an act of partitioning a two-digit numeral such as 53 symbolized for her the conceptual act of partitioning the number it signified (i.e., 53 partitioned into 50 and 3 rather than 5 and 3).

Despite these reasonably sophisticated conceptions, Holly counted on the hundreds board exclusively by ones during the first six small-group sessions. Further, in contrast to Katy, counting by tens and ones on the hundreds board did not become a routine solution method for her. Although she was observed counting in this way on three different occasions, each seemed to involve a local, situation-specific advance. Significantly, there was no indication that Holly could create abstract composite units of ten in an image-independent manner when she gave number words and numerals counting-based meanings.

Holly was first observed counting on the hundreds board by tens and ones on February 2, when she worked with Michael and Katy. Throughout the session, Katy routinely counted by tens and ones. As it so happened, Holly had drawn three collections of 10 tally marks and 7 individual marks to solve a task corresponding to $37 + 25 = \_$ when she saw Katy count on the hundreds board: "25 plus 10 makes 35, plus 10 makes 45, plus 10 makes 55. 1, 2, 3, 4, 5, 6, 7 . . . 62." The way in which she subsequently began to solve tasks by counting by tens and ones on the hundreds board indicates that she interpreted Katy's solution against the backdrop of her own image-supported creation of composites of ten. For example, she explained to Michael that she had solved a task corresponding to $37 + 57 = \_$ as follows: (Points on the hundreds board) "I started at 57 and I counted down 10, 20, 30, and then I counted these 1, 2, 3, 4, 5, 6, 7, and I came up with 94." In contrast, she counted exclusively by ones when she used the hundreds board the following day (February 3). It would, therefore, seem that the advance she made was specific to her interactions with Katy and, further, that her drawing of groups of 10 tally marks was crucial.

The other two occasions when Holly was observed counting by tens and ones on the hundreds board also seemed to be situation specific. The first of these solutions occurred on March 2, when she and Michael solved a sequence of story problems involving money. This solution, in turn, seemed to build on a prior advance that she made while solving the task, "Krista has this much money. [A picture shows coins whose total value is 37¢.] She has 25¢ less than Larry. How much money does Larry have?" In this situation, she explained her solution to Michael as follows: "Add 10 cents, that makes 47, then add another 10, 57, then 58, 59, 60, 61, 62." Her explicit reference to coin values ("add 10 cents") suggests that situation-specific imagery may have supported her conceptualization of the 25¢ as two units of 10 and 5 more.

Later in this same session, she used a hundreds board to solve the task, "Scott has this much money. [A picture shows coins with a total value of 42¢.] He has 30¢ more than Susan. How much money does Susan have?" To do so, she first counted the values of the pictured coins and then reasoned: "Then take away, 42, 41, 40, 39, 38, 37, 36, 35, 34.... I can't do this. We need the hundreds board (picks one up). 42 (points to 42 on the board and then moves her finger up three rows)—12." Given the nature of her solution to the preceding money word problem, it seems reasonable to infer that the composite units she counted when she moved up three rows on the hundreds board were image supported.

The final occasion on which Holly counted by tens and ones on the hundreds board occurred on March 23, when she solved a balance task corresponding to $16 + 16 + 16 = \_$. As noted previously, Holly had used her partitioning algorithm routinely to solve tasks of this type since the beginning of January. If she interpreted this task in this way, she would have created composite units of ten only because all the addends happened to be in the teens. In other words, if one of the addends had been, say, 36, she would have created 30 and 6 rather than 3 units of 10 and 6. Given that this solution was the only occasion during the session when she counted by tens and ones on the hundreds board, it seems reasonable to suggest that she did create units of ten in this situation-specific way. Her counting acts would, then, be expressions of the tens and sixes she created before she began to count.

In summary, there was no indication that Holly's use of the hundreds board played a significant role in supporting the temporary conceptual advances she made. On those occasions when she did count by tens and ones, she seemed to create abstract composite units of ten before she began to count. Thus, as was the case with Katy, counting by tens and ones on the hundreds board appeared to be a consequence of rather than the cause of her task interpretations. In contrast to Katy, solutions of this type did not become routine for Holly, presumably because she could not create composite units of ten in an image-independent manner when she gave number words and numerals counting-based meanings.

Ryan. During his interview, Katy's partner, Ryan, created abstract composite units of ten by relying on situation-specific imagery. Further, during his interview, he was asked how far he would have to count if he started at 43 and went to 53. He gave 10 as his answer and explained that he knew this because of the hundreds board. Nonetheless, he was never observed spontaneously counting by tens and ones on the hundreds board despite Katy's repeated attempts to explain her solutions to him. For example, on January 27, Ryan counted by ones on the hundreds board to solve the sentence $38 + \_ = 52$.

1	Ryan:	I'll start at 38.
2	Katy:	Plus ten makes 48.
3	Ryan:	No, look (attempts to count on the hundreds board by ones from 38 to 52)—16.

Here, in saying "Plus 10 makes 48," Katy seemed to be suggesting that Ryan curtail counting by ones. The way in which he flatly rejected this proposal indicates that it had nothing to do with solving the task from his point of view. Further, as was documented in his case study, the possibility that Katy created units of ten when she counted efficiently on the hundreds board did not occur to him.

There were only two occasions when Ryan was observed doing anything other than counting by ones when he used the hundreds board. The first occurred in this same session on January 27, when Katy persisted in explaining how she had solved the sentence 39 + 19 = _. Ryan, for his part, argued that the answer was 59.

1 Katy: (Points on the hundreds board) See, you have 39, and plus 10 more makes 49—50, 51, 52 . . . 58, that's 19, so it has to be 58.
2 Ryan: Okay, (points on the hundreds board), 10 plus that, 49, and then 50, 51, 52 . . . 59.

In this case, Ryan seemed to accommodate Katy's repeated explanations in an attempt to justify his answer of 59. The fact that he solved all subsequent tasks in the session by counting by ones suggests that he did not conceptualize 19 as a composite unit of 10 and 9 more. Instead, he seemed to imitate her solution by curtailing his count to 59.

The second occasion when Ryan attempted to count by tens and ones occurred 7 weeks later on March 16. In this instance, he initially solved a task corresponding to 17 + _ = 62 by counting by ones from 17 to 62, whereas Katy counted on the hundreds board, "10, 20, 30, 40—1, 2, 3, 4, 5—45". A researcher then asked Ryan if Katy had used "an okay method." He replied, "Yes, it's all right." Further, he reenacted her solution successfully at the second attempt when asked to do so by the researcher. However, as on January 27, he then proceeded to solve all subsequent tasks by counting by ones. This exchange indicates that he knew how she counted but did not understand why she did so. In this regard, it should be noted he never appeared to create abstract composite units of ten in the absence of situation-specific imagery when he worked with Katy. This inability to create units of ten when interpreting tasks would explain why his technical know-how with the hundreds board was insufficient. It would also explain why Katy's concerted attempts to explain and demonstrate how she used the hundreds board were unsuccessful.

Michael. Michael's performance during his individual interview indicated that the creation of abstract composite units of ten was not within the realm of his developmental possibilities. Further, unlike Ryan, Michael counted by ones when asked how far he would count if he started at 43 and went to 53. He did, however, know immediately that he would perform 10 counting acts to get from 40 to 50 and referred to the hundreds board to explain his answer.

Throughout the time that he worked with Holly, Michael was never observed successfully counting by tens and ones on the hundreds board, even when he attempted to imitate one of her solutions. Nonetheless, the fact that he did attempt to solve tasks in this way on one occasion indicates that he was aware that there was a more efficient way of counting on the hundreds board. This attempt occurred on March 5, when he began to count by ones on the hundreds board to solve the task, "How many do you add to 37 to get 91?" He suddenly stopped counting, saying "I've got a faster way." He then counted "1, 2, 3 . . . 9" as he pointed on the hundreds board to the squares 38, 39, 40, 50, 60, 70, 80, 90, 91. Thus, he seemed to be creating a path on the board from 37 to 91. Next, he retraced this path by counting (and pointing to squares) as follows: "1 (38), 2 (39), 10 (40), 20 (50), 30 (60), 40 (70), 50 (80), 60 (90)." Finally, he decided to check this result by counting by ones and arrived at an answer of 54.

This sequence of solution attempts suggests that, in contrast to Ryan, Michael was yet to learn how to count by tens and ones on the hundreds board. For Ryan, moving down a column on the hundreds board from, say, 37 to 47, seemed to signify a curtailment of counting the 10 intervening squares by one. In contrast, it seemed that, for Michael, only the column 10, 20, 30 . . . signified the curtailment of counting by ones.

A solution attempt that Michael made on March 23 reveals much about his understanding of counting on the hundreds board. On that day, he and Holly attempted to solve a task corresponding to $16 + 16 + 16 = \_$. Holly partitioned the 16s into 10 and 6 and attempted to count by tens and ones. Michael, for his part, counted by ones, but his answer conflicted with that which Holly proposed. He continued, "I think I can settle this. 6 and 6 is 12 (then attempts to count another six on the hundreds board)—19. 19 plus 3 ones, 20, 21, 22, so it must be 22."

Michael seemed to realize that Holly had partitioned the 16s. However, as the act of partitioning a two-digit numeral such as 16 did not symbolize the conceptual act of partitioning the number it signified, the result was 1 and 6 rather than 10 and 6. His count of 3 sixes and 3 ones on the hundreds board was an expression of the task as he understood it. Significantly, his use of the hundreds board did not help him to conceptualize 16 as 10 and 6 when he interpreted Holly's solution.

Conceptually Restructuring the Hundreds Board. The analyses just presented indicate that the four children's use of the hundreds board did not support their construction of increasingly sophisticated conceptions of ten. This conclusion is consistent with the problem-solving activity of the other four children included in the case studies. None of these children counted on it effectively by tens and ones to solve tasks. It instead appears that the children's efficient use of the hundreds board was made possible by their construction of increasingly sophisticated place–value conceptions. A comparison of the four children's case studies leads to the identification of two steps in this developmental process. The first results in knowing how to count by tens and ones on the hundreds board, and the second culminates in the ability to use the hundreds board in this way to solve a wide range of tasks. The contrast between Michael's and Ryan's problem-solving efforts clarifies what is involved in making the first of these two developmental steps.

Recall that Ryan seemed to understand how Katy counted on the hundreds board. This understanding seemed to be based on his realization that moving down a column from, say, 37 to 47 curtailed counting the 10 intervening squares. He therefore gave numerical significance to regularities in the numerals (e.g., 7, 17, 27, 37 . . .) and, by so doing, structured the hundreds board in terms of numerical composites of 10 units of one. It seems reasonable to assume that Michael could also abstract these regularities from the numerals on the hundreds board. During his individual interview, for example, he produced the number word sequence "4, 14, 24 . . . 94" when the interviewer first put down 4 squares and then repeatedly put down a strip of 10 squares. Nonetheless, the hundreds board did not seem to be structured into numerical composites of ten for him in the way that it was for Ryan. Instead, only the column 10, 20, 30 . . . seemed to signify the curtailment of counting by ones.

Michael's relatively unstructured interpretation of the hundreds board can be accounted for by first noting that a break or pause in the rhythm of counting by ones occurs at the end of each row. These breaks in sensory-motor counting activity might facilitate the isolation of the count of an intact row (e.g., 51, 52 . . . 60) as a numerical composite of 10 units of one. Counting down the column 10, 20, 30 . . . would then signify the curtailment of counting along each row by ones. The possibility that Michael had made this construction is supported by his reference to the hundreds board during his individual interview to explain why he would perform 10 counting acts when going from 40 to 50.[2] In contrast, there are no similar breaks or pauses that might facilitate the isolation of, say, 38, 39, 40 . . . 47 as a composite of ten. Instead, a pause occurs in the middle of this sequence of counting acts. Consequently, the process of giving numerical significance to a sequence such as 7, 17, 27 . . . requires the child to transcend the constraints of sensory-motor counting activity. Michael appeared unable to create numerical composites of ten in this image-independent manner, whereas Ryan was able to do so.

The second developmental step in learning to use the hundreds board as an efficient problem-solving tool can be clarified by considering why Holly learned from Katy whereas Ryan did not. As has been documented, Holly attempted to solve the sentence 37 + 25 = _ by drawing three groups of 10 tally marks and 7 individual marks. This made it possible for her to create abstract composite units of ten and thus make sense of Katy's counting activity on the hundreds board. As a consequence, she knew why Katy had counted on 3 tens and 7 ones from 25. Ryan, however, seemed to misinterpret Katy's counts by tens and ones. Recall, for example, that he assumed she had misread the sentence 39 + 19 = _ when she said "39 plus 10." From his point of view, adding 10 to 39 had nothing to do with solving the task. Although he structured the hundreds board by creating numerical composites of 10 units of one, there was no indication that these composites were single entities or units for him. Further, he did not appear to create abstract composite units of ten in an image-independent manner when he interpreted tasks. Consequently, the composite units that Katy expressed by counting on the hundreds board by tens and ones did not exist for him.

---

[2] Even in this case, the hundreds board did not serve as a carrier of meaning. Michael had to abstract from his sensory-motor counting activity to come to this realization. The most that can be said is that counting by ones on the hundreds board was the experiential situation in which he made this abstraction.

We saw that Holly could only create composite units of ten in an image-supported manner and that the advances she made in using the hundreds board were situation specific. In contrast, there was strong evidence that Katy created the composite units she routinely expressed by counting in an image-independent manner. Her use of the hundreds board, in turn, enabled her to reflect on and monitor her activity of counting by tens and ones. It seems reasonable to infer that, for her, the hundreds board was structured in terms of abstract composite units of ten rather than numerical composites of 10 ones. The units of ten and one that she expressed by counting were then simply there in the hundreds board as objects of reflection for her. Thus, her ability to create abstract composite units of ten in an image-independent manner made it possible for her to solve a variety of tasks by counting efficiently on the hundreds board.

In general, the case studies do not support Rogoff's speculation that simple demonstration and explanation might be sufficient when children learn to use a culturally developed tool such as the conventional positional numeral system for thinking. The three requirements for productive intellectual interchanges identified by Piaget, therefore, seem relevant when considering the types of social interaction that might facilitate children's attempts to use cultural tools, such as mathematical symbols, effectively.

Cultural Tools and Mathematical Activity. Thus far, the discussion has emphasized that children's use of a cultural tool such as conventional positional notation is constrained by their current interpretive possibilities. Lest this account seem one-sided, it should also be stressed that children's use of cultural tools can facilitate their subsequent mathematical development. For example, Katy used the numeral patterns on the hundreds board when she reflected on and monitored her activity of counting by tens and ones. Because she could create abstract composite units of ten in an image-independent manner, the units of ten and of one that she expressed by counting were simply there in the hundreds board as objects of reflection for her. Another instance in which the children's use of a cultural tool played an enabling role can be culled from the case studies. This concerns the balance format used to present tasks.

The analyses of Ryan's and Holly's mathematical activity indicate that both children's use of the partitioning algorithm was initially limited to balance tasks. This suggests that both children gave different meanings to two-digit numerals when they interpreted balance tasks and, say, number sentences. In particular, it would seem that they established numbers as arithmetical objects that could be conceptually partitioned only when they interpreted the notational scheme used to present balance tasks. In this regard, it should be noted that one metaphor that was generally taken as shared by the classroom community was that of a number in a box. For example, the teacher and students often spoke of the goal of a balance task to be that of finding the number that went in an empty box. Holly, in fact, used this metaphor on January 13, when she explained her interpretation of a task to Michael: "This is a take-away box, this is a take-away."

One characteristic of the box metaphor is that its use implies the bounding of whatever is placed inside it (Johnson, 1987). Consequently, Holly's and Ryan's interpretation of tasks in terms of this taken-as-shared metaphor may have supported their construction of numbers as discrete, bounded entities that they could act on conceptually. To the extent that this was the case, their creation of relatively sophisticated arithmetical entities was facilitated by a task-specific convention of interpretation. This explanation does not, of course, imply that a taken-as-shared metaphor carries with it a mathematical meaning that is self-evident to the adult. The analysis of Michael's mathematical activity indicated that numerals in boxes did not signify discrete arithmetical objects for him. Instead, it seems that a metaphor such as that of a number in a box can support some students' active, image-supported construction of experientially real arithmetical objects. The mathematical significance of the metaphor is therefore relative to a student's developmental possibilities.

Following Saxe (1991), this discussion can be summarized by saying that the notational scheme used to present balance tasks was deeply interwoven with Holly's and Ryan's cognitive activity. A similar point can be made with regard to the conventional positional numeral system. It was noted, for example, that Katy's sophisticated counting solutions were based in part on regularities she abstracted from the numerals on the hundreds board. Further, as was the case with the number-in-a-box metaphor, the significance of these regularities for the children was relative to their conceptual possibilities. Katy's mathematical reality was such that the numeral pattern 7, 17, 27 ... meant 10 more. Thus, her use of the hundreds board transcended the separation of syntax and semantics (cf. Winograd & Flores, 1986). The two were reflexively related in her activity in that the numeral patterns both constrained and enabled her numerical interpretations, and her numerical understanding made the numeral patterns significant when she used the hundreds board to solve tasks. In contrast, the numeral patterns that Michael had abstracted did not carry numerical significance and, as a consequence, did not involve the integration of syntax and semantics.

A crucial difference between the number-in-a-box metaphor and the numeral patterns is, of course, that the former was specific to the classroom community, whereas the latter are taken as shared by wider society. In this regard, when Michael's activity with the hundreds board is interpreted in terms of his mathematical enculturation, it is reasonable to say that he had abstracted *notational regularities* that are consensual with those constructed by members of wider society. Katy, in contrast, had constructed a *symbolic numerical practice* consensual with that of wider society. Here, the term *symbolic* of course implies that syntax and semantics were interrelated reflexively. More generally, this conception of symbolic mathematical activity touches on a central aspect of mathematical experience discussed by Kaput (1991):

> One can draw an analogy between the way the architecture of a building organizes our experience, especially our physical experience, and the way the architecture of our mathematical notation system organizes our mathematical experience. As the physical level architecture constrains and supports our actions in ways that we are often unaware of, so do mathematical notation systems. (p. 55)

In proposing this analogy, Kaput acknowledged Vygotsky's contention that our mathematical activity is influenced by the sociohistorically developed tools that we use. The issue at hand concerns the nature of this influence. If it is accepted that syntax and semantics are reflexively related, then it is not mathematical notation per se that constrains our thought; rather, it is our enculturation into consensual symbolic mathematical activities. To the extent that notational regularities are given mathematical significance, the process of learning to use a cultural tool such as a mathematical notation system involves the construction of an experiential mathematical reality that is consensual with those of others who can be said to know mathematics. Consequently, when we use conventional notation individually or when communicating with others, we participate in the continual regeneration of a consensual mathematical reality, and that consensual reality both constrains and enables our individual ways of thinking and notating. Both Mehan and Wood (1975) and Bruner (1986) drew attention to the more general phenomenon that we create the world about which we speak as we speak about it. The reflexivity of linguistic activity in general, and of symbolic mathematical activity in particular, indicates that Kaput's analogy is appropriate. We see the abstract mathematical reality we create symbolically as we look though the cultural tools we use.

Vygotskian and Emergent Approaches. The analysis presented of the children's conceptual restructuring of the hundreds board is reminiscent of Vygotsky's (1982) discussion of how a chessboard is perceived by people with differing levels of expertise:

> The one who does not know how to play chess perceives the structure of the pieces from the perspective of their external characteristics. The meaning of the pieces and their position on the board fall completely outside of his [or her] perspective. The very same chessboard provides a different structure to the person who knows the meanings of the pieces and the moves. For him [or her], some parts of the table become the background, others become the focus. Yet differently will the medium-level and excellent chess players view of the board. Something like that takes place in the development of the child's perception. (pp. 277—278)

This passage illustrates that Vygotsky was interested in qualitative developmental change. The primary difference between his sociocultural perspective and the approach exemplified by the case studies concerns the way in which these changes are accounted for.

In Vygotskian terms:

> Human thinking—and higher psychological processes in general—are primarily overt acts conducted in terms of the objective materials of the common culture, and only secondarily a private matter. The origin of all, specifically human, higher psychological processes, therefore, cannot be found in the mind or brain of an individual person but rather should be sought in the social "extracerebral" sign systems a culture provides. (van der Veer & Valsiner, 1991, p. 222)

In contrast, the analysis I have presented focuses on individual children's cognitive construction of increasingly sophisticated place–value numeration concepts. In this combined constructivist and interactionist approach, an account of the origin of psychological processes involves analyses of both the conceptual constructions of individual minds and the evolution of the local social worlds in which those minds participate.

It is tempting to respond to the conflicting assumptions of the two approaches by claiming that one side or the other has got things right. However, in my view, they have evolved to address different problems and issues. Considering first the sociocultural perspective, Vygotsky appeared to have developed his theoretical orientation while addressing problems of cultural diversity and change. In particular, he seemed to have been committed to the notion of "the new socialist man" and to have viewed education as a primary means of bringing about this change in Soviet society (Kozulin, 1990; van der Veer & Valsiner, 1991). Further, much of his empirical research focused on differences in the activity of members of different cultural groups in the Soviet Union.

The vantage point from which he analyzed psychological development appears to be that of an observer located outside the cultural group. From this point of view, thought and activity within a cultural group appear to be relatively homogeneous when compared with the differences across groups. An issue that immediately comes to the fore is, then, that of accounting for the social and cultural basis of personal experience. Vygotsky's sociohistorical theory of development can, in fact, be interpreted as a response to this problem (Kozulin, 1990; Wertsch, 1985). In addressing this issue, it is reasonable to treat learning as primarily a process of enculturation, and to emphasize the crucial role played by both children's interactions with more knowledgeable others and their mastery of tools that are specific to the culture. From this perspective, place–value numeration might be viewed as a culturally organized way of thinking whose appropriation involves mastering a culturally specific tool, our numeration system. An analysis conducted from this perspective might, therefore, investigate how and to what extent children's use of instruction devices such as the hundreds board facilitates this appropriation process.

In contrast to Vygotsky's interest in cultural diversity and change, constructivists and interactionists (henceforth called *emergent theorists*) typically focus on qualitative differences in individuals' activity. Thus, whereas Vygotsky emphasized homogeneity within a cultural group, emergent theorists emphasize the diversity of group members' activity. The position from which they analyze activity is, therefore, that of an observer located inside the group. An issue that comes to the fore is that of accounting for the diversity in individual learning by investigating how individuals reorganize their activity as they participate in the practices of a local community, such as that formed by the teacher and students in a classroom. Consequently, whereas analyses conducted in Vygotskian terms investigate the influence that mastery of a cultural tool has on individual thought, analyses conducted from the emergent perspective focus on the individual conceptual constructions involved in learning to use a cultural tool appropriately. From this latter perspective, it is the practices of the local community rather than those of wider society that are taken as a point of reference. With regard to the analysis of the children's use of the hundreds board, for example, Katy can be seen to have contributed to the establishment of counting by tens and ones as a classroom mathematical practice once she could create composite units of ten in an image-independent manner. In this account, her mathematical activity contributed to the establishment of local mathematical practices that both enabled and constrained her individual activity. Thus, in contrast to Vygotsky's focus on the social and cultural basis of personal experience, this combined constructivist and interactionist analysis highlights the contributions that actively interpreting individuals make to the development of local social and cultural processes (cf. Cobb, 1989).

Given these differences between the two theoretical approaches, the interesting challenge for the future is that of exploring ways in which they might complement each other by delineating situations in which one or the other might be more appropriate. This task seems worth pursuing, given that together they span the actively cognizing student, the local social situation of development, and the established mathematical practices of the wider community.

Implications of the Case Studies

Some of the conclusions reached while conducting the case studies were genuinely surprising to the American researchers, even though they had visited the classroom on an almost daily basis during the teaching experiment. As a consequence, the initial impressions about the productiveness of particular groups' interactions had to be revised. For example, during the teaching experiment, Michael and Holly's relationship was thought to be reasonably productive. Holly consistently attempted to help Michael understand her more sophisticated solutions, and few interpersonal conflicts arose as they worked together. Conversely, given the continual tension in Ryan and Katy's interactions, there were doubts about the productiveness of their relationship. As it transpired, the relative harmony of a group's interactions was not a good indicator of the occurrence of learning opportunities. Instead, two aspects of children's social relationships seem critical for their learning in a classroom in which an inquiry mathematics microculture has been established and the children can conserve their claims and conjectures. These are:

1. The development of a taken-as-shared basis for mathematical communication.
2. The routine engagement in interactions in which neither child is the mathematical authority, namely those involving multivocal explanation.

In this regard, Steffe and Wiegel (1992) argued that cooperation among children

> should involve give-and-take and a genuine exchange of ideas, information, and viewpoints. To engage in such cooperation, a child must be able to assimilate the activity of the other, and in an attempt to understand what the other is doing, modify the assimilated activity as a result of interactive communication until there is an agreement about what the activity means. (p. 459)

In other words, the children should have developed a taken-as-shared basis for communication. However, this is not

> enough to ensure genuine cooperation. Cooperative learning should not be reduced to one child attempting to understand the activity of the other. Rather, each child also has the obligation to become involved in his or her productive activity that might yield a result, and then be willing to correlate the result with the result of the other. Cooperation is possible only if the involved children are striving to reach a common goal. (Steffe & Wiegel, 1992, pp. 459–460)

Thus, neither child should be the mathematical authority in the group. The only caveat to add to this argument is that it might be premature to dismiss as unproductive interactions involving univocal explanation in which one child is an active participatory listener. As noted, none of the four groups described in the case studies engaged in interactions of this type.

On the surface, the two criteria listed previously indicate that the differences in the children's conceptual possibilities should be relatively small. Such a conclusion, of course, leads to difficulties in that homogeneous grouping clashes with a variety of other agendas that many teachers rightly consider important, including those that pertain to issues of equity and diversity. Further, it can be noted that strongest predictor of children's eventual reading ability is the reading group to which they are assigned in first grade. At the very least, it should be clear that general rules or prescriptions for organizing small-group activity cannot be derived from the case studies. Lest this outcome seem unsatisfactory, it should be noted that general principles of conduct are almost invariably too extreme and must be balanced by conflicting principles (Billig, 1987). This is clearly the case in a complex activity such as teaching, which is fraught with dilemmas, tensions, and uncertainties (Clark, 1988). Within the context of such activity, the proposed criteria suggest ways in which teachers might look at and interpret small-group interactions when making professional judgments. In this regard, it appears that neither harmonious, on-task activity nor the mere occurrence of explanations are good indicators of interactions that are productive for mathematical learning. Instead, the criteria indicate that teachers should monitor the extent to which children engage in genuine argumentation when they solve tasks and discuss their solutions. Further, teachers should intervene as necessary to guide the development of small-group norms that make genuine argumentation possible (cf. Wood & Yackel, 1990).

The comments made thus far take the general organization of the teaching experiment classroom as a given and assume that the teacher should assign children to groups. In this regard, Murray (1992) described the interactions that occurred in a third-grade classroom in which children initiated their solutions independently but were then encouraged to collaborate in groups of two or three. Although the teacher did intervene, she left the final choice of whom to collaborate with to the students. In such an arrangement, the groups were not stable but broke up and reformed according to the needs and interests of different students.

The children in Murray's classroom had more responsibility for and control over their social interactions than the children in the second-grade teaching experiment classroom had. Nonetheless, the teacher with whom Murray worked continued to act as an authority and attempted to guide the establishment of two essential social norms. First, a student could only be helped with a solution strategy that he or she had already initiated. Second, a problem-solving episode could not be terminated unless each student had solved the problem successfully by means of a strategy that he or she had initiated. Further, it was the group's rather than the teacher's responsibility to ensure that this occurred.

Murray (1992) reported that the students engaged in intimate, reflexive discussions as they interacted with chosen peers. It would, therefore, seem that they were able to establish taken-as-shared bases for mathematical activity. Murray also noted that the less conceptually advanced students objected to cooperating with their more mathematically sophisticated peers. As one child put it, "This means that they start telling me what they think, and then I don't have the time to think for myself" (p. 8). The same child went on to say that he did not actually need direct suggestions about possible solution strategies from other group members, but that working with them would definitely enable both him and them to solve problems. In the terminology of the case studies, this child appears to be describing interactions that involve indirect collaboration and multivocal explanation. Murray argued that these children intuitively realized that they could not establish a taken-as-shared basis for mathematical communication with their more conceptually advanced peers. Instead, they believed that their development of solutions was aided by collaboration with true peers. This view is consistent with the findings of the case studies if, by *true peers*, Murray meant those who do not attempt to act as mathematical authorities.

In light of Murray's experiences, teachers might want to experiment with alternatives to the social arrangements established in the teaching experiment classroom. For example, approaches that require students to initiate their own solutions seem worthy of further investigation, given that the children in the teaching experiment classroom frequently experienced a tension between the obligations of developing their own solutions and explaining their thinking to the partner. Further, although teachers should clearly initiate and guide the development of classroom social norms, they might also want to go beyond the case studies by investigating whether and to what extent students can construct the social structures they need for optimal collaborative learning when they are encouraged to engage in voluntary social interactions with their peers.

ACKNOWLEDGMENTS

The development of these case studies was supported, in part, by the National Science Foundation (Grant No. RED 9353587), and by a Sir Alan Sewell Visiting Fellowship awarded by Griffith University in Brisbane, Australia. The author thanks George Booker for his warm hospitality.

REFERENCES

Atkinson, P., Delamont, S., & Hammersley, M. (1988). Qualitative research traditions: A British response to Jacob. *Review of Educational Research, 58,* 231–250.

Balacheff, N. (1991). The benefits and limits of social interaction: The case of mathematical proof. In A. J. Bishop (Ed.), *Mathematical knowledge: Its growth through teaching* (pp. 175–192). Dordrecht, Netherlands: Kluwer.

Barnes, D., & Todd, F. (1977). *Communicating and learning in small-groups.* London: Routledge & Kegan Paul.

Bauersfeld, H. (1990). *Activity theory and radical constructivism—What do they have in common and how do they differ?* (Occasional Paper 121). Bielefeld, Germany: University of Bielefeld, Institut für Didaktik der Mathematik.

Bauersfeld, H. (1992). A professional self-portrait. *Journal for Research in Mathematics Education, 23,* 483–494.

Billig, M. (1987). *Arguing and thinking: A rhetorical approach to social psychology.* Cambridge, England: Cambridge University Press.

Bruner, J. (1986). *Actual minds, possible worlds.* Cambridge, MA: Harvard University Press.

Clark, C. M. (1988). Asking the right questions about teacher preparation: Contributions of research on teacher thinking. *Educational Researcher, 17*(2), 5–12.

Cobb, P. (1986). Clinical interviewing in the context of research programs. In G. Lappan & R. Even (Eds.), *Proceedings of the eighth annual meeting of PME-NA: Plenary speeches and symposium.* East Lansing: Michigan State University.

Cobb, P. (1989). Experiential, cognitive, and anthropological perspectives in mathematics education. *For the Learning of Mathematics, 9*(2), 32–42.

Cobb, P., & Wheatley, G. (1988). Children's initial understandings of ten. *Focus on Learning Problems in Mathematics, 10*(3), 1–28.

Cobb, P., Wood, T., & Yackel, E. (1992). Learning and interaction in classroom situations. *Educational Studies in Mathematics, 23,* 99–122.

Cole, M. (1985). The zone of proximal development: Where culture and cognition create each other. In J. V. Wertsch (Ed.), *Culture, communication, and cognition: Vygotskian perspectives* (pp. 141–161). New York: Cambridge University Press.

Davidson, N. (1985). Small-group learning and teaching in mathematics: A selective review of the research. In R. Slavin, S. Sharan, S. Kagan, R. Hertz-Lazarowitz, C. Webb, & R. Schmuck (Eds.), *Learning to cooperate, cooperating to learn* (pp. 211–230). New York: Plenum Press.

Dörfler, W. (1991). Image schemata and protocols. In F. Furinghetti (Ed.), *Proceedings of the Fifteenth Conference of the International Group for the Psychology of Mathematics Education* (pp. 17–32). Genoa, Italy: Program Committee of the 15th PME Conference.

Erickson, F. (1986). Qualitative methods in research on teaching. In M. C. Wittrock (Ed.), *The handbook of research on teaching* (3rd ed., pp. 119–161). New York: Macmillan.

Forman, E. A. (1992). Discourse, intersubjectivity and the development of peer collaboration: A Vygotskian approach. In L. T. Winegar & J. Valsiner (Eds.), *Children's development within social contexts: Vol. 2: Metatheoretical, theoretical and methodological issues.* (pp. 143–159). Hillsdale, NJ: Lawrence Erlbaum Associates.

Forman, E. A., & Cazden, C. B. (1985). Exploring Vygotskian perspectives in education: The cognitive value of peer interaction. In J. V. Wertsch (Ed.), *Culture, communication, and cognition* (pp. 323–347). Cambridge, MA: Cambridge University Press.

Forman, E. A., & McPhail, J. (1993). A Vygotskian perspective on children's collaborative problem-solving activities. In E. A. Forman, N. Minick, & C. A. Stone (Eds.), *Education and mind: The interaction of institutional, social, and developmental processes* (pp. 213–229). New York: Oxford University Press.

Gale, J., & Newfield, N. (1992). A conversation analysis of a solution-focused marital therapy session. *Journal of Marital and Family Therapy, 18*, 153–165.

Glaser, B. G., & Strauss, A. L. (1967). *The discovery of grounded theory: Strategies for qualitative research*. New York: Aldine.

Good, T. L., Mulryan, C., & McCaslin, M. (1992). Grouping for instruction in mathematics: A call for programmatic research on small-group process. In D. A. Grouws (Ed.), *Handbook of research on mathematics teaching and learning* (pp. 165–196). New York: Macmillan.

Halliday, M. A. K., & Hasan, R. (1985). *Language, context, and text: Aspects of language in a social semiotic perspective*. Oxford, England: Oxford University Press.

Johnson, M. (1987). *The body in the mind: The bodily basis of reason and imagination*. Chicago: University of Chicago Press.

Kaput, J. J. (1991). Notations and representations as mediators of constructive processes. In E. von Glasersfeld (Ed.), *Radical constructivism in mathematics education* (pp. 53–74). Dordrecht, Netherlands: Kluwer.

Kozulin, A. (1990). *Vygotsky's psychology. A biography of ideas*. Brighton, England: Harvester Wheatsheaf.

Lakatos, I. (1976). *Proofs and refutations*. Cambridge, MA: Cambridge University Press.

Lampert, M. (1990). When the problem is not the question and the solution is not the answer: Mathematical knowing and teaching. *American Educational Research Journal, 27*, 29–63.

Lave, J., & Wenger, E. (1991). *Situated learning: Legitimate peripheral participation*. Cambridge, England: Cambridge University Press.

Leont'ev, A. N. (1981). The problem of activity in psychology. In J. V. Wertsch (Ed.), *The concept of activity in Soviet psychology* (pp. 31–37). Armonk, NY: Sharpe.

Linn, M. C. (1992). The computer as learning partner: Can computer tools teach science? In L. Roberts, K. Sheingold, & S. Malcolm (Eds.), *This year in school science 1991*. Washington, DC: American Association for the Advancement of Science.

Mehan, H., & Wood, H. (1975). *The reality of ethnomethodology*. New York: Wiley.

Mishler, E. G. (1986). *Research interviewing: Context and narrative*. Cambridge, MA: Harvard University Press.

Murray, H. (1992, August). *Learning mathematics through social interaction*. Paper presented at the International Congress on Mathematics Education, Québec City, Canada.

Neuman, D. (1987). *The origin of arithmetic skills: A phenomenographic approach* (Goteborg Studies in Educational Sciences 62). Goteborg, Sweden: Acta Universitatis Gothoburgensis.

Newman, D., Griffin, P., & Cole, M. (1989). *The construction zone: Working for cognitive change in school*. Cambridge, England: Cambridge University Press.

Noddings, N. (1985). Small groups as a setting for research on mathematical problem solving. In E. A. Silver (Ed.), *Teaching and learning mathematical problem solving: Multiple research perspectives* (pp. 345–359). Hillsdale, NJ: Lawrence Erlbaum Associates.

Piaget, J. (1967). Les operations logiques et la vie sociale [Logical operations and the social world]. In *Etudes Sociologiques [Sociological Studies]*. Geneva: Librairie Droz.

Presmeg, N. C. (1992). Prototypes, metaphors, metonymies and imaginative reality in high school mathematics. *Educational Studies in Mathematics, 23*, 595–610.

Richards, J. (1991). Mathematical discussions. In E. von Glasersfeld (Ed.), *Radical constructivism in mathematics education* (pp. 13–52). Dordrecht: Kluwer.

Rogoff, B. (1990). *Apprenticeship in thinking: Cognitive development in social context*. Oxford, England: Oxford University Press.

Saxe, G. B. (1991). *Culture and cognitive development: Studies in mathematical understanding*. Hillsdale, NJ: Lawrence Erlbaum Associates.

Schroeder, T., Gooya, Z., & Lin, G. (1993). Mathematical problem solving in cooperative small-groups: How to ensure that two heads will be better than one? In I. Hirabayashi, N. Nohda, K. Shigematsu, & F.-L. Lin (Eds.), *Proceedings of the Seventeenth Conference of the International Group for the*

Psychology of Mathematics Education (pp. 65–72). University of Tsukuba, Japan: Program Committee of 17th PME Conference.

Sfard, A. (1994). Reification as the birth of metaphor. *For the Learning of Mathematics, 14*(1), 44–55.

Shimizu, Y. (1993). The development of collaborative dialogue in paired mathematical investigation. In I. Hirabayashi, N. Nohda, K. Shigematsu, & F.-L. Lin (Eds.), *Proceedings of the Seventeenth Conference of the International Group for the Psychology of Mathematics Education* (pp. 72–80). University of Tsukuba, Japan: Program Committee of 17th PME Conference.

Smith, E., & Confrey, J. (1991, April). *Understanding collaborative learning: Small-group work on contextual problems using a multi-representational software tool.* Paper presented at the annual meeting of the American Educational Research Association, Chicago.

Steffe, L. P., Cobb, P., & von Glasersfeld, E. (1988). *Construction of arithmetic meanings and strategies.* New York: Springer-Verlag.

Steffe, L. P., von Glasersfeld, E., Richards, J., & Cobb, P. (1983). *Children's counting types: Philosophy, theory, and application.* New York: Praeger Scientific.

Steffe, L. P., & Wiegel, H. G. (1992). On reforming practice in mathematics education. *Educational Studies in Mathematics, 23*, 445–466.

Taylor, S. J., & Bogdan, R. (1984). *Introduction to qualitative research methods* (2nd ed.). New York: Wiley.

Thompson, P. (1994). Images of rate and operational understanding of the fundamental theorem of calculus. *Educational Studies in Mathematics. 26*(2–3), 229–274.

van der Veer, R., & Valsiner, J. (1991). *Understanding Vygotsky: A quest for synthesis.* Cambridge, MA: Blackwell.

Voigt, J. (1992, August). *Negotiation of mathematical meaning in classroom processes.* Paper presented at the International Congress on Mathematics Education, Québec City, Canada.

Vygotsky, L. S. (1982). Forward. In L. S. Vygotsky, *Sobranie sochinenij. Tom 1. Voprosy teorii i istorii psikhologii* [Collected Works, Volume 1. Issues regarding the theory and history of psychology] (pp. 238–290). Moscow: Pedagogika. (Original work published 1934)

Vygotsky, L. S. (1960). *Razvitie vysshikh psikhicheskikh tunktsii* [The development of the higher mental functions]. Moscow: Akad. Ped. Nauk. RSFSR.

Webb, N. (1989). Peer interaction and learning in small-groups. *International Journal of Educational Research, 13*, 21–39.

Webb, N. M. (1982). Student interaction and learning in small groups. *Review of Educational Research, 52*, 421–445.

Wertsch, J. V. (1985). *Vygotsky and the social formation of mind.* Cambridge, MA: Harvard University Press.

Winograd, T., & Flores, F. (1986). *Understanding computers and cognition: A new foundation for design.* Norwood, NJ: Ablex.

Wood, T., & Yackel, E. (1990). The development of collaborative dialogue within small-group interactions. In L. P. Steffe & T. Wood (Eds.), *Transforming children's mathematics education. International perspectives* (pp. 244–252). Hillsdale, NJ: Lawrence Erlbaum Associates.

Yackel, E., & Cobb, P. (1993, April). *Sociomath norms, argumentation, and autonomy in mathematics.* Paper presented at the annual meeting of the American Educational Research Association, Atlanta.

Yackel, E., Cobb, P., & Wood, T. (1991). Small group interactions as a source of learning opportunities in second grade mathematics. *Journal for Research in Mathematics Education, 22*, 390–408.

4

Children's Talk in Inquiry Mathematics Classrooms

Erna Yackel
Purdue University Calumet

This chapter was motivated by a desire to investigate the ways in which children talked about and explained their mathematical solutions and thinking in the project classroom where mathematics instruction followed an inquiry tradition. In this tradition, instruction is designed to engage children in meaningful mathematical activity, including explaining and justifying their thinking to others (Cobb, Wood, Yackel, & McNeal, 1992). Other chapters in this book deal with languaging more generally (Bauersfeld, chap. 8, this volume) and with the development of argumentation (Krummheuer, chap. 7, this volume).

In the past several years there have been a number of attempts to modify the participation structure in mathematics classrooms by organizing instruction so that students actively engage in doing and talking about mathematics (e.g., Ball, 1993; Cobb, Wood, & Yackel, 1991; Kamii, 1989; Koch, 1992; Lampert, 1990; Murray, 1992; Richards, 1991). Such instruction contrasts with teacher-dominated instruction, in which students listen to teacher explanations, respond to teacher directives, and develop expertise using pregiven solution procedures. In the United States, much of the impetus for this change comes from the National Council of Teachers of Mathematics Standard "Mathematics as Communication." The *Standards* (National Council of Teachers of Mathematics, 1989) call for children to "talk mathematics" to help them construct knowledge, learn other ways to think about ideas, reflect on and clarify their own thinking, develop convincing arguments, and eventually extend such arguments to deductive proofs. Others have argued earlier for the benefits of dialogue in classrooms. For example, Barnes and Todd (1977) described the value of collaborative dialogue as follows:

> [Collaborative] moves are mutually supportive: by taking the trouble to elicit an opinion from someone else, or by utilizing what has been said by extending it further, the group members ascribe meaningfulness to one another's attempts to make sense of the world. This helps them to continue, however hesitantly, with the attempts to shape their own understanding by talking, and contrasts sharply with any schooling which reduces the learner to a receiver of authoritative knowledge. (p. 36)

From the student's point of view, the purposes for talking in the classroom are to explain their thinking to others and to challenge and question the thinking of others. From the teacher's point of view, talking serves the additional function of giving the teacher information about the students' progress. Such information is useful for both evaluating individual students and making instructional decisions.

This emphasis on verbal communication in the mathematics classroom underscores the importance of studying students' ways of talking about their mathematical activity. This chapter analyzes and discusses differences in students' explanations and in the solutions they develop and offer in various classroom situations. The analysis has several purposes. One is explanatory: Classroom observations clearly indicate that the solution methods children discuss and the ways in which they talk about their activity change in different situations. The investigation that formed the basis for this chapter attempted to document and account for these differences. As the investigation proceeded, it became apparent that to account for observed differences it was necessary to consider more than the social setting—an analysis that took children's interpretations into account was needed. The symbolic interactionist perspective proved useful for this purpose, because one of its central tenets is that meaning develops out of interaction and interpretation. Consequently, another purpose of this chapter is to exemplify how a sociological perspective can clarify an activity, such as explaining, that is typically thought of primarily in cognitive terms.

This chapter reports on the analyses that were conducted in the attempt to explore differences in the children's talk and explanations across situations in the project classroom. Throughout, the symbolic interactionist perspective is emphasized as a means of interpreting the data and accounting for participants' communication activity. Two major themes emerged from the analysis. The first is that the critical feature in determining the nature of children's talk and activity is the situation as it is interactively constituted as a social event rather than the social setting. For example, the analyses indicated that there are no systematic differences between the students' explanations in small-group work and in subsequent whole-class discussion. However, qualitative differences in the children's explanations and in the solutions they develop and offer become apparent when the analysis focuses on their interpretations of social events rather than social arrangements.

In the analyses presented in this chapter, the explanations a child gives for the same problem to his or her partner, the teacher, a researcher during small-group work, and to the class during subsequent whole-class discussion, are related to the child's expectations and interpretations of his or her obligations in the immediate situation. It is shown that despite well-established social norms for small-group work and whole-class discussions, the expectations and obligations are not stable across groups or across class discussions. Across groups there are dramatic differences in the way the children interact and in their interpretations of their obligations to each other. The critical role of students' interpretations indicates that it is the context rather than the setting that distinguishes qualitative differences in students' explanations. Further, the relationship between the quality of an explanation and the social situation in which it is developed is reflexive. The student participates in the interactive constitution of the social situation as a social event through his or her activity, and the social event, in turn, constrains the student's activity.

A second important theme in the chapter is the teacher's role in supporting students' attempts to explain their solutions. The analysis clarifies the distinction between the teacher's attempts to help children develop viable solutions and her attempts to make sense of students' explanations. Finally, the analysis indicates the tension that exists for the teacher as she attempts to fulfill her obligations to help students understand explanations given by their peers and, at the same time, help them develop an understanding of what constitutes mathematical explanation.

In the final section of the chapter, implications for instruction that arise from these analyses are discussed.

THE PROJECT CLASSROOM

The project classroom was described in chapter 2 of this volume. Additional details that are particularly relevant to this chapter are included here. The typical lesson in the project classroom consisted of a brief introduction, small-group work, and subsequent whole-class discussion. The purpose of the introduction was to clarify the intention of the tasks to be completed during small-group work. These problem tasks were typically posed in written format. Consequently, the introduction necessarily included clarifying symbolic conventions of the written formats. For example, some tasks were posed in the format of a pan balance. To interpret these tasks, students needed to understand that balance is achieved when the amounts on both sides of the balance are the same. In this case, the introduction to the lesson included a clarification of this basic principle of a balance. Following the introduction to the lesson, students worked in pairs for approximately 20 minutes to solve problems. After small-group work, children discussed their problem interpretations and solution attempts in a whole-class setting.

The social norms that were negotiated in the classroom are particularly relevant to discussions about students' explanations of their solution methods. The project classroom followed an inquiry mathematics tradition. In this tradition, children engage in meaningful mathematical activity, including explaining and justifying their thinking to others (Cobb, Wood, Yackel, & McNeal, 1992). When students work in small groups, they are expected to work collaboratively, develop personally meaningful solutions, explain their thinking to their partner(s), and listen to and try to make sense of their partner's explanations. During whole-class discussions, students are expected to explain their solution methods to the class, listen to and try to make sense of the explanations of others, and pose questions and challenges when appropriate.

The teacher plays a significant role in that he or she initiates and guides the development of the social norms as just described (Cobb, Yackel, & Wood, 1989; Yackel, Cobb, & Wood, 1991), facilitates children's developing ability to engage in collaborative dialogue (Wood & Yackel, 1990), and facilitates their evolving understanding of what constitutes an adequate explanation (Yackel, 1992). Additionally, the teacher initiates and guides the development of norms that are specific to mathematical aspects of the children's activity. These sociomathematical norms are critical to the children's understanding of what constitutes mathematical difference, mathematical sophistication, and mathematical elegance (cf. Voigt, chap. 5, this volume; Yackel & Cobb, 1993), and consequently influence the children's mathematical activity.

DATA

Videotape data and available transcripts were examined to identify cases in which children, who had been video recorded as they attempted to solve a problem in the small-group setting, described their solutions during the subsequent whole-class discussion. All relevant episodes were analyzed. In addition, selected episodes of the same type from a second teaching experiment conducted in a different school setting were analyzed. Several examples from the latter group are included in this chapter to exemplify different types of interactions in both the small-group setting and the whole-class discussion setting.

THEORETICAL CONSTRUCTS

Meaning as Interactively Constituted

Critical to any investigation of communication and "talk" is one's view of *meaning*. The theoretical framework underlying social interaction as used here is derived largely from Blumer (1969). The view that meaning is not intrinsic but develops out of interaction and interpretation is central to his position. According to Blumer (1969), meaning arises:

> in the process of interaction between people. The meaning of a thing for a person grows out of the ways in which other persons act toward the person with regard to the thing. . . . Thus, symbolic interactionism sees meaning as social products, as creations that are formed in and through the defining activities of people as they interact. (pp. 4–5)

4. CHILDREN'S TALK

Symbolic interactionism emphasizes the interpretative process involved in the development of meaning as an individual responds to, rather than simply reacts to, another's actions. The response is based on the meaning an individual ascribes to another's actions. It is the emphasis on interpretation that distinguishes symbolic interactionism.

The symbolic interactionist position regarding the constitution of meaning is compatible with the constructivist perspective as stated by von Glasersfeld (1983): "For communication to be considered satisfactory and to lead to what we call understanding, it is quite sufficient that the communicators' representations be compatible in the sense that they do not manifestly clash with the situational context or the speaker's expectations" (p. 53). Implicit in this position is the view that participants in the interaction have expectations and intentions that guide their activity. A participant's interpretation involves taking account of the intentions of another, at least implicitly. Accordingly, the analysis of children's activity and talk focuses on the children's intentions.

Context

In this chapter, context is used in the sense of Bateson (1972) to refer to the internal (cognitive) state of the child rather than to external conditions, such as who the participants are or the format in which a problem is presented. In this sense, context is individual and indefinite in scope. The external conditions form the setting. This distinction between setting and context is consistent with Cobb's use, as illustrated by the following example. Cobb (1986) found that some children interpreted the task 16 + 9 presented in horizontal format using plastic numerals as a different task from 16 + 9 written vertically on a traditional workbook page. One child solved each task differently and arrived at different answers without being puzzled by the fact that his answers did not match. For him, they represented different contexts. On the other hand, for those children who immediately commented that the second problem is "just the same" as the first, they represented the same context.

In this chapter, the use of context rather than social setting indicates the importance of taking social interaction into account. The analysis will show that a child's interpretation of his or her obligations and his or her expectations of other participants in the immediate situation are critical to the nature of his or her activity and talk. Thus, the same social setting (e.g., small-group work) may represent completely different contexts for different individuals or even different contexts for the same individual from one occasion to another. These differences are explored in this chapter.

Reflexivity

The ethnomethodologists' concept of reflexivity is useful in accounting for the way in which situations that occur in mathematics classrooms emerge. Leiter (1980) described reflexivity as:

> A property of social phenomena which . . . makes social facts the product of interpretation. . . . Accounts and settings, then, mutually elaborate each other. The account makes observable features of the setting—which, in turn, depend on the setting for their specific sense. The features of a setting that are revealed by descriptive accounts and behavior do not just explicate the setting; they, in turn, are explicated by the setting. (p. 138)

There is a reflexive relationship between context and a child's actions; the child's interpretation helps to define a situation to be as he or she interpreted it. For example, if a student's interpretation of his or her obligation in the immediate situation is that he or she is to clarify which of several answers is correct, then by fulfilling that obligation, the student, in effect, participates in creating the situation as one that requires clarifying the answer. The teacher's interpretations contribute along with the children's interpretations to the constitution of the situation. In this sense, situations as well as meanings are interactively constituted.

DISCUSSION AND ANALYSIS OF CASES

An overview of the results is given in this paragraph as an orientation to the subsequent analysis of cases. The analysis of the targeted episodes that follows shows that for children the critical feature is not the social setting, such as small-group problem solving or whole-class discussion, but instead is the way in which the situation is realized as a social event. That is, it is the context that is relevant, not the social setting. The presence of the teacher or a researcher during small-group work results in different interaction patterns than when the children are alone. Children may alter their explanations and sometimes their solution methods due to their differing expectations, their perceptions of their obligations, and their own personal agendas. For example, in the presence of the teacher a child's goal may be to establish that he or she had the correct answer rather than to elaborate on difficulties he or she had with a particular interpretation of the problem. On the other hand, the child's goal might be to elicit the teacher's assistance to resolve a conflict between two differing interpretations. When children are interacting during small-group work, with no adult present, they may be attempting to convince each other of the viability of their own personal interpretation, or they may be attempting to take each other's perspective. When children talk during whole-class discussion, it may be to report (and explain) a previously developed solution method or it may be to enter into the ongoing discussion using their prior small-group activity as a basis for their contribution. In some cases, they may abandon their prior small-group work and develop a solution that addresses the questions and challenges of the immediate discussion. The child's interpretation of the immediate situation is critical to the nature of the activity (and talk) in which the child engages.

4. CHILDREN'S TALK

Qualitative Differences in Student's Explanations

To illustrate children's differing activity and talk in various situations that they interpret as different social events, we consider the following example.

Example 1—February 2. Jack and Jamie are working together on a problem in small-group work, first alone, then in the presence of a researcher, and later in the presence of the teacher. Finally, they participate in the subsequent whole-class discussion of the problem. Although these four situations can be viewed as different social settings, the analysis indicates that it was not the setting per se, but the individual children's expectations and their interpretations of their obligations, that were the critical factors in their activity and developing explanations.

The problem under discussion is the last problem on the activity page shown in Fig. 4.1, involving 38 and 38. In each case, the task is to figure out what number to put in the empty box to make the left and right sides balance.

Because both boys related the last task to previous tasks on the same page, their solutions for Tasks 2 and 3 are described. Jamie solved the second task by adding 40 and 40 to get 80, adding 6 to 80 to get 86, and then adding 6 more to get 92, and the third task by relating it to the second.

1   Jamie: 48 and 48 make, just make 2 more—4 more than that (pointing to previous task). Just 4 more than that. What's that? That makes 96.

For his part, Jack simply indicated that he agreed with Jamie's solution of the second task. He proposed to solve the third task by adding 8 and 8 and 80, but there is no evidence that he completed the calculation.

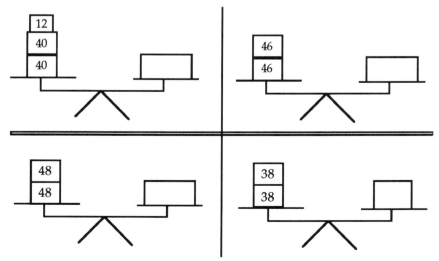

FIG. 4.1. Student page of balance tasks.

The following dialogue ensued when the boys attempted to solve the last task on the page.

1	Jamie:	Thirty . . .
2	Jack:	38, 38.
3	Jamie:	Just take away . . .
4	Jack:	8 from that (points to 38).
5	Jamie:	20, 20, take away 20. That would be uh, 76, 76.
6	Jack:	Take away 10. Take away 10.
7	Jamie:	No, take away 20.
8	Jack:	Take away 10. Take away 10 (points at 48 and 38). Look, 40, 30. Not 20.
9	Jamie:	I know, oh—Yeah.
10	Jack:	Take away 10.
11	Jamie:	20. 70—so that should be 76.
12	Jack:	No. Take 10 away from that answer (points to previous task), that answer (points to current task) . . . 86.
13	Jamie:	Wait. 40 take away—so take away 4 more. Take away 20, makes 70.
14	Jack:	No, take away 10. Look. It's 40, 30 (points to current and then previous task).
15	Jamie:	80 take away, 80 take away 20. 80 take away 20 makes what?
16	Jack:	We're not supposed to be taking away 20.
17	Jamie:	But look . . .
18	Jack:	Because there is 48 and 43. That's just taking away 10. So take away 10 from that (points to 96).
19	Jamie:	What?
20	Jack:	See. Right there (points to 48). That is 10 higher than that (points to 38). So take away 10 from that (points to 96).
21	Jamie:	Um-um [no]. It's supposed to be 70. (Writes down his answer of 76.)
22	Jack:	I don't agree with you. (Jamie fetches another activity page. Jack writes 86 as his answer.)

The dialogue shows that both boys proposed to solve the task by relating it to the previous task. However, they had differing ideas about how to do so. Because there is a dispute, it seems appropriate from the observer's perspective that each child would clarify his interpretation of the problem and provide a rationale for his proposal. The first evidence of an attempt to do so was given by Jack in Line 8, when he said, "Take away 10. Take away 10 (points at 48 and 38). Look, 40, 30. Not 20."

Jamie's reply in Line 9, "I know, oh—Yeah," on the other hand, gave no rationale. Further, this remark might have been an indication that although he may have been aware of the discrepancy in their solution procedures, he might have been unaware of the discrepancy between his and Jack's interpretations of the problem. He appeared to agree with Jack, acknowledging that this problem involved 30 whereas the previous problem involved 40. Nevertheless, from the observer's perspective, they had differing interpretations of the problem that were not explicitly addressed in the subsequent verbal interaction. Jamie simply repeatedly restated the need to "take away 20," and reiterated the result as being 76 or in the 70s, whereas Jack repeated "take away 10." Jack attempted to provide a rationale on three occasions, first in Line 8 when he said, "Look, 40, 30. Not 20"; later in Line 14 when he said "No, take away 10. Look. It's 40, 30 (pointing to current and then previous task)"; and still later in Line 18, "Because there is 48 and 43. That's just taking away 10. So take away 10 from that (points to 96)." (Jack's repeated references to the 40 and the 30 lead us to interpret the "43" as misspeaking.) This last remark gave explicit evidence that Jack's intention was to give a rationale for the solution method he was advocating. Here, he was relating his interpretation of the task to his suggested solution procedure. However, his explanation failed to explicate the differences between his and Jamie's interpretations.

There was a distinct change in the nature of Jamie's "talk" when a researcher entered the scene a few moments later. At that point, both boys gave a rationale for their solution activity in response to the researcher's request, "How did you do it?"

32	Jamie:	Well, take away 10, 10 from each of these (points to the previous task).
33	Researcher:	Yeah.
34	Jamie:	Take away 10 from each of these make 20, take away 20. He put 80. I think that's—I think it's 76.
37	Researcher:	How did you do it? (to Jamie)
38	Jack:	See, that's just 10 more—10 less than that.
39	Researcher:	Yeah.
40	Jack:	So the answer has to be 10 less.

Jamie's reply to the researcher elaborated his interpretation of the problem and explained the origin of the "20" that was to be taken away. His previous comments to Jack did not provide this clarification. His expectation may have been that Jack shared his interpretation of the task. (There is some support for this hypothesis from the previous history of this pair; see Yackel, Cobb, & Wood, 1993.) Another possibility is that Jamie may not have felt obliged to give Jack an explanation, although he did feel obliged to give one to the researcher. The researcher had previously established his interest in how the children thought about the tasks. In any event, Jamie fulfilled his obligation to explain his thinking to the researcher. Further, his reply indicated that he understood which aspects of his thinking were critical to providing an explanation that would be adequate for the researcher.

For his part, Jack's reply to the researcher differed from his prior comments to Jamie in that he then went on to verbally relate his observation, "just 10 more—10 less than that," to the answer, "So the answer has to be 10 less," whereas previously he had only pointed to the answer. If either child had said to the other what he said to the researcher, they may have resolved their disagreement. The difference in the explanations given by each child and in the nature of their "talk" indicates that each of them interpreted the presence of the researcher as a different context than that of working as partners. Both boys understood that in the presence of the researcher they were obligated to explain their solutions. In this situation, Jamie clarified both his problem interpretation and his solution in detail, whereas in his previous interaction with Jack he gave no rationale for his suggestion to "take away 20."

When the teacher entered the group later, the boys immediately told her about their conflict over the problem, even though they were already working on another activity page. Because of their previous experience in this class, they might have expected the teacher to help them resolve their disagreement by discussing their solution procedures with them. However, the situation evolved differently. Two solution methods were discussed with the teacher, both of which differed from the methods the boys were attempting to use previously.

The teacher's initial comment signaled a relationship to the previous problem. Jack accepted the teacher's implicit suggestion to relate the problem to the previous one, but Jamie did not. Instead, Jamie solved the problem by adding 30 and 30 and 8 and 8, which is the same method he used to solve the second task on the page.

 1 Teacher: You know that 48 and 48 make 96, so 38, 38.
 2 Jamie: Make ...
 3 Jack: 10 less than that.
 4 Jamie: 30, 60 ...
 5 Teacher: Not quite 10.
 6 Jamie: 68, 69, 70 ... 76 (puts up fingers as he counts the last 8), 76.
 7 Teacher: He still gets 76. Okay, there is another way you can figure this. What's 30 and 30?
 8 Jamie: 60.
 9 Teacher: And what's 8 and 8?
10 Jamie: 16.
11 Teacher: You got a 60 and a 16.
12 Jamie: That makes 76.

The teacher's initial move was to implicitly suggest a solution procedure that she anticipated the boys might use, given her knowledge of the solution methods they had previously used and her knowledge of their cognitive capabilities. However, she listened to their proposals and explanations before making further suggestions. When Jamie explained his solution, the focus shifted from relating the two tasks to the result of the computation. The children's expectation was that the teacher would follow up on their solution attempts. In this case, the teacher chose to follow up on Jamie's explanation by suggesting an alternative way to figure out the sum of 30, 30, 8 and 8, namely adding the two 30s, the two 8s and then adding the partial sums. The episode concluded without an opportunity for Jack to explain his thinking.

The solutions discussed in detail in the presence of the teacher were not the same as those the children attempted when working as partners or in the brief interchange with the researcher, even though the teacher implicitly suggested that they might relate the current and prior tasks when she first entered the group. Jack's response indicated his expectation that the discussion would be about their prior activity. However, Jamie's introduction of a different solution method indicated that he did not share Jack's expectation. One possible interpretation is that in the teacher's presence, Jamie's intention was to establish that his answer was correct. By using a different method, he was able to do so without referring to the disagreement he and Jack had about the problem. In this way, he initiated the interactive constitution of the activity to be that of developing a viable solution, rather than a discussion of his and Jack's prior solution activity. The teacher contributed to this interactive constitution with her proposal for an alternative method of computing the sum. Her earlier reply to Jack, "Not quite 10," signaled that Jack's interpretation was not appropriate. However, there was no indication of why and the discussion did not return to this issue. There was a potential situation for Jack to give an explanation, but it did not develop, because the teacher and Jamie interactively constituted the activity to be one of developing a solution. Because the teacher did not pursue her attempts to interpret Jack's comment, and because Jamie offered a different solution from the one he was advocating earlier in his discussion with Jack, the episode concluded without a confrontation over the two solution methods and the two interpretations that were the subject of the boys' prior small-group activity.

In the subsequent whole-class discussion, Jamie first explained that he and Jack had disagreed on the interpretation of the problem. Then he gave the variation of his solution method that the teacher had suggested when she intervened in their small-group work, thereby clarifying that the correct answer was 76. Finally, Jamie shifted the topic of discussion to his and Jack's disagreement in small-group work.

16 Jamie: Well, um—Jack was a little mixed up cause he thought it was only 48 take away—and no 48. He thought just 48 and no other 48 and 38 and no other 38. And just take away 10 would be 86. But we had to take away 20.

Here, Jamie directly addressed the difference between his and Jack's interpretations of the problem. He explained what he inferred to be Jack's interpretation, "He thought . . . ," and gave an explicit rationale for what Jack did based on that inference. He then went on to explain his own interpretation, "But we had to take away 20." However, he stopped short of explicating the origin of the 20 as he did when he said to the researcher, "Well, take away 10, 10 from each of these—Take away 10 from each of these to make 20. Take away 20." These children understood that in interactions with the researcher, they would be asked to explain how they thought about the tasks, whereas in whole-class discussion the teacher would typically assist them with their explanations. The children's obligations to explain their solutions were different in the two cases. As the episode continued, the teacher supported the children's expectation that she would provide assistance.

17	Teacher:	Aha! Why did you pick 20?
18	Jamie:	Cause 48 and 48, just take away—take—48 take away 10 makes 38.

Jamie's response still provided no differentiation between his thinking and Jack's, who repeatedly said, "Take away 10," during their small-group attempt to solve the problem. It was the teacher who continued and completed the explanation.

19	Teacher:	Oh. That's what I was trying to get you to say. All right. Do you see that, boys and girls? He said here is 48. Here is 38 (pointing to 48 and 38) and I know that's 10 less. And 48 and 38 is 10 less (pointing to the other 48 and 38). So how's that going to affect your answer?

The episode concluded with another child from the class answering the teacher's question with, "Two tens less."

In summary, in this example, the solution processes the boys described and the way in which they explained and talked about their thinking differed, depending not on the social setting, but instead on their individual contexts. Even though the classroom social norms included the expectation that children should explain their thinking, in only two instances did either boy explicitly attempt to give what would count as a rationale for us. Jack attempted to give a rationale to Jamie during small-group work, and Jamie gave a rationale to the researcher. Neither child gave a rationale to the teacher or during the whole-class discussion. The children were aware that during class discussion the teacher frequently intervened to rephrase and elaborate on their solution descriptions for the benefit of other students.

The example also showed that the two boys' interpretations of the purpose of interacting with the teacher during small-group work were not the same. For Jack, the occasion was one of discussing prior solution attempts and, presumably, resolving their differences. For Jamie, it was one of establishing the correct answer. The same social setting represented different contexts for them.

A second example complements and contrasts with the first. In this example, as in the first, two children discussed the same problem in different situations. The children engaged in small-group work, both with and without the presence of a researcher. Later, they explained their solution in the class discussion. This example, like the first, illustrates the relevance of children's interpretations of the situation in accounting for differences in their explanations. The additional supporting evidence that this second example provides is important because it demonstrates that not all pairs of children interpreted situations in the same way. In particular, unlike the pair in Example 1, these children attempted to fulfill an obligation to explain when working in small groups as well as in a whole-class discussion. They genuinely sought to convince each other of the viability of their interpretations and solution procedures by providing extended supporting arguments throughout their small-group work.

An additional purpose of this example is to illustrate differences between children's descriptions and explanations of activity in progress and those of

4. CHILDREN'S TALK

already-completed activity. As this example demonstrates, descriptions and explanations were not limited to small-group work for activity in progress and subsequent whole-class discussion for already-completed activity. Rather, explanations of both types occurred in each of the social settings. This suggests that the potential of whole-class discussion following small-group work extends well beyond that of reporting back.

Example 2—January 26. In this example Ryan and Katy are working together to solve the tasks shown in Fig. 4.2. During the introductory phase of the lesson, which was designed to clarify the intention of the tasks, the teacher called

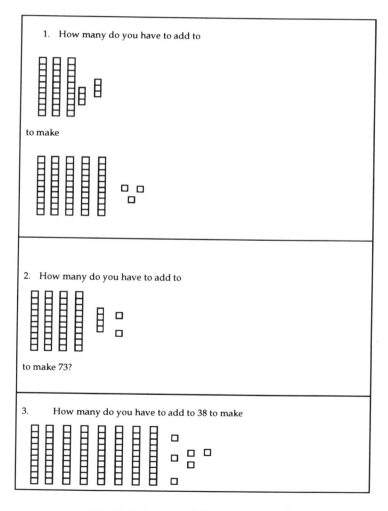

FIG. 4.2. Student page of strips-and-squares tasks.

on several children to explain their interpretations. In the course of these explanations, various solution methods were proposed and discussed, including methods that were primarily image based and others in which the pictorial material (diagram) was given numerical significance.

Katy and Ryan's solution activity during the first two tasks is described to document their interpretation of small-group work as an occasion for explaining. The two children solved the first two tasks using differing approaches. Katy used a numerical approach. She ascribed numerical significance to the pictorial material. She first figured out what number was represented by the picture and then counted on, using her fingers, to determine the answer. Ryan, on the other hand, used an image-based approach. He used multilinks to symbolize the picture shown in the problem, and reasoned from the multilinks. Ryan began the first problem by putting out 3 ten-bars and 6 cubes (ones). He attempted to solve the problem by putting down 2 additional ten-bars and removing 3 cubes (ones). The two children engaged in a lengthy discussion. Each child attempted to convince the other of the validity of their method. Unlike the previous example, in which the two boys gave little or no rationale to each other for their interpretations and suggestions, Katy and Ryan each attempted to explain to each other in detail how they were interpreting the (first) problem. This difference in the interaction between these two pairs is consistent with the findings of Cobb's case studies (chap. 3, this volume). In this case, both children accepted the obligation to explain their thinking to their partner, including their interpretation of the problem and their solution attempts.

Katy repeated her solution to the first task in its entirety at least four different times and clearly indicated her intention to be one of explaining how she figured out her solution, as indicated by the following statements excerpted from the extensive (and repetitive) dialogue.

27 Katy: Come here, come here. I think you're not getting this right. All right, you have that many numbers. And this makes that, right? Makes 6 (starts counting on her fingers) 7, 8. That makes 36.

31 Katy: Here, I'll explain to you how I got the number. We have that many numbers. We have that many numbers. 36. Add 10 more, 46. Forty-six, 47, . . . , 53.

35 Katy: I'll show you how I got my number. See, uh, 36, and then 10 more (holding up both hands) makes 46. 46, 47, 48, 49, . . . , 53 (putting up one finger with each number word and stopping when she says 53—she has 7 fingers up).

Ryan also attempted to give convincing explanations, but he had not formulated a solution that resulted in a numerical answer. His first solution relied on the imagery of the cubes and the ten-bars and was based on describing how many cubes and ten-bars to put down and remove.

4. CHILDREN'S TALK 145

28 Ryan: Look, look, look. Well, see this [picture on the activity page] is 36, but we have to take away 1 of these things [group of 3 cubes]. And then you add 2 more of these [bars of 10] and that makes 53. That's how I get it.

Later, he suggested an alternative solution method, again an image-based approach. Here again there was clear evidence of his intent to explain to Katy.

34 Ryan: Katy, look, this is how I got it. Well, look. You have to take away a 10. Now that you do this, you take this [1 group of 3] and put it over here (on top of the other group of 3) and then add 4 more, and then put another 3. Work it that way.

There is no doubt that in this case the children were attempting to clarify their interpretations of the problem and their solution activity for each other. For them, working as partners carried this obligation.

The next portion of the episode is particularly relevant for demonstrating differences in the way children talk about their solution activity in progress and about completed solution activity. As the episode shows, Katy took the perspective of the researcher into account when she described her solution after its completion. Her account was different than when she first carried out the solution. Ryan, on the other hand, gave essentially the same description while solving the problem during small-group work and in reporting to the whole class later.

First Katy's explanations are discussed. The task was the third problem on the activity page. Katy gave her initial explanation to a researcher who was standing close by. The researcher's responsibility was to make sure that the children's task interpretations and solution methods were documented on the videotape for future reference. Therefore, it was the researcher's responsibility to ask questions when deemed appropriate to encourage the children to verbalize their thinking. In this episode, Katy accommodated the researcher by talking aloud as she solved the problem. Her talk included disjointed segments that made sense only to someone who was simultaneously observing her solution activity as she worked with the diagram on the activity page. Katy seemed to assume that the researcher shared her interpretation of the task and of her activity. An intervening "Mm-hmm" from the researcher provided her with some reassurance that this assumption was valid.

5 Katy: It's a minus.
6 Researcher: It's a minus.
7 Katy: See there's 10, 20, 30 (counting 3 ten-strips at the left) and here's 1, 2, 3, 4, 5, 6, (counting 6 ones).
8 Researcher: Mm-hmm.
9 Katy: And you have 6, 7, 8, (covering up the next 5 ten-strips but counting 2 ones at the top of the ten-strip farthest to the right). Then you minus 1, 2, 3, 4 (counting ten-strips not previously

counted). 8 (referring to the 8 remaining ones of the ten-strip from which she took 2 ones). I think it's 48, minus."

Katy then decided to make a mark through each of the 48 she counted, 4 ten-strips and 8 ones of a fifth ten-strip. The researcher stepped back and Katy repeated her solution to herself to verify her answer.

The interaction between the two children began when Katy turned to attend to Ryan's activity. Ryan had been working to develop an image-based solution using multilinks (for a further discussion of image-based solutions see Cobb, chap. 3, this volume). He had miscounted and had put out 7 ten-bars instead of 8. Katy corrected the error and explained what she had already figured out, "Take away 48."

While Ryan seemed to be puzzling over the solution, Katy addressed the researcher. She explained her solution in some detail. This time she gave special attention to explaining the marks she made on the paper, as if she anticipated that this aspect of her explanation might be misunderstood. A condensed version of her explanation follows.

39–49 Katy: If we have 30—38 and we're trying to get 38 and we have, we have 86 we have to take away and I decided to take away 30, 40—8. All these marks mean I don't, I don't take them. I just take the rest. Yeah, I took these, 10, 20, 30. And there have to be 8 more. So I counted 1, 2, 3, 4, . . . 8. See, and then and then and then I marked them other ones out.

The researcher completed the argument with "and you counted them." Katy may have assumed that the researcher would take it for granted that the final step in her solution was to "count them" and, therefore, felt it was unnecessary to say so. She had already indicated that her answer was 48.

The dialogue illustrates explicitly how Katy's interpretation influenced what she said. In this explanation, Katy first clarified her interpretation of the problem by saying, "If we have 86 and we're trying to get to 38 we have to take away." Her use of "we" in this part of the explanation indicates that she assumed that her interpretation was shared by others as well. In clarifying her interpretation, she gave a rationale for her procedure, namely "taking away." She changed from saying "we" to saying "I," signaling that she was now describing her own solution activity. She explained the marks she made, "All these marks mean I don't, I don't take them." She continued to explain how she used the marks to get her result, "I just take the rest. So I counted . . ." Katy's language here indicates an explanation after the completion of activity. Her intention was not to clarify her thinking for herself but for another, and in doing so she provided elaboration as if to rule out potential misinterpretations that she anticipated the listener may have made. As Bruner noted:

> In simple declarative speech acts, ones does not indicate by overt speech that which one can take for granted as an element of knowledge or experience in the interlocutor's mind. I do not say "This room has walls" to my partner in conversation unless the

matter is in question. . . . One uses a negative declarative "naturally" only under conditions of plausible denial. . . . The use of negation presupposes a context that merits plausible denial. (Bruner, 1986, p. 84)

Katy's interpretation of the situation was that she was obliged to clarify her solution. However, for her, clarifying her solution included much more than trying to explain her thinking; it also included explicitly attempting to take into account the point of view of the listener. To do so, Katy had to make judgments about which aspects of her thinking and which aspects of her solution procedure to explicate and elaborate. Her comment, "I don't take them," indicates that she was aware that the listener might misinterpret what she meant by marking out some of the squares. Taking the perspective of another required her to take her own explanation as an object of reflection and, in this sense, contributed to her learning (Cobb, chap. 3, this volume; Yackel, Cobb, & Wood, 1991).

Ryan's activity provided a contrast to Katy's activity. Whereas her descriptions changed to having the character of an after-the-fact description indicative of reflection when she gave her final explanation to the researcher, Ryan's descriptions did not do so.

In the ensuing exchange with the researcher, Ryan described a solution based on Katy's diagram rather than on his multilinks display. A condensed version of his explanation follows.

| 53–61 | Ryan: | Well, I say it's 48 because—You take away 50 and then, take away 50 and then you have 6. Look, you have to take away the—you have to take off 2—from the ten-bar and then add to here and then that makes 38. |

This solution, like his solution to the first task, was image based. His explanation did not clarify how his process resulted in the answer of 48, leaving it unclear the extent to which he understood Katy's solution.

When the teacher called on Katy and Ryan during whole-class discussion, Ryan immediately reported their answer to be 48.

1	Ryan:	48.
2	Teacher:	48. Do you say the answer is 48?
3	Ryan:	Yes.
4	Teacher:	All right. Now, how did you figure this out? How did you figure it out? You explain it to me.
5–8		[inaudible]
9	Ryan:	We took away 50 and then, we have 30 left and then there is 5 over here so we knew that wouldn't work so . . .
10	Teacher:	Right!
11	Ryan:	We have to take 2 off of the ten-bars—add to there and then add it to the 30 and that makes, and that, and that would make all up—and that would make 48.

The solution method Ryan described was the same one he was attempting to use in small-group work. His reporting of the answer as 48 suggests that he may have been influenced by Katy's explanation, because he had not arrived at a numerical answer.

The language Ryan used in the whole-class setting indicates that he was aware that the expressed intent of the whole-class discussion was to "talk about" your solutions and give a rationale for your activity rather than simply repeat (or recreate) it. Initially he used the past tense, indicating a description of prior activity, "We took away 50 and then, ... so we knew that wouldn't work ...," then switched to describing current activity, "so ... we have to take 2 off of the ten-bars ... and that, and that would make all up." The last phrase is further indication that he was attempting to figure out the result, but he abandoned the attempt (he had not explained this in prior small-group work, either) and simply stated the result, "and that would make 48."

This example illustrates that qualitative differences in the way children explain their activity are not determined by the social setting of small-group work or whole-class discussion. In each of these two social settings there are times when the children described their ongoing activity and when they described prior activity. In the final instance, Ryan's activity in the whole-class discussion was characterized by a shift from reporting about prior activity to currently engaging in mathematical activity. Nevertheless, throughout the episode both children attempted to fulfill their obligation to give explanations of their problem interpretations and solution attempts. In so doing, they were actively involved in the interactive constitution of the situation as a social event that warranted explanation.

Whole-Class Discussion as an Occasion for Reconceptualizing a Problem

In project classrooms, whole-class discussion following small-group problem solving was initially established as a time for children to report on their small-group work. Students would explain their interpretations of problems and their solution processes. However, as it was interactively constituted, the discussion became much more than a report-back session. The example in this section is from a second project classroom in a different school setting. In this classroom, the teacher regularly solicited solutions different from those already discussed and, as the year progressed, encouraged children to ask questions of each other when they did not understand or when they wished to challenge a solution method or explanation. Whole-class discussion then became dynamic. Children had to go beyond trying to make sense of the explanations given to formulating appropriate questions and challenges. In addition, they made comparative judgments about various solutions to decide whether their own was sufficiently different to be offered (Yackel, Cobb, & Wood, in press). As a result, children sometimes proposed solution methods during class discussion that were markedly different from those they developed during small-group work, indicating a reconceptualization of the problem. The

4. CHILDREN'S TALK

following example, which took place in the middle of the school year, is an illustration.

Example 3—February 2. Travonda and Rick are small-group partners. Travonda read the problem and developed a solution to which Rick agreed. The problem was: "Susan had 63 baskets in her shop. She sold 33 of them. How many does Susan have now?"

 1 Travonda: Susan had 63 baskets in her shop. (She places her finger on 63 on her hundreds board.) She sold 33 of them. How many does Susan have now? 10, 20 (moving her hand up the hundreds board by rows of 10). I'll start from there. (She starts over pointing to 63 and counting up by tens, then by ones on the hundreds board.) 10, 20, 30, 1, 2, 3. She had 30 left (pointing to 30 on the hundreds board).

Because her method was clarified visually, Rick had no need to question or disagree. The method of counting by tens on the hundreds board was one that both children used routinely.

During the subsequent whole-class discussion, the first student called on, Shaun, gave 33 as the answer. When he was challenged by another student to explain how he got 33, Shaun explained that he mistook the 63 for 66. He gave no further explanation. Travonda then raised a question in the form of a remark.

20 Travonda: I got a question. Susan had 63 baskets in her shop. Um, you got to add some to the 33 to make 63.

The teacher acknowledged this question as a "good one" and said, "She wanted to know what did you do? Is that what you did? How did you do it?" Here, the teacher reinterpreted Travonda's remark as indicating that Shaun had not clarified his interpretation of the problem, and specifically asked if the interpretation Travonda proposed was the one Shaun intended. Shaun responded by again explaining that he and his partner thought the 3 in 63 was a 6. At this point in the episode the teacher called on Travonda to explain her solution.

25 Teacher: Would you like to tell us how you got yours?
26 Travonda: We got 30 because we said, we said 30 plus 33 that would equal 63. We said the um, 3 tens to the 30 that would make it 60, um and we already had a 3 over there so we said got have to put a 0. That's how we got 63.

This solution was not the same as the one Travonda developed during small-group work. However, it was consistent with the interpretation of the problem she implicitly suggested in her first remark in Line 20. Without explicitly acknowledging that she was doing so, she undertook the task of demonstrating how to figure out what to add to 33 to make 63. She did so using a combination of a collections approach and the partitioning algorithm. Interestingly, she implied that the solution had been developed during small-group work. In this way she was able to give the illusion of meeting the implied obligation of describing the solution she developed during prior small-group work. At the same time, she was able to engage in the current discussion by responding to her own "question" to Shaun. It may be that when Travonda heard Shaun's incorrect answer to the problem, she attempted to verify it by adding. That could account for her reconceptualization of the problem as a missing addend task, whereas she had treated it as a subtraction problem earlier. At least for Travonda, whole-class discussion was an opportunity to engage in an ongoing discussion and was not restricted to reporting back from small-group work. To participate as Travonda did required that she be able to develop a solution, on the spot, consistent with her alternative interpretation of the problem.

Teacher's Role in Supporting Students' Attempts to Explain

A second theme that emerged from the analysis of the data is the teacher's role in supporting students' attempts to explain. This theme came to the fore especially in those situations that might be characterized as involving breakdowns in communication. A breakdown in communication occurs when there are differences in the participants' taken-as-shared interpretations and the discrepancies are not resolved.

The data that formed the basis for this chapter included episodes in which children were unsuccessful in their attempts to communicate either with each other or with the teacher during small-group work. The analysis indicated various reasons for breakdowns in communication. The first is that participants may have had differing interpretations and were unaware of the differences. In this case, there was no apparent need for them to elaborate and clarify their thinking. The initial interaction between Jack and Jamie in Example 1 is an illustration. A second cause of breakdown in communication indicated by the analysis was the incompatibility of the children's interpretations of the immediate task and of what constitutes mathematical activity more generally. In this case, children's genuine attempts to explain were unsuccessful, in part, because their notions of what constitutes an explanation differed substantively. A third cause of the breakdown in communication was the children's inability to articulate their thinking in a way that others could understand.

The notion of *situation for explanation*, as described previously by Cobb, Wood, Yackel, and McNeal (1992), proved useful for this aspect of the analysis. Teachers and students give explanations in an attempt to clarify aspects of their thinking that they deem to be not readily apparent to others. However, when they do so there is no guarantee that other participants in the interaction attempt to interpret and make sense of the explanation. The distinguishing feature of a situation for explanation is that it is interactively constituted by the participants in the interaction and requires that someone attempt to interpret an explanation that is offered. From this

4. CHILDREN'S TALK 151

point of view, explaining is not an individual but a collective activity. The usefulness of this construct for analyzing situations of communication breakdown is that it permits distinctions based on the intentions of the listener as well as of the speaker.

This section discusses examples in which communication broke down during small-group work. In the first example, the children had differing views of what constituted mathematical activity. In the second example, taken from a different project classroom, one child was unable to articulate his thinking to his partner or the teacher. In each case the classroom teacher joined the group, but with differing results. The discussion of the examples includes a description of the children's explanation attempts and of the teacher's interventions, including the role those served in facilitating the establishment of situations for explanation.

Example 4—February 3. Andrea and Andy were working together in the small-group setting to solve the problem shown in Fig. 4.3.

As was documented in previous discussions of the project classroom (Cobb, Wood, Yackel, & McNeal, 1992) and in Cobb's case studies (chap. 3, this volume), Andrea had an instrumental view of mathematics learning. For her, mathematical activity consisted of following procedural instructions. By contrast, for Andy, mathematical activity consisted of developing personally meaningful ways of solving problems.

Andrea and Andy spent much time on this problem during small-group work, after an initial intervention from the teacher in which she helped them figure out that they might think of the diagram as the number 35. Andy symbolized 74 by drawing 7 ten-strips and 4 squares and used it to reason that you could get the diagram shown in the problem from his by taking away 4 tens (strips) and adding 1 (square). For the entire 5 minutes that the children worked on this task without the teacher's presence, Andrea did not attempt to develop a solution of her own. During this time, Andy repeatedly attempted to explain his solution to Andrea, who showed no evidence of understanding him. His solution was image based and did not result in a numerical answer.

Andrea's failure to understand is consistent with her procedural approach to solving mathematical tasks. She typically used developed procedures for solving

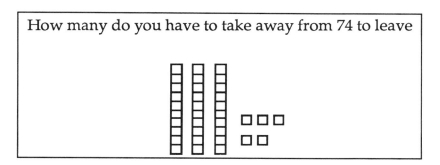

FIG. 4.3. Strips-and-squares task.

was possible for her to interpret a task as an addition task. The procedure she had figured out for solving tasks presented using the strips-and-squares format, as was the task posed here, applied only to those where the largest number in the problem was represented pictorially. Presumably she could have used her method to solve the problem by using Andy's diagram, but her procedural approach was incompatible with the explanation Andy developed from his diagram. Further, her interpretation of this task was that the answer to the problem should be numerical. Andy's explanations, like Ryan's in Example 2, were image based and resulted in descriptions of how to move strips of ten and squares (ones), but did not result in a numerical answer. Andy and Andrea had no basis for developing a taken-as-shared interpretation of Andy's diagram or of his activity. Neither one could conceptualize the task in terms of abstract units of ten that were composed of 10 ones (Cobb, chap. 3, this volume). Nonetheless, Andy did attempt to explain his image-based method to Andrea several times.

34	Andy:	Listen, listen. You leave 30, right? (pointing to his diagram) You take away 4 from 70, and you add a 1 . . .
35	Andrea:	That would be 75.
36	Andy:	No 3, 4, (4 squares in his diagram). It's 4, there is already 4 there! There is already a 4 there, see? And then you would add a 1. 74! So you take the 4 [ten-strips] from your 74 . . .
48	Andy:	Okay, take away 4 [ten-strips], take away these. You take away 4 [ten-strips] and you add 1 [square]. (Points to his strips-and-squares representation.) That will be 5 just like over here, just like 5. (Points to the activity page.)

Later, the teacher rejoined the group at Andy's request. Andy explained his solution:

| 65–67 | Andy: | What I did was, like I draw 70 and 4 more. You add a 1 and you take away 4 tens. |

He further elaborated on his image-based solution. From Andy's perspective this constituted a complete explanation. His interpretation of the task was that he should figure out how to begin with a pictorial symbolization of 74 and modify it to make it look like the pictorial symbolization of 35 shown in the problem. The explanation Andy gave to the teacher was much like those he gave to Andrea. From his perspective, he was attempting to clarify his activity. The teacher then developed an alternative solution with Andrea, based on starting with the diagram on the activity page.

80	Teacher:	(To Andrea) All right. Draw a picture down here—Andrea, you do it. Make a stack of 45, another 10, just make one.
81	Andrea:	(Draws a strip.)
82	Teacher:	45, OK. Go to the next one.
83	Andrea:	(Draws a strip.)
84	Teacher:	55.

85	Andrea:	(Draws a strip.)
86	Teacher:	65, OK. Now 75, but we only need 74. So what are we going to do? How many are going to be in the last stack?

Andrea drew another strip, so now there were 75.

94	Teacher:	OK. Look, wait a minute 10, 20, 30, 40, 50, 60, 70, 71, 72—10, 20, 30, 40, 50, 60, 70, 71, 72, 73, 74, 75 (counting on the diagram). But we need only 74, so what are we going to do about this?

After another prompt by the teacher, Andrea said, "Just take away one!" In Lines 105 and 107 the teacher continued to develop the solution with Andrea, ignoring Andy's intervening comment in Line 106.

105	Teacher:	Take away one (from 1 of the added ten-strips), you know you have 10 in each of these [strips], right? 1, 2, 3, 4, 5, 6, 7 . . .
106	Andy:	(Interrupts) What I did is I just put 4 like 74 like it says, and then I added 1 like I said over here, and I took away these 4 tens.
107	Teacher:	(Marks 10 squares on each of the 4 strips added and marks off 1 square.) Okay, so now that's a 9 instead of a 10 there. That would give you what you need. 74. All right, now what is this number that you just drew? (Andrea counts to get 39 and writes that down as the answer.)

An essential feature of the solution developed by the teacher and Andrea is that the modification made to the pictorial symbolization was given numerical significance. In this sense, the solution, unlike Andy's, was compatible with Andrea's interpretation that the problem should have a numerical answer. Andrea participated in the development of the solution only to the extent of responding to teacher directions. The teacher provided the rationale for the procedure. By doing so, she inadvertently contributed to Andrea's interpretation of her actions as procedural instructions. As in any communication attempt, the participants interpreted the statements and actions of others in terms of their own understanding and perspective. As was noted earlier, context is individual. Andrea's view of mathematical activity as carrying out procedures influenced how she interpreted the teacher's comments and actions. Even though the teacher attempted to involve Andrea in the development of the solution method and its rationale, the dialogue shows that the teacher provided both the method and the rationale herself. Thus, despite her best intentions, the teacher contributed to the interactive constitution of a situation that did not challenge Andrea's view of mathematical activity.

Meanwhile, Andy was attempting to get the teacher to attend to his solution. He kept interrupting, presumably in an attempt to compare his activity to the teacher's. He admitted he did not understand her approach.

120	Andy:	I don't understand what *you* are doing now.

126 Andy: I know I don't agree.

Throughout this small-group episode, both Andy and the teacher attempted to provide clarification for their thinking and activity. However, their attempts were unsuccessful. On the other hand, at no point did Andrea attempt to explain; she merely carried out the teacher's directives. Both Andrea and Andy explicitly said that they did not understand. Andrea did not understand Andy, and Andy did not understand the teacher.

During the subsequent whole-class discussion, one child reported a solution similar to that developed by Andrea and the teacher during small-group work. However, the report was cryptic. The teacher noticed the similarity and offered to reenact the solution activity she and Andrea had developed together.

14 Teacher: Okay. This is what Andrea and Andy did one time, too. Didn't we do that one? We extended the picture and how many did you have then? Okay. Let's do that. We know there are 75 and I will draw in how many more? 35? Is that what you did? 45 (draws in 1 more strip).
15 Andrea: 3 more [ten-strips] than that and 1 out and then we get the 38.
16 Teacher: Okay. Now that gives us 40 [from 35 to 75] but we don't want 40. 10, 20, 30, for—we only need these many more. So, I'm gonna block out 1. Now let's see what we have left.

In this reenactment of the drawing procedure, Andrea and the teacher essentially reversed roles. In this case, the teacher drew the strips at Andrea's direction, whereas previously Andrea drew the strips at the teacher's direction. However, as in the small-group work, Andrea did not participate in explaining the procedure—the teacher assumed that responsibility. This is consistent with our contention that Andrea interpreted the teacher's actions during small-group work as procedural instructions. There is no evidence that Andrea understood the rationale for using the procedure. In fact, when the teacher later called for an explanation from the group Andrea deferred to Andy.

34 Andrea: Well, Andy will tell you. He is the one who knows how to draw.

Andy reported that they had solved the problem in two separate ways, and proceeded to describe his way.

37 Andy: We took these (points to 5 squares) and we had 5 of 'em and what I did I put 4 more right there like that . . .
38 Teacher: Okay.
39 Andy: And I added (pause) 1 little one right there. I added more to make 5 out of these and this were all gone.
40 Teacher: Okay.
41 Andy: And I came up with 35. I erased all those 4.

The teacher responded by talking about what she perceived to be an error in her original diagram for the problem. Neither she nor any of the children reacted in any way to Andy's explanation.

In this example, Andy was consistent, giving essentially the same explanation to Andrea, to the teacher when she intervened in the group, and in the whole-class discussion. In each case, the language he used was similar. He gave an account of the actions he performed with the strips and squares giving confirming evidence that his interpretation of the task was image based. There was no need for explanation involving numerical operations, because the appropriateness of his actions could be checked visually. Nevertheless, he gave a rationale for his activity and continued to attempt to make sense of the task and of the activity of Andrea and the teacher, consistent with his general view of mathematical activity as developing personally meaningful solutions to problems.

Several important issues arise from this example. First, a potential situation for explanation arose in that initially Andrea did not understand the problem, whereas Andy had developed an interpretation of it. However, there is no evidence that Andrea attempted to interpret the explanation that Andy offered. Presumably the teacher attempted to interpret Andy's explanation when she joined the group. However, she quickly dismissed it to develop an alternative solution with Andrea. Therefore, there was no possibility that Andy could succeed with his explanation attempts during small-group work. No one was attempting to make sense of them. Consequently, a situation for explanation was not constituted. Likewise, in the subsequent whole-class discussion, a potential situation for explanation arose when Andy described his solution procedure. However, once again there is no confirming evidence that anyone was attempting to interpret his explanation.

Second, an illusion of understanding can be created when a teacher and student develop a solution together, with the student following teacher directives. Gregg (1992) referred to this as an *illusion of competence*. However, by analyzing the dialogue it is possible to determine whether one of the participants, in this case the student, is attempting to communicate aspects of his or her mathematical thinking that the student considers to be not readily apparent to others. In both small-group interventions and in whole-class discussions, the teacher can inadvertently guide the interactive constitution of what, from the observer's perspective, is the illusion of competence. In the preceding example, during the whole-class discussion the teacher and Andrea interactively recreated the illusionary situation that they had first developed in the small-group intervention.

The next example illustrates an instance in which neither the teacher nor another participating student understood a child's explanation attempts, even though they made a serious effort to do so. This example is significant, because the solution attempt was relatively sophisticated and involved complex relationships. Consequently, it was important that the student attempting to explain not conclude that his use of sophisticated procedures was inappropriate. In essence, a distinction needed to be made between the adequacy of the explanation and the viability of the solution procedure. The adequacy of an explanation is dependent not only on what is said, but also on the interpretations of those attempting to make sense of what is said (Yackel, 1992). Consequently, clarity could develop only when the participants in the interaction resolved their incompatibilities. Prior analyses of small-group interactions indicate that one way in which children negotiate mathematical meaning when they have a dispute is to make explicit what they have previously only taken as shared (Cobb, Yackel, & Wood, 1992; Yackel, Cobb, & Wood, 1993). The following example illustrates that even when incompatibilities remain and mutual understanding does not develop, there can be some benefits from the interaction. In this instance, even though the teacher did not understand the explanation attempt, he facilitated the development of a situation for explanation. As a result, the student's thinking and reasoning were in no way diminished.

Example 5—October 23. This episode among Ray, Jameel, and the teacher occurred 2 months into the school year, in a project classroom different from the one that is the primary subject of this book. The problem under consideration was: "Roberto had twelve pennies. After his grandmother gave him some more he had twenty-five pennies. How many pennies did Roberto's grandmother give him?"

Ray and Jameel spent considerable time working on this problem. Jameel first suggested a thinking strategies approach.

5 Jameel: 12 and 11 is 23. 14 no, 15.

Ray acknowledged that he understood Jameel's thinking with, "12 and 7 is 19 . . . what could make 25?" but pursued a solution using unifix cubes. He put out a collection of 12 cubes using a ten-bar and 2 ones and eventually added 13 using another ten-bar and 3 ones. His final collection of cubes consisted of the two collections side by side. In the meantime, Jameel also got out a collection of 25 cubes, but had his arranged in a ten-bar of one color and a long strip of 15 of another color.

In the extended discussion that ensued, the boys talked about whether or not both unifix cubes representations were appropriate. Jameel insisted that his 10 and 15 were "the same thing" as Ray's 12 and 13. Ray disagreed. From the social interactionist perspective, we might say that Ray's interpretation was that Jameel had violated two "perceived" sociomathematical norms. First, a representation with cubes should model the problem as stated, and second, when solving a problem one should not "change the task," such as, in this case, to discuss (arbitrary) number relationships. Ten and 15 have the same sum as 12 and 13, but the task at hand was to figure out how many more are needed to get from 12 to 25.

When the teacher entered the group the boys both acknowledged their disagreement. After discussing the interpretation of the problem and clarifying the goal as figuring out how many to add to 12 to get 25, the teacher attempted to get Jameel to explain his thinking. A lengthy discussion followed. At (only) one point Jameel explicitly indicated his thinking strategy approach and why the 10 and 15 were useful to him.

122 Jameel: Well, we *play* like he had 10, and we, and uh—
123 Teacher: Where did you get the 10 from?

124 Jameel: I said, we could *play* like we had 10 and um, and just—all you gotta do is just say, um, 13, 14, 15, and it's 3.

From the observing researcher's perspective, Jameel was using the compensation strategy, that is, he was reasoning that because 10 and 15 are 25, then 12 and 13 are 25. However, Jameel's explanation was indirect. He argued that to figure out how much one needs to add to 12 to get 25, simply take the last three numbers in the counting sequence, 13, 14, 15, and add those three to the 10 to get 13. Jameel's inarticulate attempt to explicate the compensation strategy was met with, "I don't quite understand," from the teacher.

In his further attempt to clarify his thinking to the teacher, Jameel removed 3 cubes from the stack of 15, placed them off to the side, and indicated to the teacher that the remaining 12 cubes represented the number of pennies Roberto originally had, whereas the stack of 10 and the 3 cubes off to the side represented the 13 additional pennies his grandmother gave him. To make sense of Jameel's second explanation, it is necessary to think of the 12 pennies Roberto already had as a subset of the stack of 15. Apparently Jameel's reasoning now was that if 15 and 10 make 25, then 12 and 13 make 25. Because 12 indicates the number of pennies Roberto had (and 13 the number his grandmother gave him), the analogy implies that in the 15–10 combination, the 15 should also represent the number of pennies Roberto had (and 10 the number his grandmother gave him). Yet in Jameel's first explanation, his statement, "*play* like he had 10" indicated the opposite. A further confusion is that he used 13, 14, and 15 in two different ways. On the one hand, each number is associated with one object in the totality, as though each object is "named" with a number name in counting. On the other hand, each number represents the cumulative total counted so far. The teacher left the group without understanding Jameel's explanation. Finally, Jameel counted on from 12 to 25 using his fingers and determined that he had counted 13 fingers. This explanation was not challenged by Ray, and they went on to the next problem.

During the discussion Jameel altered his explanation several times but, after his initial remark, "*play* like he had 10," he never addressed the critical question of where the 10 or the 15 came from, or what they represented, even though the teacher explicitly told Jameel repeatedly that this was what was not shared in their interpretation. Critically, even though the teacher did not understand Jameel's explanation attempts, he continued to try to make sense of them. Thus, a situation for explanation continued to occur. The teacher never indicated to Jameel that he was wrong, only that he, the teacher, did not understand. Had he implied that Jameel was using inappropriate thinking, the incident might have had the effect of discouraging Jameel from using relational thinking to develop solutions in the future. As it stood, Jameel could conclude that his explanation was inadequate, but he had no reason to conclude from the interaction with the teacher that his thinking was erroneous.

In the whole-class discussion of the problem, a number of solution methods were offered. Jameel participated in the discussion of the solutions offered insofar as he challenged the teacher's reference to the "1" in 12 and the "1" in 13 as "one" and "one." Jameel insisted that they each represented ten, not a single one. Given that he had been unsuccessful in explaining his method to the teacher during small-group work, and that he had repeatedly asked if he could go on to explain the next problem, it is not surprising that when the teacher called on Jameel to explain the current problem he demonstrated the unquestioned counting-on solution with which he and Ray had concluded their small-group work on the problem. He gave no hint of his thinking strategy approach, even though none of the other solutions given were based on a thinking strategies approach and different solutions were encouraged, nor did the teacher refer to the discussion he had with Jameel during small-group work.

Although Jameel was left to draw his own conclusions about the appropriateness of his activity during small-group work, the teacher's actions encouraged rather than stifled his explanation attempts. The fact that the teacher had not rejected his explanation but had only indicated that he did not understand it, coupled with the fact that the teacher accepted Jameel's challenge that the "ones" in 12 and 13 should be called "tens" indicated that he respected Jameel's thinking.

IMPLICATIONS FOR THE CLASSROOM

As the commonly accepted mode of instruction in mathematics shifts from the traditional approach to an inquiry approach in which students develop personally meaningful solutions and explain their thinking to their peers, it is imperative that teachers (and researchers) understand the features of children's talk, their explanations of their thinking and their solution methods. More important, it is necessary for teachers to understand that students' activity is reflexively related to their individual contexts, and that the teacher contributes, as do the children, to the interactive constitution of the immediate situation as a social event. Analyses such as those discussed in this chapter are potentially useful in this regard.

In this section, we return to each of the issues that emerged from the analysis to reemphasize them and to discuss implications for classroom mathematics instruction. The first is the importance of children's interpretations of social events rather than social settings as the critical feature in determining qualitative differences in their explanations and descriptions of their solutions. The second is that the teacher plays an important role in facilitating children's attempts to explain. On the one hand, the teacher may assist children in saying what they are trying to say to provide clarity for others. On the other hand, the teacher may serve the function of maintaining a situation for explanation by persisting in trying to help a child make sense of his or her own explanation attempts. As the data show, teachers must be cautious in their attempts to help children develop viable solutions. The teacher may unintentionally inhibit children's attempts to explain, either by constituting the situation as one of achieving a viable solution or by relieving the children of the responsibility of figuring out for themselves which aspects of their explanations require further elaboration and explication.

The examples demonstrate that children's interpretations of their obligations in the presence of a teacher, or in the presence of a researcher during small-group work, can result in children giving different explanations and solutions than when they are working only with each other. In the example with Jack and Jamie, the boys changed their explanations and, in some cases, their solutions. These changes were based on their differing expectations, goals, and intentions. For example, our interpretation is that in the presence of the teacher Jamie felt obliged to establish that he had the correct answer, whereas in the presence of the researcher he felt obliged to explain how he had been attempting to solve the problem. An implication for the classroom is that the teacher cannot assume that children interpret the obligation to explain their solution attempts uniformly, or that their explanations are describing prior activity. Further, the teacher must be aware of the possibility that, without the teacher's assistance, one partner in the pair may not have an opportunity to pursue issues that remained unresolved in the group. This implies that teachers must exercise caution when working with partners or calling on them during class discussion. Because the explanations a child gives and the solutions discussed may represent the thinking of only one child in the pair, teachers must take care to attend to both partners. Second, as we have seen with the examples of Jamie during the small-group intervention of the teacher and of Travonda during the whole-class discussion, explanations that are not reports of prior activity can be beneficial and can represent powerful thinking. The teacher's recognition of the creative nature of such activity can contribute to fostering the development of alternative solution methods and reconceptualizations of a problem.

The examples also demonstrate that the teacher must be aware of the children's potential interpretations of his or her interventions, including those that are counter to the teacher's intentions. In the case of the interventions with Jamie and Andrea during small-group work, the teacher intervened to encourage the children to develop a viable or an alternative solution method. Nevertheless, in each of these cases, the teacher contributed to the expectation that developing a solution and determining the correct numerical answer were her purposes. Further, in both cases the solution method that these children reported during whole-class discussion was the one suggested by the teacher during small-group work, indicating that the children may have interpreted the teacher's suggested method as preferential.

The analyses also provide teachers and researchers with some indication of the variation of activity that may occur during small-group work. Children may take it for granted that their partner shares their interpretation and understands their thinking, as in the example with Jack and Jamie, or they may actively seek to convince their partner of their interpretation and solution method, as in the case with Katy and Ryan. In some cases they may make sense of their partner's activity, whereas in others, such as with Andrea and Andy, the children may not have a basis for communicating even though one or both of them may attempt to do so.

The teacher's role in establishing situations for explanation emerged as an important theme. As the examples make clear, constituting a situation for explanation neither requires nor implies the development of a viable solution. Conversely, ensuring the development of a viable solution does not imply that a situation for explanation is constituted. In Example 1 (Jack and Jamie), Example 4 (Andrea and Andy), and Example 5 (Jameel and Ray), the teacher intervened in the small-group work. For Jack and for Andy, potential situations for explanation did not develop, because in each case the teacher and another student in the group interactively constituted the topic of discussion to be that of developing a viable solution. Neither Jack nor Andy succeeded in getting the teacher to interpret their explanations. (In the case with Andy, it could be argued that the teacher did understand Andy's explanation but was attempting to negotiate an interpretation of the problem that went beyond that of a purely image-based task. Either she chose to ignore Andy's solution or she did not know how to incorporate it into a numerical interpretation, which she felt responsible to encourage.) In each of these two cases, however, a viable solution was developed with the assistance of the teacher.

The critical feature of a situation for explanation is that there are participants in the interaction who attempt to make sense of the explanation that is offered. The interaction between the teacher and Jameel in Example 5 exemplifies this feature. This interaction was qualitatively different from those in the previous examples. In this case the teacher was intent on making sense of Jameel's explanation, even though he was unsuccessful in doing so. Consequently, we can say that a situation for explanation was interactively constituted by Jameel and the teacher. However, a viable solution (that everyone in the interaction understood) was not developed in the presence of the teacher. Only after the teacher left did Ray and Jameel apparently reach consensus on a solution, one that Ray had proposed even before the teacher entered the group. For completeness' sake it should be noted that while the teacher was present with the group, a situation for explanation did not emerge for Ray. At one point the teacher asked Ray to explain his thinking but, before he finished doing so, the teacher interrupted him and asked Jameel to continue with his explanation. It appears that the teacher was not listening to Ray's explanation.

As these examples have illustrated, teachers must be aware of the distinction between assisting students in developing solutions and attempting to make sense of their solution attempts. Situations for explanation are likely to arise for students when the teacher attempts to make sense of their solution methods, but are not likely to arise when the teacher introduces a solution method of his or her own.

The role of teacher interventions in assisting children to give explanations emerged as another important issue in the analysis. When the teacher intervenes to help clarify an explanation that a child is giving during whole-class discussion, the teacher's actions can serve both to assist the children in understanding what constitutes an explanation and to interfere with the children in developing their own explanations. They can assist the children to understand inasmuch as the teacher repeats, elaborates, and so forth, based on her understanding of the other children's potential for making sense of what is being said. They can interfere in that they relieve the students of the obligation of figuring out for themselves what repetitions and elaboration might be useful. On the one hand, it is the teacher's responsibility to help students learn how to describe and talk about their mathematical thinking. On the other hand, it is the teacher's responsibility to help them learn what constitutes an acceptable explanation. The latter involves judging the fit between what is being said and what is taken as shared by the class. Rather than taking the responsibility for judging this fit him- or herself, the teacher can ask children if they understand and encourage them to ask questions and request clarification. In this way, the teacher contributes not only to children's developing understanding of what constitutes an acceptable explanation, but also to the interactive constitution of the obligation to listen to and try to make sense of the explanation attempts of others.

REFERENCES

Ball, D. L. (1993). Halves, pieces, and twoths: Constructing and using representational contexts in teaching fractions. In T. P. Carpenter, E. Fennema, & T. A. Romberg (Eds.), *Rational numbers: An integration of research* (pp. 157–195). Hillsdale, NJ: Lawrence Erlbaum Associates.

Barnes, D., & Todd, F. (1977). *Communicating and learning in small groups*. London: Routledge & Kegan Paul.

Bateson, G. (1972). *Steps to an ecology of mind*. London: Paladin.

Blumer, H. (1969). *Symbolic interactionism*. Englewood Cliffs, NJ: Prentice-Hall.

Bruner, J. (1986). *Actual minds, possible worlds*. Cambridge, MA: Harvard University Press.

Cobb, P. (1986) Contexts, goals, beliefs, and learning mathematics. *For the Learning of Mathematics*, 6(2), 2–9.

Cobb, P., Wood, T., & Yackel, T. (1991). A constructivist approach to second grade mathematics. In E. von Glasersfeld (Ed.), *Radical constructivism in mathematics education* (pp. 157–176). Dordrecht: Kluwer.

Cobb, P., Wood, T., Yackel, E., & McNeal, B. (1992). Characteristics of classroom mathematics traditions: An interactional analysis. *American Educational Research Journal*, 29, 573–604.

Cobb, P., Yackel, E., & Wood, T. (1989). Young children's emotional acts while doing mathematical problem solving. In D. B. McLeod & V. M. Adams (Eds.), *Affect and mathematical problem solving: A new perspective* (pp. 117–148). New York: Springer-Verlag.

Cobb, P., Yackel, E., & Wood, T. (1992). Interaction and learning in mathematics classroom situations. *Educational Studies in Mathematics*, 23, 99–122.

Gregg, J. (1992). *The acculturation of a beginning high school mathematics teacher into the school mathematics tradition*. Unpublished doctoral dissertation, Purdue University, West Lafayette, IN.

Kamii, C. (1989). *Young children continue to reinvent arithmetic, 2nd grade: Implications of Piaget's theory*. New York: Teachers College Press.

Koch, L. C. (1992, August). *Developing mathematical voices*. Paper presented at the Seventh International Congress on Mathematical Education, Québec City, Canada.

Lampert, M. (1990). When the problem is not the question and the solution is not the answer. Mathematical knowing and teaching. *American Educational Research Journal*, 27, 29–63.

Leiter, K. (1980). *A primer on ethnomethodology*. New York: Oxford University Press.

Murray, H. (1992, August). *Learning mathematics through social interaction*. Paper presented at the Seventh International Congress on Mathematical Education, Québec City, Canada.

National Council of Teachers of Mathematics. (1989). *Curriculum and evaluation standards for school mathematics*. Reston, VA: National Council of Teachers of Mathematics.

Richards, J. (1991). Mathematical discussions. In E. von Glasersfeld (Ed.), *Radical constructivism in mathematics education* (pp. 13–51). Dordrecht: Kluwer.

von Glasersfeld, E. (1983). Learning as a constructive activity. In J. C. Bergeron & N. Herscovics (Eds.) *Proceedings of the Fifth Annual Meeting of the North American Chapter of the International Group for the Psychology of Mathematics Education* (Vol. 1, pp. 411–469). Montreal: PME.

Wood, T., & Yackel, E. (1990). The development of collaborative dialogue within small group interactions. In L. P. Steffe & T. Wood (Eds.), *Transforming children's mathematics education: International perspectives* (pp. 244–252). Hillsdale, NJ: Lawrence Erlbaum Associates.

Yackel, E. (1992, August). *The evolution of second grade children's understanding of what constitutes an explanation in a mathematics class*. Paper presented at the Seventh International Congress on Mathematical Education, Québec City, Canada.

Yackel, E., & Cobb, P. (1993, April). *Sociomathematical norms, argumentation, and autonomy in mathematics*. Paper presented at the Annual Meeting of the American Educational Research Association, Atlanta, GA.

Yackel, E., Cobb, P., & Wood, T. (1991). Small group interactions as a source of learning opportunities in second grade mathematics. *Journal for Research in Mathematics Education, 22*, 390–408.

Yackel, E., Cobb, P., & Wood, T. (1993). Developing a basic for mathematical communication within small groups. In T. Wood, P. Cobb, E. Yackel, & D. Dillon (Eds.), *Rethinking elementary school mathematics: Insights and issues* (Journal for Research in Mathematics Education Monograph Series No. 6, pp. 33–44). Reston, VA: National Council of Teachers of Mathematics.

Yackel, E., Cobb, P., & Wood, T. (in press). The interactive constitution of mathematical meaning in one second grade classroom: An illustrative example. *Journal of Mathematical Behavior*.

5

Thematic Patterns of Interaction and Sociomathematical Norms

Jörg Voigt
University of Hamburg

The underlying motive of the study discussed in this chapter is to understand how intersubjectivity is established in mathematics classrooms, if the radical constructivist perspective on the learner's subjectivity is taken as a starting point. The popular answer, that the student could discover mathematical truth on his or her own, does not seem sufficient because the student (at least the elementary school student) has not fully learned the process of mathematical reasoning, and is not yet a member of a mathematical community. This study searches for an answer by analyzing interactions (i.e., mutual influences) between the teacher and the students within the classroom microculture.

Although the investigation focuses on theoretical concepts developed within a theoretical network, the concepts are based on the analyses of actual classroom episodes. Thus, classroom episodes in the text will be used to illustrate theoretical considerations and empirical findings.

In the first section, an interactionist approach is presented that mediates between individualism and collectivism. This contrast juxtaposes a psychological perspective that stresses the learner's autonomy and his or her cognitive development, and a collectivist perspective that criticizes the "child-centered ideology" and views mathematics learning as the student's socialization into a pregiven culture. The comparison of these perspectives is made in order to present contrasting ideas and to outline an interactionist approach. This interactionist approach views negotiation of meaning as the mediator between cognition and culture.

In the second section, several basic ideas of the interactionist perspective are presented. A main assumption is that the objects of the classroom discourse are ambiguous, that is, open to various interpretations. The participants gain a taken-as-shared understanding of the objects when they negotiate mathematical meanings. In these processes, the students and teacher achieve a thematic coherence in their discourse. Interactively, they constitute a mathematical theme that, on the one hand depends on the participants' contributions, whereas on the other hand cannot be sufficiently explained by the thoughts and intentions of any one person alone.

Whereas the second section focuses on the dynamics of classroom situations, the third section attempts to explain the regularities of the classroom microculture. Thematic patterns of interaction that contribute to the stability of the mathematics discourses are reconstructed. In particular, this section argues that the evolution of mathematical themes, described as improvisations of thematic patterns of interaction, seems to correspond to the students' cognitive development.

The fourth section stresses the relationship between the students' mathematical goals and values that emerge in the negotiation of mathematical meanings between the teacher and the students. That is, the students direct their mathematical activities toward their own mathematical goals, and these goals are indirectly influenced by the teacher. In the classroom analyzed, the teacher and the students mutually constitute sociomathematical norms, such as what counts as an elegant mathematical explanation. This analysis demonstrates how the teacher can positively influence the students' mathematical thinking and cognitive development in an indirect and subtle way.

This study does not offer direct suggestions for improving mathematics instruction. The primary interest is to develop theoretical concepts that make classroom processes understandable. But what is the relevance of such theoretical work for the improvement of mathematics education? We cannot improve the classroom microculture in the same way that we can change the mathematical curriculum or the classroom macroculture characterized by general principles and teaching strategies. The microculture lives its own life, and its characteristics depend on hidden patterns, conventions, and norms that, like the students' attitudes and the teaching style, are difficult to change. Therefore, we should conceptualize the change of a microculture as an evolution rather than as a rearrangement. In order to influence and direct that evolution, it is helpful to understand the regularities and dynamics of the processes within the classroom life.

RELEVANCE OF THE INTERACTIONIST APPROACH

A basic assumption of the interactionist point of view is that cultural and social dimensions are not merely peripheral conditions of learning mathematics but are, in fact, intrinsic to the learning of mathematics. This view has to be justified, because mathematics learning is often considered to be a process that occurs within the mind of a lonely learner who is free of social dynamics and cultural influences.

Philosophies like Platonism or intuitionism assume that mathematics expresses eternal relationships between objects that are intuitive as well as objective. Tymoczko (1986) called such theories "private theories" because, in the ideal situation, a single isolated mathematician discovers or creates mathematical knowledge. Diverging from these private theories, other philosophers, such as Lakatos (1976) and Wittgenstein (1967), considered mathematics to be a product of social processes. Using the literary form of classroom discussions, Lakatos described how mathematical concepts are stabilized or changed over time through the processes of agreement and refutation. Wittgenstein understood mathematics as invented (not discovered), and thus as humanmade. Wittgenstein explained the inexorableness of mathematical argumentation in our experience through his concept of the "language game." According to this view, the firmness of numbers and figures and the rigor of proving procedures are effects of the rigid language games invented by mathematicians. (For a further discussion of language games, see Bauersfeld, chap. 8, this volume.)

In mathematics education, there exists another obstacle to the interactionist point of view: the prominence of child-centered conceptions that stress the autonomy of the learner. Several educators who refer to Piaget's genetic epistemology or to Bruner's early educational claims of "discovery learning" view a child's mathematical knowledge as the product of his or her conceptual operations. However, although the sociological aspects of Piaget's research are not fully developed, Piaget himself stated that "Social interaction is a necessary condition for the development of logic" (quoted by Doise & Mugny, 1984, p. 19). Similarly, Bruner (1986) remarked on a development in his own work: "I have come increasingly to recognize that most learning in most settings is a communal activity, a sharing of the culture" (p. 127). Quite early, Bruner (1966) doubted that the student could discover all school knowledge: "Culture, thus, is not discovered; it is passed on or forgotten. All this suggests to me that we had better be cautious in talking about the method of discovery, or discovery as the principal vehicle of education" (p. 101).

Nevertheless, I also see the danger of overemphasizing the focus on social and cultural aspects of mathematics learning. Edwards and Mercer (1987, p. 168) claimed that the "child-centered ideology" needs to be replaced (with another ideology?). Solomon (1989) insisted that in mathematics lessons, "Learning is the initiation into a social tradition" (p. 150), and that "Knowledge gain is a question of entering a social existence" (p. 86). She also maintained that "Development is the progressive socialization of the child's judgments, their turning in with those of the adult by means of the appropriate public criteria" (p. 118). Solomon illustrated her point with regard to arithmetic: "Thus knowing about number can be described as entering into the social practices of number use, and coming to know as a process of initiation into a social understanding of when and how to act in particular situations" (p. 8).

In these works, the culture of the mathematics classroom seems to be pregiven. That is, the student is regarded as an object of the teacher's activities or as a rather passive participant of the classroom processes. The student's unusual and unexpected actions may be explained as mere deviations, and differences among individuals' developments are often unexplained (Cobb, 1990).

The risk of defining learning mathematics as "learning how to participate in 'social practices'" (Solomon, 1989) is that a smoothly proceeding classroom discourse may be interpreted as an indication of successful learning. In fact, several studies analyzing discourse processes in detail have concluded that students' attempts to participate in the constitution of traditional patterns of interaction can be an obstacle for learning mathematics (Bauersfeld, 1988; Jungwirth, 1991; McNeal, 1991; Steinbring, 1989; Voigt, 1985, 1989). In usual lessons, the classroom discourse often tends to degenerate into rituals that are constituted step by step. Surprisingly, the teachers are unaware of these regularities in the microprocesses, and they misinterpret the students' participation in the classroom discourse. Therefore, we should resist the temptation to identify learning mathematics with learning how to participate successfully in patterns of interaction. We can, however, analytically distinguish between the dynamics of learning processes and the dynamics of interaction processes. This view will enable us to investigate the relationships between mathematical learning and social interaction.

In view of these problems, the interactionist approach allows the researcher to consider social aspects, and, at the same time, avoids the danger of overemphasizing the cultural aspects. This approach focuses on individuals' sense-making processes and on the ways in which they interactively constitute and stabilize mathematical meanings. The interactionist approach is based on microsociology, and is particularly influenced by symbolic interactionism (Blumer, 1969; Mead, 1934) and ethnomethodology (Garfinkel, 1967; Mehan, 1979).

Bauersfeld, Krummheuer, and I have modified the sociological concepts in order to deal with the specifics of teaching and learning mathematics (Bauersfeld, 1980; Bauersfeld, Krummheuer, & Voigt, 1988; Krummheuer, 1983; Voigt, 1984). Of course, the interactionist point of view is not sufficient for understanding classroom processes as a whole; it clarifies the dynamics and regularities of the microculture in mathematics classrooms. Specifically, it focuses our attention on the negotiation of mathematical meanings in the local events of classroom life.

In closing this section, some general methodological comments are given. Because microsociology is based on the "interpretative paradigm" (in contrast to the normative one; cf. Wilson, 1970), interactionist studies prefer microethnographical methods (Erickson, 1986; Voigt, 1990). These methods involve the detailed description and interpretation of transcripts in the context of case studies. The transcripts are analyzed as documents describing processes. For example, instead of asking what the mathematical meaning of something really is or how often the meaning is constituted, we reconstruct how a mathematical meaning emerges and how it is stabilized as the classroom life evolves. Mathematical meanings are considered to be products of interaction processes and to be specific to the microculture studied.

When analyzing transcripts, the interpreter attempts to reconstruct the constructions of the observed students and teachers (Schütz, 1971). The basic relativistic assumption of Schütz's social phenomenology as well as of von Glasersfeld's radical constructivism is that the actor's experiential reality is the reality as he or she interprets it. This assumption is valid for the interpreter as well. Therefore, every observed and documented classroom episode is open to different interpretations. In order to justify the researcher's particular interpretations of an episode, the researcher should explicate empirical background knowledge (i.e., products of

5. THEMATIC PATTERNS OF INTERACTION

earlier interpretations) and theoretical concepts. Therefore, in contrast to quantitative positivist studies, ethnographical studies have to involve "thick descriptions" (Denzin, 1989) and the elaboration of theoretical concepts.

NEGOTIATION OF MATHEMATICAL MEANINGS

Ambiguity and Interpretation

According to folk beliefs, the tasks, questions, symbols, and so on of mathematics lessons have definite, clear-cut meanings. In order to realize the relevance of the concept of negotiation, these beliefs have to be challenged. If one looks at microprocesses in the classroom carefully, the tasks and symbols are ambiguous and call for interpretation (with regard to the ambiguity of tools, such as of the hundreds board, see Cobb, chap. 3, this volume). What is the meaning of "5" to young children in a specific situation? Is the meaning bound to concrete things (e.g., "the little finger of my left hand")? Does the sign remind the student of previous activities (e.g., "a difficult number to write")? Does it evoke specific emotions (e.g., "my favorite number")? Is its meaning related to other numbers (e.g., "equal to 2 + 3, 1 + 4, 0 + 5"), and so on?

The problem shown in Fig. 5.1 is taken from the project classroom. There are several different ways to interpret this problem. For example:

• The problem can be interpreted as a practical one. Multilinks have to be manually connected to each other. In this case, the number of the additional squares has to be counted. There is no necessity to figure out the first number or to count tens.

• The problem can give rise to operations with bars and squares in one's imagination. The additional bars and the surplus squares have to be counted in order to get seven bars and three squares.

• The problem can be interpreted as a problem of calculation with abstract units of numbers. The first representation of a number has to be enumerated. Then the

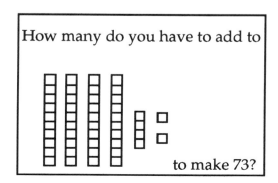

FIG. 5.1. Strips-and-squares task corresponding to 46 + __ = 73.

difference between the two numbers has to be figured out according to rules of the numeration system.

The meaning of ten differs in these options for understanding. Ten can be taken as the quantity of several objects, or as the name of a standard bar, or as a unit. Using cognitive concepts, Cobb (chap. 3, this volume) discusses the possible interpretations of the tasks in detail.

According to one of the ethnomethodological assumptions, every object or event in human interaction is plurisemantic (i.e., "indexical," Leiter, 1980). In order to make sense of an object, the subject uses background knowledge and forms a context for interpretation. If the background understanding is taken for granted, then the subject does not necessarily experience the ambiguous object as plurisemantic but rather as factual. For example, some students solved the problems by counting the missing bars and surplus squares separately. Carol, one of the less conceptually advanced students in the classroom, solved the problem shown in Fig. 5.2 this way:

1 Teacher: How are we going to figure this out? We are going to do this together. What are we going to add to this number to go to this number? How are we going to do this? What are we going to have to do? (She waits until several students raise hands.) Carol.
2 Carol: (She goes to the board.) Add 2 ten-bars take away 3. (She points to the top parts of the problem).
3 Teacher: Add 2 ten-bars take away 3. Hm, OK. So she says (writes on the overhead projector Carol's solution) add 2 tens and take away 3 ones. All right, that's one solution. I don't know if it is right or wrong, but that is what she says might work. What about Michael?

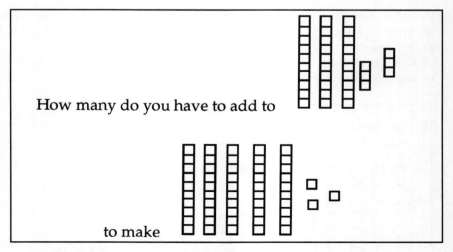

FIG. 5.2. Strips-and-squares task corresponding to, 36 + __ = 53.

5. THEMATIC PATTERNS OF INTERACTION 169

Interpreting the problem as a practical one, Carol had no reason to combine the amounts of the missing bars and surplus squares to make one number. Moreover, the observers assumed that Carol had a conceptual understanding of ten as a numerical composite rather than an abstract composite unit (for the cognitive concepts see Cobb, chap. 3, this volume, or Cobb & Wheatley, 1988). Therefore, we can infer that Carol, by herself, could not evaluate her solution as a provisional one. Indeed, for the practical purpose of combining cubes and bars, Carol's solution was sufficient.

The ambiguity of objects is not only a peculiarity of special scenes or problems; it comes out as a long-term characteristic of classroom discourse if the teacher and the students interpret the objects in systematically different ways. Referring to Goffman's frame analysis (1974), Krummheuer (1983) reconstructed different background understandings ("framings") between the teacher and the students of an algebra class over a longer period of time. Comparing the interactions during collaborative learning with the interactions during frontal class teaching, he demonstrated that "misunderstandings" between the teacher and the students are quite usual. Even if the participants use the same words, they can think differently. Using epistemological arguments, Steinbring (1991) explained why the disparity between the teacher's and the students' background understandings exists. New mathematical concepts can implicate new understandings of what was known before. In order to learn the new mathematical concepts that the teacher introduces, the students must change their background understandings toward the teacher's understanding. This disparity, or nonreciprocity of perspectives, can give rise to learning opportunities in social interaction.

The mathematical problems posed in the project classroom can be solved in different ways according to different conceptual possibilities. Further, the classroom participants are obliged to understand and compare their different interpretations of the problems. During processes of interaction, they negotiate the mathematical meanings and try to come to an agreement about which results and which arguments could be taken as mathematical solutions and appropriate mathematical explanations, respectively.

Social Interaction

From the view of symbolic interactionism, interaction is more than a sequence of actions and reactions. One participant of an interaction monitors his or her action in accordance with what he or she assumes to be the other participants' background understandings, expectations, and so forth. At the same time, the other participants make sense of the action by adopting what they believe to be the actor's background understandings, intentions, and so forth. The subsequent actions of the other participants are interpreted by the former actor with regard to his or her expectations and can prompt a reconsideration, and so on. For example, the student can interpret the teacher's reaction as a specific evaluation of his or her own thinking even though the teacher's reaction might only have been an expression of amusement at a particular moment during the student's action. Using his or her background knowledge of the teacher's supposed emotion and consequently feeling shame, the student might search for more advanced ways to interpret the problem at hand and to solve it. Another student experiencing a similar situation might cope by pleasing the teacher superficially. Of course, the teacher could react differently according to his or her background knowledge of the two students' different dispositions. Analyzing the interaction between the teacher and a student, Jameel, Yackel (chap. 4, this volume) presents a concrete example of such mutual interpretations and expectations.

Of course, the subject's background understanding is not fixed and stored knowledge that is unchangeable during the interaction processes; at any time during these processes, the subject may realize various background understandings. The result may be that the realized background understanding will be confirmed or altered. However, in the context of learning mathematics, the situations of interest are those in which the students cannot be expected to understand the classroom objects as intended by the teacher. In the next two episodes, a student offered an advanced solution that the teacher did not expect from all students. Therefore, she adjusted her reaction to her expectation of the conceptual possibilities of the other students:

The problem $5 \times 7 = \_\_$ is discussed after the problem $5 \times 5 = \_\_$ was solved:

1	Teacher:	You're ready for the next one. All right. It says $5 \times 4 = 20$, the next one says $5 \times 5 = 25$. If we know that, can we figure out 5×7? Alex?
2	Alex:	Um 3 more 10 more.
3	Teacher:	10 more?
4	Alex:	Well 10 more . . . from 25 [inaudible].
5	Teacher:	Jack, I'm going to ask you for the next answer. You and Jamie be ready.
6	Jack:	OK.
7	Teacher:	Go ahead (to Alex).
8	Alex:	It's 10 more than this [$5 \times 5 = 25$] because the 7 is 2 more than 5.
9	Teacher:	So you think the answer would be 10 bigger?
10	Alex:	Yeah.

The problem $5 \times 7 = \_\_$ is ambiguous because there are different ways to think about it and to solve it. Alex's way of solving can be considered advanced when compared with other students' ways of solving. If we assume that Alex pursued this way because he anticipated that the teacher would like it, we can understand that Alex's activities were influenced by the social interaction with the teacher. That is, Alex could have interpreted the teacher's first turn as an invitation to compare the problem in question with the former one. Conversely, the teacher's activities might have been influenced by her interpretation of how Alex's activities indicated his conceptual possibilities. During prior lessons, Alex offered similar comparisons between problems and his answers indicated that he took the structural relationships

5. THEMATIC PATTERNS OF INTERACTION

into account correctly. Perhaps the teacher addressed Alex first because she expected him to offer a similar method again.

Moreover, the teacher's following reaction to Alex's solution may have been influenced by her interpretation of the other students' competencies:

11	Teacher:	Do you see that boys and girls? He's saying $5 \times 5 = 25$.
12	Students:	Oh yeah.
13	Teacher:	This is 2 bigger. It's 2 fives bigger, not 2 ones.
14	Students:	Oh yeah.
15	Teacher:	But 2 fives . . . and so 25 and 10 make . . .
16	Students:	35, 35.
17	Teacher:	If we put 5 on a finger and count 7 of them we have 5, 10, 15, 20, 25, 30, 35. So that's another way to figure it.

On the one hand, the teacher stressed Alex's solution and explained the mathematical links between the two problems. On the other hand, she also presented a rather plain way of solving. Presumably, the teacher wanted to highlight Alex's way, but she anticipated that not all students would understand it and that she would overtax some of them. For example, 20 minutes before, another student, Peter, had inferred from $3 \times 7 = \underline{21}$ that 3×8 has to be 22 (1 more) or that it has to be 28 (7 more).

The next classroom episode also illustrates interaction processes that involve different background understandings.

The participants have discussed fractions as the parts of the area of circles. Now, the teacher uses pieces of apples as an embodiment of fractions:

1	Teacher:	If we have 20 kids and 25 apples; how are we going to split it up so everyone gets the same amount . . . equal pieces?
2	Student:	1.
3	Student:	We can give everybody 1 and what are we going to do with the other 5 apples?
4	Student:	Throw them away.
5	Teacher:	We're going to throw them away. Well we can throw them away, but that's kind of wasteful.
6	Student:	Split them in half.
7	Teacher:	[simultaneously] What are we going to do? Split them in half.
8	Students:	Split them in fourths, split them in fourths.
9	Michael:	Split them into 20 more pieces.
10	Teacher:	Right. Split them into 20 more pieces. So sometimes a fraction is not only a whole thing or a group it has . . . it has extra pieces.
11	Alex:	You split the apples into 5 pieces because . . .
12	Bonnie:	No, into fourths.
13	Teacher:	Wait a minute. Sh [to rest of class], OK.
14	Bonnie:	5 apples 5×4 is 20.
15	Teacher:	5×4.
16	Bonnie:	It would be fourths. Split the apples into fourths.

17	Teacher:	We would split the apples into fourths. And so how much would everybody get?
18	Bke:	One and a fourth.
19	Teacher:	One and a fourth apples. Or we would get 5/4ths. (She circles 5/4 on the overhead projector.) Wouldn't we?

On the one hand, the story can be understood as a pragmatic problem so that the distribution of only 20 apples is reasonable. At first, the students suggested that every child would get one apple, leaving five remaining. In real life, it is not necessary to distribute the remaining apples. On the other hand, the story can be taken as a representation of how fractions can be handled. Acting from this perspective, the teacher had to cope with the pragmatic solution, the first perspective. The teacher spoke of wastefulness. If a student thought of the problem as true to life, he or she had to change his or her perspective in order to understand the intellectual problem intended by the teacher. By negotiation of meaning, the participants arrived at a provisional agreement. In the interaction process, two different background understandings became adjusted so that a joint solution was achieved.

Mathematical Meanings Taken as Shared

It is not only important that the teacher and the students attempt to understand each other. A third thing, the accomplishment of intersubjective meanings taken as mathematical ones, is essential in mathematics teaching and learning. The point is not that teacher and students "share knowledge"; from the symbolic interactionist and the radical constructivist points of view, mathematical meanings are only taken as shared when they are produced through negotiation. Krummheuer (1983, with reference to Goffman) and Cobb (1990, with reference to von Glasersfeld) used the terms *working interim* and *consensual domain*, respectively, to describe this situation. The participants interact as if they interpret the mathematical topic of their discourse as the same, although they cannot actually be certain that their subjective background understandings are consistent with those of the other participants. "We have then a kind of interactional 'modus vivendi.' Together the participants contribute to a single over-all definition of the situation which involves not so much a real agreement as to what exists but rather a real agreement as to whose claims concerning what issues will be temporarily honored" (Goffman, 1959, p. 9f). One can never be sure that two persons are thinking similarly if they collaborate without conflict, especially if they agree about formal statements and processes. One of the characteristics of formal mathematics is that people can coordinate their actions smoothly although they are actually ascribing different meanings to the objects. For example, in the earlier scene, the teacher and the students speak of "fourths." They act as if they are thinking of the same thing, but several interpretations such as a part of an apple, or a part of five apples, or a number could be assumed.

The term *taken as shared* describes the participants' conviction that meanings are shared, or the participants' willingness to neglect doubts in view of inevitable

5. THEMATIC PATTERNS OF INTERACTION 173

ambiguities, or the presumption that the meanings will be shared if the others will "read between the lines." The participants' conviction that meanings are shared, fixed, and definite is described by ethnomethodological studies (Garfinkel, 1967; Leiter, 1980) that use the concept of "commonsense knowledge" referring to Schütz and Luckmann (1973). In his ethnomethodological writings, Cicourel (1970) used the concept of the "et-cetera-rule" to hint at the participants' confidence that they could interpret the discourse with further clarifications but the provisional understandings are sufficient for the purposes at hand.

In learning situations, students participate in processes of mathematical argumentations without ensuring that the mathematical meanings are immediately clarified. In this episode, the students, Jack and Jamie, are solving the following problems:

1. $50 - 9 = \underline{41}$
2. $60 - 9 = \underline{51}$
3. $60 - 19 = \underline{41}$
4. $41 + 19 = \underline{60}$
5. $31 + 29 = \underline{60}$
6. $31 + 19 = \underline{50}$
7. $32 + 18 = \underline{}$

We do not know by direct observation how the two students solved the first six tasks because the videotape begins when the students are solving the seventh task. In the classroom, several ways of solving were allowed and encouraged; for example, counting by ones, using the hundreds board, or relating a new problem to a previously solved one (Wood & Yackel, 1990).

The videotape begins when Jack makes a statement:

1	Jack:	Uh-huh, that's 18, not 19.
2	Jamie:	Yeah, but that's 32 not 31.
3	Jack:	Oh yeah!
4	Jamie:	They're the same thing.
5	Researcher:	What's the same thing?
6	Jamie:	These two. [He points to $31 + 19 = 50$ and $32 + 18 = \underline{}$.]
7	Researcher:	Hang on, I was asking Jack. Which ones? We've got 31 and 19.
8	Jack:	Makes 50.
9	Researcher:	Yeah.
10	Jack:	And, look 32 and 18. See, it's just one more than that [points to the tasks], and that's one higher than that.

Later on in the interview, Jack and Jamie indicated that they solved the fifth task by adding the tens and the ones $31 + 29 = 30 + 20 + 1 + 9 = 50 + 10 = 60$. The sixth task was solved by comparing the second addends of the sixth and of the fifth tasks. The students were able to combine two tasks that did not differ in one addend. But the fifth task, which differs in both addends from the preceding task, was solved as an isolated one.

At this point, when the students solve the seventh task, a new way of solving emerged in the course of interaction. We do not know whether Jamie had solved the problem by a compensation strategy, such as comparing the sixth and the seventh tasks before the episode starts. The analysis of the interview with Jamie indicates that he would be able to do so. Therefore, it may be that Jamie had offered this solution to Jack who, at the beginning of the episode, resisted and had to be convinced by Jamie.

Nevertheless, it is also possible to assume that the compensation emerges "between" the students without one student thinking of it ahead of the negotiation. This interpretation illustrates the interactive constitution of mathematical meaning from the interactionist point of view: The meaning of the task is interactively constituted as one in which the increase of one addend compensates the decrease of the other. At the beginning, each of the students focused his attention on the change of one addend and points to it. Each then altered his attention because he was stimulated by the other and accepted the other's statement. The students gained a tacit agreement without checking whether they did in fact "share" a common knowledge. Through this negotiation, a meaning of the task was constituted. Without this negotiation, we cannot assume that one student would have constructed this meaning on his own or that one student would have taken over the responsibility for the solution alone.

In the classroom, the participants constitute taken-as-shared mathematical meanings in an interactive way. What is referred to as "meaning taken as shared" emerges during processes of negotiation. From the observer's point of view, the meaning taken as shared is not a partial match of individuals' constructions, nor is it a cognitive element. Instead, it exists at the level of interaction. "Symbolic interactionism views meaning . . . as arising in the process of interaction between people. The meaning of a thing grows out of the ways in which other persons act toward the person with regard to the thing. . . . Symbolic interactionism sees meanings as social products" (Blumer, 1969, p. 5). In the scene just described, the meaning of the seventh task emerged when the students interact. The task was taken as the "same thing" compared with the sixth task. When the researcher and the students interacted, this meaning was stabilized between them.

Interactive Constitution of a Theme

In the course of negotiation, the teacher and the students (or the students among themselves) accomplish relationships of mathematical meanings that are taken as shared. From the observer's point of view, I call these relationships of meanings a mathematical *theme*. In the episode last described, the theme is the comparison of two tasks because the participants seem to pay attention to differences and correspondences of the tasks.

Because "the teacher is not safeguarded against the students' creativity" (Bauersfeld, personal communication), the students can make original contributions to the theme so that it may not represent the content that the teacher intended to establish. Realizing his or her intentions, the teacher is dependent on the students' indications of understanding, just as the students are dependent on the teacher's acceptance of their contributions. So the theme is not a fixed body of knowledge, but, as the topic of discourse, it is interactively constituted and it changes through the negotiation of meaning.

In the next episode, a student makes a statement that, presumably, the teacher did not expect. His statement further extends the preceding theme.

This excerpt picks up after a discussion in which a story problem was solved, and the solutions of the story problem were discussed. The first problem is as follows: Jay has some baseballs. He lost 19 of them. He had 29 left. How many baseballs did he have before he lost some? At this point of the discussion, the solutions of the next story problem are discussed: After Mary gave away 29 stickers to Julie, she had 19 left. How many stickers did Mary have to begin with?

1	Teacher:	Joy.
2	Joy:	48.
3	Teacher:	48, how did you get that answer?
4	Joy:	Well, I just used . . . well, we used the top one. (The teacher smiles and nods.) And it equals 29.
5	Jack:	(He interrupts.) They're the same numbers.
6	Teacher:	Ah-hah, Jack, did you want to add something to that?
7	Jack:	Yes, they're just different words. (A general hub-bub of comments as the teacher speaks, as if to indicate the insight is shared. Among this Linda speaks.)
8	Linda:	They're cousins.
9	Teacher:	They're cousins, that's a nice way to explain it, they're cousins.

At the beginning, the participants compared the two story problems. Joy and Jack focused their attention on the correspondence between the numbers. The other students' excitement indicated that they saw the correspondence. Should we be satisfied if pure numbers and their relations form the theme? Do the students know why they have to add the numbers? We do not know whether they realized a correspondence at the surface level of numerals or at a semantic level. If the students did not pay attention to the story because they were tempted to look for signal words without realizing the intended semantic structure of the story problem, then they were only realizing a surface correspondence.

In light of this danger, Linda's utterance, "They're cousins," should not be judged as only a joke. Her contribution reintroduced the story into the theme, so that her statement could be taken as a starting point for the discussion of the story. In the discussion, the intended semantic structure of the problem could be derived.

If the teacher does not direct and evaluate the students in a rigid, step-by-step way, and if a dialogue consisting of students' original contributions is established, then the theme can be described as a river that produces its own bed. The result of the dialogue is not clear in advance. In one of the episodes presented earlier, Jamie and Jack were not forced to solve the task by a compensation strategy. But, if they had solved several tasks in a similar way and if they had successfully justified that way, the observer might expect that the comparison of tasks would be a common theme between Jack and Jamie in future problems.

Themes gain stability because people are obliged to take care of the thematic coherence of discourse (except small talk), and because mathematical discussions are constrained by specific, rigid obligations. In cases of conflict, the participants clarify what is taken as the theme. If a participant is accused of straying from the theme, he or she may be forced to justify the relevance of his or her divergent contribution. The theme may also be changed through meta-communicative remarks or markers (Luhmann, 1972; see also Krummheuer, chap. 7, this volume).

When the interactional concept of theme is applied to teacher–student interactions, it mediates between two theoretical perspectives. One perspective stresses the experiential situation as a person subjectively constructs it. The individual's conceptual operations are of interest in this perspective. The other perspective views the global cultural context, as stabilized and institutionalized by a community (of mathematics teachers and other persons) over a longer period of time. Because the theme is interactively constituted, it would not exist without the teacher's and the students' contributions. It is related to the students' individual thinking processes as well as to the mathematical and educational claims of a global context that the teacher represents.

The theoretical perspective of the second section can be now summarized. In the classroom situation, objects are ambiguous. Because of the differences between their situated interpretations and their background understandings, respectively, the teacher and the students negotiate mathematical meanings. The students and the teacher arrive at mathematical meanings that are taken as shared. The relationship between these meanings forms the theme of the discourse. The theme unites students' contributions and those of the teacher, who is particularly concerned that the theme fits with the intended mathematics.

In elaborating these concepts, I attempt to take two theoretical demands that were mentioned in the first section into account. On the one hand, the concepts are not restricted to issues of individual psychology. The individual's sense-making processes are assumed to be influenced by social affairs. On the other hand, the individual's actions and thoughts are viewed as originally influencing the social processes.

(THEMATIC) PATTERNS OF INTERACTION

The Accomplishment of the Microculture

In time, the negotiation of meaning forms commitments between the participants and stable expectations for each individual. Through mutual accommodations, the participants form the impression that they know what mathematics teaching and learning is. When the interaction proceeds smoothly, a "common sense knowledge"

(Schütz, 1971) is taken as shared. What was initially constituted explicitly, now remains tacit. Cobb called this process the *institutionalization of knowledge* (1990). Studying everyday life, ethnomethodologists point out that the knowledge is confirmed to be shared and to be given by descriptive "accounting practices":

> The stories that people are continually telling are descriptive accounts.... To construct an account is to make an object or event (past or present) observable and understandable to oneself or to someone else. To make an object or event observable or understandable is to endow it with the status of an intersubjective object. (Leiter, 1980, pp. 161–162)

Accordingly, from the interactionist point of view, we understand an explanation as a social process as opposed to an individual act. The participants' indications that something is mathematically evidential (or standing in need of proof) make it taken as mathematically evidential (or standing in need of proof). In the episode described earlier, Jack and Jamie made the compensation observable and understandable between themselves; in the end of their negotiation, it was taken for granted that the results of the sixth and the seventh tasks are the same. The boys endowed their solution with the status of intersubjectivity, which has to be established once more when the researcher joins them.

The ambiguity of a single object is reduced by relating its meaning to a context that is taken as shared. At the same time, the context is confirmed by constituting the meaning of the single object. The context and the singular meaning elaborate each other. From that ethnomethodological point of view, meanings are not given by a cultural context that would exist independently of the negotiation of meaning, but the cultural context is continually and simultaneously constituted. Ethnomethodologists uses the term *reflexivity* in order to describe such a relationship in which two parts are mutually dependent (Leiter, 1980; see also Cobb, chap. 3; Krummheuer, chap. 7; and Yackel, chap. 4, all this volume). For example, first graders realize that colored blocks, chips, and so forth are used differently in the mathematics classroom than at home. With regard to the subject "math," the members of the classroom ascribe mathematical meanings to the blocks. At the same time, the meaning of what is called "math" becomes clearer to the first graders: "One has to count chips, to compare blocks, and so on."

If the observer looks at the classroom life like an ethnographer who investigates a strange culture, the observer might be surprised by what is taken for granted by the participants: The use of fingers is taken as an explanation, a picture is taken as a calculation task, and so on. However, in the treadmill of everyday school life, the participants would say that they know what mathematics or the mathematics classroom really is. In everyday classroom situations, the teacher and the students often constitute the context routinely without being aware of this ongoing accomplishment, so that the context seems to be pregiven.

For example, in one of the previous episodes the students realized that the split of apples should be taken as the representation of a fraction. We can expect that, if given more apple problems, the students and the teacher would constitute the context "fractions" routinely, without conflicts.

In everyday classroom practice, the teacher and the students suppose that the characteristics of their microculture are definite. In fact, they are taken as shared, diffuse, and vague. Further, students' familiarity with the microculture is merely assumed. The microculture can be constituted without the participants talking about it explicitly, and it can be accomplished indirectly, by way of mathematical activities.

For example, describing a specific method of solving as "simple" does not only ascribe meaning to the method, it also gives meaning to the context of mathematical argumentation. In the classroom observed, the participants use the term *simple* in the sense of "mathematically elegant" (and not as cognitively easy). In this way, Jack's and Jamie's compensation strategy described earlier would be evaluated as simple. This positive evaluation supports the students in orienting their activities toward that advanced mathematical argumentation. The students are given confirmation that the construction and use of complex thinking strategies and abstract considerations are characteristics of the inquiry mathematics classroom. Hence, the microculture of a classroom can be constituted during interaction processes without the necessity of having the participants explain the characteristics of the culture.

From a distance, an ethnographer can think of other classrooms with different characteristics. For example, in many mathematics classrooms teachers do not usually expect a variety of ways of solving but rather a specific one (the "natural" one, or the "appropriate" one). During frontal teaching, they elicit and control the classroom discourse step by step (Voigt, 1985, 1989). Hence, comparing different classrooms we can view different microcultures depending on the participants' activities, even though the participants themselves could view the microculture they establish as natural and pregiven.

Patterns of Interaction

Because of the ambiguity and different background understandings in the classroom, the negotiation of meaning in the microsituation is fragile. Even though a context is taken as shared, there is a permanent risk of a collapse and disorganization of the interactive process. *Patterns of interaction* function to minimize this risk (Voigt, 1984, 1985).

The patterns of interaction are considered as regularities that are interactively constituted by the teacher and the students: "What is presented is a level on which processes remain processes and do not coagulate into entities, to which the very process from which they were abstracted is assigned to as effect" (Falk & Steinert, 1973, p. 20).

For example, the *elicitation pattern* (Voigt, 1985) hints at the combination of two apparently contradictory claims. The idea of elicitating a clear-cut body of mathematical knowledge is juxtaposed with the claims of a liberal and child-centered classroom (cf. Maier & Voigt, 1992). In the pattern, three phases can be distinguished:

5. THEMATIC PATTERNS OF INTERACTION

- The teacher proposes an ambiguous task, and the students offer different answers and solutions that the teacher evaluates preliminarily. This phase corresponds to the claim that the students are stimulated to make varied and spontaneous analyses and discoveries according to their competence.
- If the students' contributions are too divergent, the teacher guides the students toward one definite argument, solution, and so on. Believing that it helps the students, the teacher poses small questions and elicits bits of knowledge. This phase corresponds to the idea of Socratic catechism in which the teacher elicits bits of knowledge that are associated with small steps in reasoning.
- The teacher and the students reflect and evaluate what has transpired.

The pattern of interaction is a structure that is not necessarily intended by the teacher. In the dynamics of human interaction, the pattern is constituted turn by turn. Again, the elicitation pattern might serve as an example. If the students offer statements that diverge from the teacher's expectation, the teacher can feel obliged to assist the students by small hints and questions that lead toward the teacher's expectation. Then, the students can feel the obligation to follow the teacher step by step in order to arrive at what the teacher might expect. If the students' following contributions do not fulfill the teacher's expectations, the teacher might try to guide the students more directly and explicitly so that, in effect, the option of the first phase is lost. In the end, the students produce the desired result. The teacher may doubt whether or not the students have gained any insight. Therefore, the teacher encourages the students to reflect on the preceding process. (This last phase cannot often be reconstructed in traditional classrooms.)

The following classroom episode illustrates the first two phases of the elicitation pattern. The participants discuss the balance problems shown in Fig. 5.3.

1	Teacher:	All right, let's look at the next one. 36 (writes on the overhead projector), 36, 46 (writes on the overhead projector) and this time they gave us 16 and a mystery number. Now could, oops, I'm off the screen. (She adjusts the overhead projector.) All right now, we have a 36 and a 46. I know Linda and Kathy did it. How did you figure out 36 and 46 without doing it in your head again?
2	Linda:	We both [inaudible].
3	Teacher:	Linda. How did you figure this without ever counting in your head? Come on, give me a hand here.
4	Linda:	We got 66.
5	Teacher:	How did you know that . . . what's the answer here? Betsy and Carol, what did you get?
6	Betsy:	We got 98.
7	Teacher:	You got 98 over here?
8	Betsy:	No, in the box we did.
9	Teacher:	Okay, not yet. How did you know 36? What is 36 and 46? [pause]

10	Jamie:	82.
11	Teacher:	How did you know that, Jamie?
12	Jamie:	Well, um. We knew on this one this was 82.
13	Teacher:	Exactly. If you looked right over there, you would have found the answer, to that side. 36, 46 (points to problem) 36, 46, and the last one is the same thing, 36, 46. All right. So part of the problem is already solved for you. You already know that 82 is what that equals out to—balances out to be on the one side. Now, how am I going to balance out 82 when it already has 16? How did you and your partner do it? I need some help. Duane and John, did you get it done?
14	Duane/John:	Yeah.
15	Teacher:	Okay. What's your answer? What's your solution?
16	Duane:	92.
17	Jamie:	I disagree.
18	Students:	Disagree.
19	Duane/John:	[Start talking.] Uh uh, you did it.
20	Teacher:	Let's not argue about it. No, no, we were not going to blame anybody for it. 82. You know you had 16 already and 16 is real close to 20 and it's more than 10, so it's going to be someplace in between there, because 16 is between those two numbers. How are [we going to] figure out—how we're going to balance 16? [pause] My *goodness*, everybody is just kind of.... Michael can you help me. You and Holly?
21	Holly:	We got ...

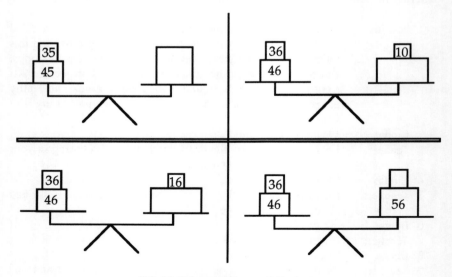

FIG. 5.3. Balance problems activity sheet.

5. THEMATIC PATTERNS OF INTERACTION 181

22	Michael:	98.
23	Teacher:	You say it's 98. 98 plus 16 is balancing out to 82.
24	Alex:	90 is higher than 80.
25	Teacher:	Right. Um?
26	Michael:	Counted on our fingers, like she did. Andrea did.
27	Teacher:	I think you went the wrong way.
28	Alex:	I think you plus instead.
29	Jamie:	I think you subtract.
30	Teacher:	Right. I think you went the wrong way. Instead of adding you should have . . . ?

The problem was ambiguous because the teacher interpreted it as a missing addend problem, whereas Betsy and Michael added all numbers given. Certainly, the task could be understood as adding the weights of all specified boxes. Although the teacher may have taken it for granted, it is only a didactical convention that the weight of the unspecified box has to be determined.

After Betsy offered the sum of all known weights, the teacher initiated the proceduralization of a specific way of solving. As a first step, she asked for the sum of 36 and 46. In this way, alternative options regarding how to solve the task were neglected (e.g., the weight 36 could have been decreased by 16 so that the solution would have been the sum of 20 and 46). The first step that the teacher had in mind (36 + 46 = 82) was established, but the hope that the next students solve the task as expected did not come true. The teacher gave several hints at her interpretation of the task as a missing addend task: "Now, how I am going to balance out 82 when it already has 16?" "16 is between those two numbers." At last, the teacher narrowed the argumentation down by a suggestive hint: "Instead of adding you should have . . . ?" Even without knowing the problem, many people presumably knew the answer that the teacher intended.

In different studies, the elicitation pattern is reconstructed as a typical one in traditional mathematics classrooms. However, in the project classroom, it is not predominant (even though an episode from this classroom is used as an illustration). Rather, a *discussion pattern of interaction* can be reconstructed in many cases of teacher–student interaction (for the teacher's role, see Wood, chap. 6, this volume):

- The students have solved the problem at hand during small-group work. Now, the teacher asks a student to report.
- The student declares a solution to the problem and explains it.
- The teacher contributes to the student's explanation by further questions, hints, reformulations, or judgments, so that a joint explanation or solution emerges and is taken as valid.
- The teacher asks other students for different ways of solving. The first phase starts again.

Two episodes illustrate this discussion pattern. In the first case, the teacher controls and proceduralizes the explanation in several steps. In the second case, a

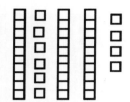

FIG. 5.4. Collection of strips-and-squares representing 40.

more complex and dynamic process of argumentation develops that involves new considerations by both a student and the teacher.

The teacher presents the strips-and-squares pattern shown in Fig. 5.4 on the overhead projector.

6	Teacher:	And now, how much do we have now? (She waits for children to solve the problem.) Rick?
7	Rick:	35.
8	Teacher:	35. How did you get that answer, Rick?
9	Rick:	Well, me and Duane were counting.
10	Teacher:	All right. Uh . . . let's hear from Kathy.
11	Kathy:	40.
12	Teacher:	40. How did you get that for an answer?
13	Kathy:	I [inaudible].
14	Teacher:	You what? You saw . . . OK, she saw the 3 first and then what?
15	Kathy:	Then the 6 there and the 4.
16	Teacher:	All right. She said and then I saw the 6 here, let me get my pen. . . . She saw this 6 and then she saw this 4 and what the 6 and 4 make?
17	Students:	10.
18	Teacher:	10. And 30 plus 10 [students simultaneously] equals (nods).
19	Students:	40.
20	Teacher:	Good.

At first, the teacher posed the problem. Then, she asked for a result. Rick gave a number that was taken as an answer by the teacher. The teacher asked Rick for a report of his process of solving. Rick labeled it as "counting," and the teacher accepted it. Next, another student was called on. At this point, the theme converted from the student's report of her process of solving into the interactive constitution of a joint way of solving. By reacting differently to Rick's and Kathy's approaches, the teacher indicated which approach was favored. (The functions of such valuations are discussed in the last section.)

In the episode just described, the teacher elicited several steps of the solution, and the students participated smoothly. In comparison, in the next episode a conflict

5. THEMATIC PATTERNS OF INTERACTION

of interpretations arises, and the teacher's and the students' positions seem to be more symmetrical:

A story problem is discussed: "Daisy Duck invited 50 children to her birthday party. Nineteen of them were girls. How many were boys?"

1	Teacher:	Okay, Alex what do you say?
2	Alex:	It's 31.
3	Teacher:	You think it's 31.
4	Alex:	Because 30 + 20 is 50, 30 + 20 is 50.
5	Teacher:	(The teacher is distracted by another pair, Andrea and Andy. She carries on a side conversation.) All right, look he said . . . 3 + 20?
6	Alex:	30 + 20.
7	Teacher:	30 + 20. Where did you get the 30?
8	Alex:	Equals 50 and that takes up the 50 children that were at the party.
9	Teacher:	But this is 19, right?
10	Alex:	I know.
11	Teacher:	All right.
12	Alex:	And so 50 − 20 would be 30.
13	Teacher:	Okay. What he is saying is instead of making the 19, I made it 20.
14	Alex:	No.
15	Teacher:	No, you didn't?
16	Alex:	50 − 30 is 20.
17	Teacher:	50 − 30.
18	Alex:	50 . . . Well, I don't know what I did. 50 − 20 is 30 . . .
19	Teacher:	Right.
20	Alex:	But it's a 19 instead of a 20 so it has to be one higher than it, because that number is one less than 20, so it's 31.
21	Teacher:	All right.

In this case, the student played his part sovereignly. Presumably, Alex profited from a compensation strategy by using the knowledge of the equation 20 + 30 = 50 and by replacing simultaneously 20 by 19 and 30 by 31. Diverging from that explanation, the teacher replaced 19 by the easier number 20, calculating 50 − 20 = 30, and then she replaced 30 by 31. The teacher, who had often been observed struggling to understand Alex's explanations, offered a way of solving while seeking Alex's participation. For all that, Alex resisted. He became doubtful, and then reorganized his explanation by using the teacher's argument of replacing 19 by 20.

Through their discussion, the student and the teacher constituted an explanation that perhaps neither would produce individually. They arrived at a knowledge taken as shared. Because Alex used the indexical term *it*, which could possibly refer to three different meanings, we cannot be sure that we or the teacher understood Alex's thoughts correctly. Nevertheless, Alex and the teacher could profit from their negotiation.

There are some important differences between the elicitation pattern and the discussion pattern that was produced in the last episode. In the elicitation pattern, the solution is the main aim; whereas in the discussion pattern, the solution is the starting point of an explanation (similar to the pattern of assertion → proof in mathematicians' communications). In the elicitation pattern, the students are forced to follow the teacher's way of solving step by step if they want to participate; whereas, in the other pattern, the argumentation profits from the students' original contributions. In the one case, the student's own competencies become hidden, in the latter case they become public. In situations in which I reconstruct the elicitation pattern, the students participate successfully if they learn how to solve problems as expected by the teacher. Conversely, in situations in which I reconstruct the discussion pattern, the students have the opportunity to learn how to argue mathematically.

If we can reconstruct the elicitation pattern, it is often in situations in which the participants come in conflict with regard to the result of the tasks. If no competing results are given, or if the students and the teacher agree on a result offered, then the discussion pattern is usually constituted. Additionally, there seem to be situation-specific effects that are contrary to the teacher's intention of discussing different ways to solve. For example, the episode that was used to illustrate the elicitation pattern was taken from the first mathematics lesson after winter break. The long pause could explain the missing adjustments between the participants' activities and the mutual expectations that would have characterized the discussion pattern. Metaphorically speaking, the teacher and the students fell back into the traditional pattern that had been routinely established before the project started.

Because the interactionist point of view stresses the sense-making processes that mediate action and reaction, we have to take into account that even the participants' pure expectation that they will run into severe conflicts when discussing divergent solutions could initiate discourses similar to the traditional elicitation pattern. Because of the permanent ambiguity, practitioners also want assurance, relief, implicit orientation, and reliability: "We like to settle down like in a familiar nest, the nest of everyday life" (duBois-Reymond & Söll, 1974, p. 13).

Thematic Patterns

The patterns of interaction presented previously are not specific to mathematics classrooms; they could be reconstructed in other classrooms as well, but "thematic patterns (procedures) of interaction" are more specific to mathematics classrooms (Voigt, 1989). A thematic pattern is produced when the teacher and the students routinely constitute a theme around some related issues. The variety of options for continuing the theme is constrained by the specific conventions for interpreting the task at hand. In the following paragraphs, several thematic patterns of interaction and their implications for learning mathematics are discussed.

5. THEMATIC PATTERNS OF INTERACTION 185

In the thematic pattern of *direct mathematization*, a story or a picture is interpreted as a specific calculation problem, whereas alternative interpretations do not become thematic. For example, in the classroom episode that illustrated the elicitation pattern, the picture of a balance could be mathematized in several different ways. Answering the question by adding the weights of all signed boxes was just as rational as answering the other question the teacher intended. One of the characteristics of the pattern of direct mathematization was that the teacher took the intended mathematization, the intended mathematical modeling, for granted. If the students did not act according to the mathematical model expected, there was a "risk" that the teacher would have forced the students to follow the teacher's way of solving step by step. Parts of the picture or the story were interpreted as fragments of calculations. For example, the term *running away* was taken as a signal for subtraction. Mathematics gets the image of a proceduralized body of knowledge, of a network of recipes one has to follow in order to solve text problems or picture problems. In this example, the teacher explicitly hinted to the empirical meaning of the picture several times:

- She pointed to the "sides" of the balance.
- She spoke of "balancing."
- Later on, she said: "Well that throws our balance off right there. So now what are we going to do? Rick and Chris, I need your help. Now what are we going to do? We are overboard. We are overweight on that side. We need to lighten up somehow."

In these cases, the teacher established an empirical argument as a test of the result presented by the students. But the students did not appear to accept these thematic suggestions. Further, none of the students thematized empirical meanings of addition or subtraction. So, in spite of the teacher's hints, the mathematical modeling was negotiated implicitly.

Another example of the pattern of direct mathematization can be reconstructed in the whole-class discussion of a story problem.

The story problem is: "Sam had 48 guppies. 17 of them died. How many were left?"

1	Teacher:	He had how many to begin with, John?
2	John:	He had 48.
3	Teacher:	48 to begin with, that's right.
4	John:	And 17 died.
5	Teacher:	And 17 died.
6	John:	[Inaudible]
7	Teacher:	Yes, go ahead, explain it.
8	John:	He had 32.
9	Teacher:	OK, he said there's 32 guppies left. (She writes 32 on the overhead.)
10	Students:	Disagree, disagree . . .

11	John:	Why don't you go and get the hundreds board to see if I'm wrong. (She has turned and is talking to other students.)
12	Teacher:	OK, Andrea, so Betsy and Carol, I haven't heard from them today.
13	Betsy:	We think it is 25.
14	Teacher:	They think it's 25. (She writes 25.)
15	Students:	Disagree, disagree . . .
16	Teacher:	Oh, my. How about Katy and Ryan?
17	Katy:	42.
18	Students:	No, no . . .
19	Ryan:	The answer isn't 42, it's 24.
20	Students:	No, disagree . . .
21	Teacher:	Well, now what are we going to do?
22	Linda:	Oh, count.
23	Teacher:	Andy and Andrea.
24	Andrea:	23.
25	Students:	Disagree . . .
26	Teacher:	23. (She writes 23.)
27	Students:	Disagree, 31, 32, disagree, disagree . . .
28	Teacher:	(She writes 32.) All right, we have several different answers. I mean, 31 and 32 are pretty close, 25 is not very close, 42 is way different, and then there's 23. So we're not even in the neighborhood with some of our answers. How are we going to figure it out?
29	Ryan:	29, 29.
30	Teacher:	Ladies. (She looks at Betsy and Carol.)
31	Student:	We don't like this, we think that it is 64.
32	Ryan:	It's 29.
33	Teacher:	64? You only have 48 to begin with.
34	Ryan:	29.
35	Student:	I disagree.
36	Teacher:	Chris, put your pencil away.
37	John:	I had an idea.
38	Teacher:	OK, John has an idea, let's hear it.
39	John:	Why don't we go get the hundreds board?
40	Teacher:	OK, we could get the hundreds board and figure it out that way. (Betsy gives teacher a hundreds board.)
41	Teacher:	Betsy has one, can you see it if I hold it up like this?
42	Students:	Yeah.
43	Teacher:	How many guppies did we have?
44	Students:	48.
45	Teacher:	How many died?
46	Students:	17.
47	Teacher:	OK, let's count back and find out.
48	Teacher:	(The students speak in unison as the teacher points on the hundreds board:) One, 2, 3 . . . 17.

5. THEMATIC PATTERNS OF INTERACTION

49 Students: 31, 31.
50 Teacher: So that means, how many are still alive?
51 Students: 31.
52 Teacher: According to our counting it was 31.

The story problem was ambiguous because there were several possible interpretations. The teacher expected that the difference between 48 and 17 would be determined. However, Sam could have been counted too, because he was also a living being; or it was possible to interpret the text in such a way that 17 guppies died before Sam had 48 guppies. In that case, the numbers given had to be added. The student who offered 64 as a result may have interpreted the story in this sense.

If we step back from our standard point of view and analyze the story problem from a distance, we begin to see that the mathematization of stories was problematic. The classroom participants who are very familiar with such stereotyped word problems may have expected that the mathematization is evident. In this case however, the participants did not explicitly negotiate the mathematical modeling. That is, they did not explicitly transform the story from everyday understanding to a problem of calculation. In the thematic pattern of direct mathematization, numbers and calculations were directly pointed out. In the end, the teacher gave the decisive technical instruction: "OK, let's count back and find out."

On the one hand, such classroom processes can be criticized, because the well-known shortcomings of students' habits of solving story problems and picture problems may be rooted and confirmed in these processes (Neth & Voigt, 1991). On the other hand, this ritual of implicit mathematical modeling of stereotyped problems may be seen as a relief to the participants. They can focus their attention on the difficulties of calculation and counting, which, as evidenced in the last episode, the students clearly had. The ambiguity, which from the ethnomethodological point of view exists (on principle), has to be restricted by the practitioners to a bearable ambiguity so that the participants can realize the contours of a theme. One of the characteristics of the project classroom is that the participants presume that calculation, counting, and related mathematical activities stand in need of explanation but mathematical modeling does not.

In the following discussion, two other thematic patterns are outlined. These patterns occur when the task contains figures that can be interpreted in two different ways: as representations of materials, or representations of numbers. The students were working on the task shown in Fig. 5.5.

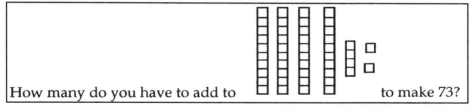

FIG. 5.5. Strips-and-squares task corresponding to 46 + __ = 73.

TABLE 5.1
Two Thematic Patterns

Thematic Pattern of Counting Materials (T_1)	*Thematic Patterns of Calculation With Two-Digit Numbers (T_2)*
Working together, some students interpret the signs as representations of concrete materials. They compare the bars and cubes separately. A typical solution would be: Add 3 bars and take away 3 ones.	Working together, some students interpret the signs as representations of numbers. The difference is calculated within the system of numbers. A typical solution would be: 46, 56, 66 are 20, and 4, and 3, it's 27.
Quantities of materials are thematized.	The application of arithmetical rules is thematized.

The students know that the strips-and-squares are representations of multilinks and that they can use the concrete materials to solve the task. After analyzing several lessons, I reconstructed two relatively stable thematic patterns, as shown in Table 5.1.

Cobb (chap. 3, this volume) and Yackel (chap. 4, this volume) make similar distinctions when analyzing cognitive processes. Cobb reconstructs image-supported and image-independent cognitions. Yackel reconstructs figural and numerical approaches. Nevertheless, the interactional analysis of thematic patterns does not claim to describe cognitive processes occurring within one individual; rather, it reconstructs themes between persons.

Two episodes are used to illustrate the pattern T_1, the thematic pattern of counting materials. The students were working on the task shown in Fig. 5.6.

1 Teacher: How are we going to figure this out? We are going to do this together. What are we going to add to this number to go to this number? How are we going to do this? What are we going to have to do? (She waits until several students raise their hands.) Carol.

2 Carol: (Goes to board.) Add 2 ten-bars, take away 3. (She points to the top parts of the problem.)

3 Teacher: Add 2 ten-bars, take away 3. Hm, OK. So she says (writes Carol's solution on the overhead projector) add 2 tens and take away 3 ones. All right that's one solution. I don't know if is right or wrong, but that is what she says might work. What about Michael?

4 Michael: (He comes to the screen.) Have 2 tens and take away... (pauses) take away these (points to 3 squares), 3, 2 of these and 1 of those.

5 Teacher: Take away 2 of those and 1 of those. (She writes on board.)

Carol did not combine the quantities of the bars and the squares (ones). The teacher accepted Carol's offer so that a solution was constituted. Nevertheless, the teacher indicated that the solution was not ratified and that it was a personal one. Michael carried out the theme that Carol had initiated. He specified which squares had to be taken away by changing the picture of the collection in detail.

5. THEMATIC PATTERNS OF INTERACTION

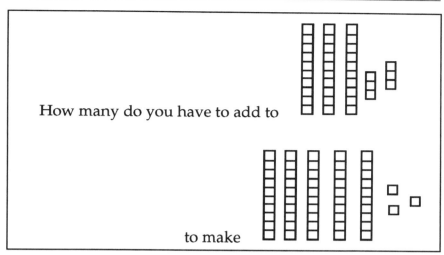

FIG. 5.6. Strips-and-squares task corresponding to 36 + __ = 53.

The analysis of an interview with Michael and the observation of Michael during small-group work indicated that Michael constructed counting-based rather than collection-based numerical meanings. Hence, the thematic pattern in which Michael participated was constrained by Michael's conceptual possibilities (see Cobb, chap. 3, this volume).

The following episode serves as the second illustration of the thematic pattern T_1.

The teacher presented the collection shown in Fig. 5.7 on the overhead projector. The participants agreed that there "is 31 altogether." Later on, the teacher changes the collection as shown in Fig. 5.8 and asks: "How did I change it?"

11	Teacher:	John has something he wants to say. John?
12	John:	You do . . . you took away 1 ten-bar.
13	Teacher:	OK, this is John. He said I took away 1 ten-bar.
14	John:	And . . . 9 . . . 9 little things right there [inaudible] [long pause]. Well, and on there before you had 11, 11 little ones and I think you took away [long pause] 2 ones.
15	Teacher:	So he says I take away a ten and 2 ones.
16	Students:	Yeah! [comments]

John compared the bars and the squares in two different steps. By interrupting John and by evaluating the first step, the teacher also structured the theme so that the thematic pattern of T_1 was interactively constituted.

The following classroom episode provides an illustration of the pattern T_2, the thematic pattern of calculation with two-digit numbers.

After a collection of strips and squares (65) had been shown on the overhead projector the teacher rearranged the strips and squares as shown in Fig. 5.9. The

FIG. 5.7. Strips-and-squares collection representing 31.

FIG. 5.8. Strips-and-squares collection representing 19.

FIG. 5.9. Strips-and-squares collection representing 54.

teacher asks: "How did I change that?" The answer "11" is given, and explained by arguments of calculation with two-digit numbers.

13	Michael:	I think it is 54. (He walks to the front.) 10, 20, 30, 40, 41 . . . 54 (counting on screen).
14	Teacher:	54! If we started out with 65 and we have how many now? How many did you say, Michael?
15	Michael:	54!
16	Teacher:	And we have 54 here already. What's the difference between 65 and 54?
17	Students:	11.
18	Teacher:	Andy?
19	Andy:	11.
20	Teacher:	11. How many think it's 11?
21	Students:	(They raise their hands.)
22	Teacher:	You got it.

At the beginning of that episode, materials or figures of materials were counted by Michael. The teacher asked a question in terms of calculation: "What's the difference between 65 and 54?" Andy answered as intended by the teacher. The calculation with two-digit numbers was thematized. The point is that the difference between 65 and 54 was stated, without reference to the rearrangement of materials.

The analysis of an interview with Andy and the analyses of other classroom situations indicate that Andy could count numerical composites, but not abstract units of ten (see Cobb, chap. 3, this volume). Andy was able to take ten as a strip

5. THEMATIC PATTERNS OF INTERACTION

of 10 ones. We can infer that, in this situation, Andy's thinking was bound to strips and squares that are visualized materials in his mind. Nevertheless, it was possible for Andy to participate in the pattern of calculation of two-digit numbers, presumably without constructing abstract units of ten.

Often, the T_2 pattern was not purely constituted. Several times the pattern was initiated, but broke off. In these situations, the teacher seemed to want the calculation with two-digit numbers to be realized, but several students did not join in. The next episode serves as an example. The students were discussing the task shown in Fig. 5.10.

The pattern T_1 was realized ("add 3 tens and take away 2 ones"), and the teacher tried to elicit the answer "27" (see one of the episodes described earlier). Then, she asked the students for a different way.

12	Teacher:	Someone do it a different way? First of all, what do you have to do?
13	Michael:	Add.
14	Teacher:	If you, you are given the problem. How many do you have to add to . . . (pointing up to the number) to make 73? What do you have to figure out first?
15	Jamie:	The answer.
16	Teacher:	Jamie, the answer, right! (She is laughing.)
17	Jamie:	40 . . . like over there, which number that was.
18	Teacher:	Right! What was the picture of, what is that the picture of?
19	Jamie:	46.
20	Teacher:	46. (She writes 46 on the overhead projector.) OK. Now, we are ready to go. Now, Holly and Michael did you do that first? Did you add up these? Did you add these, um, first?
21	Michael:	Well, we added up the 3 tens, cause I figure how many else . . . because I have most of this and I took away the 2 ones because if we still had the 2 ones it'd be 75, so I took away those 2 ones.
22	Teacher:	OK.

My interpretation is that the teacher expected the transformation of the geometrical representation into a numeral to be the first step. Presumably, Jamie's answer, "The answer," had not been expected by the teacher and her response was to laugh.

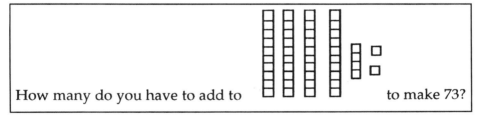

FIG. 5.10. Strips-and-squares task corresponding to 46 + __ = 73.

Then, Jamie fulfilled the teacher's expectation. The teacher asked Holly and Michael whether they "add these up first." Perhaps the teacher intended to ask whether Holly and Michael had realized the first step of T_2, that is, whether they had added the parts of the first number to arrive at a numeral. But Michael reported that they compared the bars and squares of both numbers separately. The teacher accepted this answer and, hence, T_1 was realized after all.

Two thematic patterns have been presented that differ with regard to the meaning of ten. The theme of T_1 is more strongly related to the material objects, whereas T_2 bypasses the material meaning and emphasizes the relations within the system of numerals. Although sometimes the teacher tries to initiate an argumentation according to the second pattern, the students tend to engage in the first one, causing the teacher to join in as well. The evolution of a theme that involves increasingly advanced mathematical arguments has to be supported by the students' construction of increasingly sophisticated units of ten. This cognitive construction of sophisticated mathematical concepts enables the students to contribute to corresponding themes.

Evolution of Themes

On the one hand, the individual's cognitive development contributes to the evolution of themes between individuals. On the other hand, the following considerations should emphasize that the evolution of themes contributes to the individual's cognitive development. Taking both arguments into account, we can begin to see a reflexive relationship between interaction and learning. Because the theoretical concept of reflexivity and its advantage is sometimes difficult to understand, I will develop a theoretical rationale before the analysis continues.

In the third section, reflexivity was defined as a property of the relationship between the context of a classroom situation, that is, the microculture, and particular meanings that are interactively constituted in the situation (Leiter, 1980). From this theoretical point of view, a mutual relationship exists between meanings and culture. The microculture makes the meanings in the particular interactions understandable, while at the same time the microculture exists in and through these very interactions. For example, what the students experience as inquiry mathematics is not pregiven, but it emerges in classroom interaction.

In addition, the ethnomethodological concept of reflexivity will be extended in order to describe the relationship between learning and interaction. The motive for the extension is the claim that the theory should mediate between collectivism and individualism. From both perspectives, one is tempted to assume a one-way dependence between the classroom culture and the individual's learning, as if one was only a precondition of the other. If we assume a reflexive property of learning and interaction (and negotiation, respectively), we get the scheme shown in Fig. 5.11 in which interaction mediates cognition and culture (cf. Voigt, 1990). The reflexive relationship between the meanings that are interactively constituted, and the microculture was outlined in the beginning of the third section. Further, this

5. THEMATIC PATTERNS OF INTERACTION 193

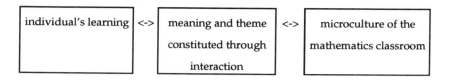

FIG. 5.11. Schematic representation of reflexivity between learning and interaction.

reconstruction of thematic patterns emphasizes the influence of the individual's learning on interactions with others. Now, in order to complete the scheme, the individual's learning will be viewed as influenced by social interactions; more precisely, as influenced by the evolution of themes.

The following analysis reconstructs the evolution of mathematical themes, and it argues that this evolution supports the students' learning. The analysis does not take the teacher's experiences into account. Wood (chap. 6, this volume) reconstructs the teacher's point of view.

At some points in the classroom discourse a student's explanation may initially assume the pattern of T_1; that is, at first, the bars are compared. But, later, the tens and ones are composed. Two episodes can serve as illustrations.

The answers, "Add 2 ten-bars take away 3 ones" and the "17" were given. Jamie combined these answers:

2	Jamie:	And just like I put 20 take away 3 more is 17.
3	Teacher:	Say that again. Did you hear what he said? (She looks at the class.) Say that again, Jamie.
4	Jamie:	20, just like if we had these 3 and these 2 together. That would be 20 altogether. Take away 3 would be 17.

Jamie was able to "construct collection-based abstract units of ten at least in situations where he could rely on task-specific figural imagery" (Cobb, chap. 3, this volume). With regard to the relationship between mathematics learning and social interaction, the interesting point is that in the previous discourse a thematic coherence was established between the separate counting (and comparing) of materials and the calculation with two-digit numbers. Students who do not have the conceptual possibilities that Jamie displayed can be stimulated to extend or change their interpretation of the task so that they can participate in the interaction of type T_2. Then, these students can feel the obligation to combine the tens and ones conceptually to one number to get the result. In the following episode, the students were working on the task shown in Fig. 5.12.

4	Teacher:	How about Katy and Ryan? What do you have to add to 38 to make 86?
5	Ryan:	47.
6	Teacher:	You say the answer is 47.

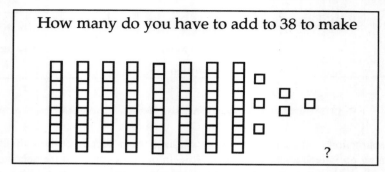

FIG. 5.12. Strips-and-squares task corresponding to 38 + __ = 86.

7	Ryan:	[inaudible]
8	Teacher:	OK.
9	Ryan:	Which one are we on?
10	Teacher:	Number 3.
11	Ryan:	Oh! 48.
12	Teacher:	48. Do you say the answer is 48?
13	Ryan:	Yes.
14	Teacher:	All right, now, how did you figure this out? How did you figure it out? You explain it to me.
15	Ryan:	Well.
16	Teacher:	(She is talking to another child.) Why don't you share this with him, so they can see?
17	Ryan:	Well, [inaudible].
18	Teacher:	But it wouldn't work.
19	Ryan:	Yeah. It would. We took away 50 and then, we have 30 left and then there is 8 over here so we knew that wouldn't work so . . .
20	Teacher:	Right!
21	Ryan:	. . . We have to take 2 off one of the ten-bars . . . and then add it to the 30 and that makes, and that, and that would make all up . . . and that would make 48.

I assume that in the beginning Ryan counted the difference between the ten-bars. The teacher's evaluation "but it wouldn't work" indicates that Ryan thematized something that was unusual or not wanted. Ryan opposed the teacher's evaluation. He related the tens and ones to each other and arrived at a numerical result.

For Ryan "the construction of composite units of ten was within the realm of his conceptual possibilities" (Cobb, chap. 3, this volume; for Ryan's activities with others, see Yackel, chap. 4, this volume). On the surface, Ryan seemed to be able to transcend the constraints of the materials, although he still relied on figural imagery. The interesting point is that, in the episode, the collections were thematized numerically as well as materially. Students in these situations who do not initially think in terms of abstract units of tens and ones because of their current

5. THEMATIC PATTERNS OF INTERACTION

conceptual possibilities might gain an "entry" to it. These students are supported by imagined materials that they can try to transcend. They may interpret the materials as representations of numerical relations, although they only have a hunch of these relations.

In the next episode, the teacher manipulates the materials in order to enact arguments that are related to the calculation of two-digit numbers.

The teacher presented the strips-and-squares collection shown in Fig. 5.13 on the overhead projector.

14 Teacher: ... Holly. What would you like to say?
15 Holly: Well, um ... (She goes to the screen.)
16 Teacher: When you saw this what did you ... how did you decide to put these numbers together? These squares and ...
17 Holly: One thing is ... you know there is 2 ten-bars that's make 20 and then you count ...
18 Teacher: Sshh (shushing a student who interrupted Holly's explanation).
19 Holly: You count these, you count these 1, 2, 3 [3 squares] 4, 5, 6, 7, 8, 9, 10 [6 squares], 11. Since there's 11 [inaudible] you see then you 1, 2, then take 10 and that makes 30. And then you put that extra 1 for 11 and you get 31.
20 Teacher: That's right. She is telling me that if I take all of these ones (moves 3 squares to 6 squares, moves 1 square to 6 squares) and squeeze them together (pushes them together and uses a strip to push them together) I'm going to make a ten-bar or a ten, and so that's how she figured out that if you take all those ones and make a ten out of it then you have a ten-bar. It's the same thing. OK, good. Now. Go ahead sit down. Thank you, Holly. That was a nice explanation.

Holly began by counting the bars and squares separately, but in her last turn, she split and composed tens and ones without speaking of concrete bars and squares. We know that Holly "routinely created and counted abstract composite units of tens when she interpreted and solved tasks that involved strips and squares" (Cobb, chap. 3, this volume).

It is interesting to note that the teacher's materialization and reenactment of Holly's account provided learning opportunities for other students. That is, the

FIG. 5.13. Strips-and-squares collection representing 31.

teacher helped those students who were dependent on activities with materials to understand Holly's explanation. This understanding encouraged the students to extend the conceptual range of their thinking beyond a level attained using less advanced methods, such as counting by ones.

The theoretical considerations of the third section can be summarized as follows. The microculture of the mathematics classroom has been described as constituted through social interaction and negotiation of mathematical meaning. The stability of the microculture is described by patterns of interaction, especially thematic patterns of interaction. The relationship between patterns of interaction and learning mathematics was illustrated in cases in which the patterns were improvised and the themes evolved. These considerations enable us to describe the process of learning mathematics as intrinsically related to the negotiation of meaning.

SOCIOMATHEMATICAL NORMS

It seems reasonable to criticize interactional regularities in which the teacher forces students to fulfill specific obligations, such as artificial conventions in school mathematics. Also, because the negotiation process between teacher and student can be risky and problematic, one might call for more "teacher-free" situations. In response to this call, this section argues that the teacher's influence on the students' mathematical activities must not be suppressed but has to become more indirect. The teacher's influence on the students' goals are described in an indirect way analogous to the manner in which Saxe (1990) reconstructed the influence of cultural settings on an individual's orientation.

Yackel, Cobb, and Wood (1991) described how social norms are constituted and stabilized in the project classroom. But "sociomathematical norms" are also constituted and stabilized in the project classroom. The terms *mathematical norm* or *sociomathematical norm* are used here to describe a criteria of values with regard to mathematical activities. Sociomathematical norms are not obligations that students have to fulfill; they facilitate the students' attempts to direct their activities in an environment providing relative freedom for interpreting and solving mathematical problems.

The students solved the following tasks:

$27 + 9 =$ ___
$37 + 9 =$ ___
$47 + 9 =$ ___
$47 + 19 =$ ___
$48 + 18 =$ ___
$49 + 17 =$ ___
$49 + 19 =$ ___
$49 + 21 =$ ___
$70 - 21 =$ ___
$60 - 21 =$ ___

5. THEMATIC PATTERNS OF INTERACTION

Many students solved the tasks as isolated problems. Their solutions included using fingers, tally marks, multilinks, or the hundreds board. Some students compared the tasks and made use of previous solutions in order to solve the next ones. Now, during whole-class discussion, different ways of solving were compared. The teacher accepted all correct explanations. For example, the teacher said: "That was one way of doing it. *Now*, who did it a different way?"

It is important to note that the teacher evaluated the explanations of correct answers differently. If the offered way of solving seemed to be cognitively demanding, she reinforced the way. For example, Holly and Michael compared three tasks:

1 Teacher: Okay, you're talking 27 + 9, 37 + 9, 47 + 9.
2 Holly: Those are about the same thing. Because they are all—they're all 7s.
3 Teacher: Right.
4 Holly: Except the ones that are going to go up higher 36, 46, 56.
5 Teacher: Okay.
6 Holly: 46.
7 Teacher: Okay. So you say should be 46 and . . . ?
8 Holly: 56.
9 Teacher: Okay. Raise your hands if you agree at your seats. Good! (Several students raise their hands.) Now, so they saw a pattern. They saw aha! All these are adding 9s. They saw that each of the 27, 37, 47 all end with 7. So they saw that. Raise your hand if you saw that happening. (Students raise their hands.) *Good.*

In the classroom observed, the teacher characterized such solutions as "insightful" ways or "simple" ways or as "discoveries." She also outlined the mathematical structure explicitly, or she asked the students to explain their methods once more while requesting the other students to listen, or she expressed emotional delight.

By her evaluations, the teacher indicated mathematical claims implicitly. It was up to the students to decide whether or not they would orient their activities toward the teacher's claims. (Usually, the students expect the teacher to represent the culture of the mathematical discipline.) In the classroom observed, several students accepted the intellectual challenge. Looking at the fourth, fifth, and sixth tasks, John said: "I made two discoveries . . . the beginning numbers go 47, 48, and 49 . . . then it was 17 the bottom one then 18 . . . 19 . . . so it's all the same answer." And Michael, who realized that the fourth solution had to be 10 more than the third one, announced: "I found a discovery."

In this case, what counts as an elegant mathematical solution is interactively constituted. The teacher did not set a standard that the students had to obey step by step. Some students, and not the teacher, offered mathematically advanced contributions. The teacher evaluated the contributions so that she represented the discipline of mathematics. Then, the students could take the teacher's evaluations as hints to more advanced explorations. In the future, the teacher was to evaluate these explorations as "insightful ones," and so on. Through these interaction processes, the quality of the mathematical discourse evolved. As a consequence, the teacher's claim and the students' goals fit together. A sociomathematical norm was interactively constituted.

Of course, the teacher did not accept incorrect results. Although the teacher encouraged the students to present their results without being afraid of embarrassment, the teacher also ensured that the students did not speculate. In her evaluations, she offered reasons for the patterns of numbers that the students suggested. The teacher had to achieve a balance. "Rigor alone is paralytic death, but imagination alone is insanity" (Bateson, 1979, p. 219).

What is the meaning of the criteria "insightful" or "simple" (in the sense of mathematically elegant)? There were no definite and explicit regulations necessary, because implicit criteria for context-dependent values were sufficient. The only way the student could gain an impression of the sociomathematical norms and orient his or her activities toward the teacher's claims was to interpret the teacher's judgments in a specific context. At the same time, the teacher had to adjust her claims to the students' possibilities. Thus, the sociomathematical norm emerged between the participants.

The students had the freedom to solve the tasks by methods they choose themselves, and they were free to interpret the teacher's judgments in the situation. Nevertheless, the teacher could be confident that the students were willing to explore mathematical relations by overstepping former standards. Also, the teacher could be confident that the students take the teacher's indications into account. The students' goals and the mathematical norms of the classroom discourse developed reflexively.

During the teaching experiment, the teacher and the project members focused their attention on the constitution of social norms in the classroom. Wood (chap. 6, this volume) describes the corresponding changes of the classroom discourse during the teaching experiment by considering the teacher's point of view. The teacher's intention was not to assess the students' contributions during the classroom discourse, but rather to encourage students to discuss their contributions among themselves. However, by analyzing the whole-class discussions from a distance, it is possible to reconstruct the teacher's indications of her mathematical claims. Her assessments contributed to the constitution of sociomathematical norms between the teacher and the students. Although the teacher did not intend the constitution of sociomathematical norms and was not conscious of them, her contributions supported the evolution of an inquiry mathematics classroom.

The contradiction between the teacher's intentions and her actions gives rise to two general statements. First, the change of the classroom culture is not restricted to intended improvements. Moreover, the unintended and unconscious side effects of reforming classroom life are no less powerful than are the intentional effects—even in a teaching experiment. As a consequence, if we want to reform classroom life, we should not only know what we want to do, but we also need the means to understand what happens in fact. Second, the teacher's indirect and unintended influences on the students' orientation might be inevitable conditions of the constitution of intersubjectivity. If we appreciate mathematics as produced by a

5. THEMATIC PATTERNS OF INTERACTION

community that, for example, determines what counts as an elegant explanation, we can view the teacher's influences in the classroom processes as a necessary condition for learning mathematics.

The interactionist approach mediates between individualism and collectivism. Roughly said, individualism tends to explain mathematics learning as the product of the laws of both cognitive development and the self-orientation of the individual who experiences a problem. Collectivism attempts to understand mathematics learning as the socialization of the individual into a pregiven culture. From the interactionist point of view, the students and the teacher influence each other. In fact, the indirect, subtle influences are especially important. The teacher and the students interactively constitute mathematical meanings and sociomathematical norms as taken as shared so that the students' learning and the microculture develop mutually.

REFERENCES

Bateson, G. (1979). *Mind and nature. A necessary unit.* New York: Dutton.
Bauersfeld, H. (1980). Hidden dimensions in the so-called reality of a mathematics classroom. *Educational Studies in Mathematics, 11*, 23–41.
Bauersfeld, H. (1988). Interaction, construction, and knowledge: Alternative perspectives for mathematics education. In T. Cooney & D. Grouws (Eds.), *Effective mathematics teaching* (pp. 27–46). Reston, VA: National Council of Teachers of Mathematics and Lawrence Erlbaum Associates.
Bauersfeld, H., Krummheuer, G., & Voigt, J. (1988). Interactional theory of learning and teaching mathematics and related microethnographical studies. In H. G. Steiner & A. Vermandel (Eds.), *Foundations and methodology of the discipline mathematics education* (pp. 174–188). Antwerp: Proceedings of the Theory of Mathematical Education (TME) Conference.
Blumer, H. (1969). *Symbolic interactionism: Perspective and method.* Englewood Cliffs, NJ: Prentice-Hall.
Bruner, J. (1966). Some elements of discovery. In C. S. Shulman & E. R. Keislar (Eds.), *Learning by discovery: A critical appraisal* (pp. 106–113). Chicago: Rand McNally.
Bruner, J. (1986). *Actual minds, possible worlds.* Cambridge, MA: Harvard University Press.
Cicourel, A. (1970). Basic and normative rules in the negotiation of status and role. In H. P. Dreitzel (Ed.), *Recent sociology, No. 2, Ptterns of communicative behavior* (pp. 4–45). New York: Macmillan.
Cobb, P. (1990). Multiple perspectives. In L. P. Steffe & T. Wood (Eds.), *Transforming children's mathematics education: International perspectives* (pp. 200–215). Hillsdale, NJ: Lawrence Erlbaum Associates.
Cobb, P., & Wheatley, G. (1988). Children's initial understandings of ten. *Focus on Learning Problems in Mathematics, 10*(3), 1–28.
Denzin, N. K. (1989). *Interpretative interactionism.* Newbury Park, CA: Sage.
Doise, W., & Mugny, G. (1984). *The social development of the intellect.* Oxford: Pergamon Press.
duBois-Reymond, M., & Söll, B. (1974). *Neuköllner Schulbuch* [Neuköllner schoolbook] (Vol. 2). Frankfurt: Suhrkamp.
Edwards, D., & Mercer, N. (1987). *Common knowledge. The development of understanding in the classroom.* London: Methuen.
Erickson, F. (1986). Qualitative methods in research of teaching. In M. C. Wittrock (Ed.), *Handbook of research on teaching* (3rd ed., pp. 119–161). New York: Macmillan.
Falk, G., & Steinert, H. (1973). Über den Soziologen als Konstrukteur von Wirklichkeit, das Wesen der sozialen Realität, die Definition sozialer Situationen und die Strategien ihrer Bewältigung [The

sociologist constructing reality, the essence of social reality, the definition of social situations, and methods for coping with these situations]. In H. Steinert (Ed.), *Symbolische interaktion. Arbeiten zu einer reflexiven Soziologie* [Symbolic Interaction: Works in Reflexive sociology] (pp. 13–46). Stuttgart: Klett.

Garfinkel, H. (1967). *Studies in ethnomethodology*. Englewood Cliffs, NJ: Prentice-Hall.

Goffman, E. (1959). *The presentation of self in everyday life*. New York: Doubleday.

Goffman, E. (1974). *Frame analysis: An essay on the organization of experience*. Cambridge, MA: Harvard University Press.

Jungwirth, H. (1991). Interaction and gender: Findings of a microethnographical approach to classroom discourse. *Educational Studies in Mathematics, 22*, 263–284.

Krummheuer, G. (1983). *Algebraische Termumformungen in der Sekundarstufe I—Abschlußbericht eines Forschungsprojektes* [Algebraic transformations in high school—Final report of a research project]. Materialien und Studien [Materials and Studies] (Vol. 31). Bielefeld, Germany: Institut für Didaktik der Mathematik.

Lakatos, J. (1976). *Proofs and refutations: The logic of mathematical discovery*. London: Cambridge University Press.

Leiter, K. (1980). *A primer on ethnomethodology*. New York: Oxford University Press.

Luhmann, N. (1972). Einfache Sozialsysteme [Simple social systems]. *Zeitschrift für Soziologie, 1*(1), 51–65.

Maier, H., & Voigt, J. (1992). Teaching styles in mathematics education. *Zentralblatt für Didaktik der Mathematik, 7*, 249–253.

McNeal, B. (1991). *The social context of mathematical development*. Unpublished doctoral dissertation, Purdue University, West Lafayette, IN.

Mead, G. H. (1934). *Mind, self and society*. Chicago: University of Chicago Press.

Mehan, H. (1979). *Learning lessons*. Cambridge, MA: Harvard University Press.

Neth, A., & Voigt, J. (1991). Lebensweltliche Inszenierungen. Die Aushandlung schulmathematischer Bedeutungen an Sachaufgaben [Staging the life-world: The negotiation of mathematical meanings of story problems]. In H. Maier & J. Voigt (Eds.), *Interpretative Unterrichtsforschung* [Interpretive research in classrooms] (pp. 79–116). Cologne, Germany: Aulis.

Saxe, G. B. (1990). *Culture and cognitive development: Studies in mathematical understanding*. Hillsdale, NJ: Lawrence Erlbaum Associates.

Schütz, A. (1971). *Gesammelte Aufsätze* [Collected papers] (Vol. I). Den Haag: Niyhoff.

Schütz, A., & Luckmann, T. (1973). *The structures of the life-world*. Evanston, IL: Northwestern University Press.

Solomon, Y. (1989). *The practice of mathematics*. London: Routledge.

Steinbring, H. (1989). Routine and meaning in the mathematics classroom. *For the Learning of Mathematics, 9*(1), 24–33.

Steinbring, H. (1991). Mathematics in teaching processes. The disparity between teacher and student knowledge. *Recherches en Didactique des Mathématiques, 11*(1), 65–108.

Tymoczko, T. (1986). Making room for mathematicians in the philosophy of mathematics. *The Mathematical Intelligencer, 8*(3), 44–50.

Voigt, J. (1984). *Interaktionsmuster und Routinen im Mathematikunterricht—Theoretische Grundlagen und mikroethnographische Falluntersuchungen* [Patterns of interaction and routines in mathematics classrooms—Theoretical foundations and empirical findings]. Weinheim, Germany: Beltz.

Voigt, J. (1985). Pattern and routines in classroom interaction. *Recherches en Didactique des Mathématiques, 6*(1), 69–118.

Voigt, J. (1989). Social functions of routines and consequences for subject matter learning. *International Journal of Educational Research, 13*(6), 647–656.

Voigt, J. (1990). The microethnographical investigation of the interactive constitution of mathematical meaning. In *Proceedings of the 2nd Bratislava International Symposium on Mathematics Education* (pp. 120–143). Bratislava, Czechoslavakia: Bratislava University Press.

Wilson, T. P. (1970). Concepts of interaction and forms of sociological explanation. *American Sociological Review, 35*, 697–710.

Wittgenstein, L. (1967). *Remarks on the foundations of mathematics*. Oxford: Blackwell.

Wood, T., & Yackel, E. (1990). The development of collaborative dialogue within small group interactions. In L. P. Steffe & T. Wood (Eds.), *Transforming children's mathematics education* (pp. 244–252). Hillsdale, NJ: Lawrence Erlbaum Associates.

Yackel, E., Cobb, P., & Wood, T. (1991). Small-group interactions as a source of learning opportunities in second-grade mathematics. *Journal for Research in Mathematics Education, 22*, 390–408.

6

An Emerging Practice of Teaching

Terry Wood
Purdue University

In the previous chapters, the focus of the discussion was on the various aspects of the classroom culture that have influenced children's learning, and of the role that children's interaction and subsequent discourse involving reasoning and argumentation has in their understanding of multiple units (including ten), place value, and the system of whole numbers. These chapters provided illustrations of the opportunities that are available for students in the classroom to validate their mathematical meanings. Complementing these previous efforts, the focus of this chapter not only continues the emphasis on the interactive process of learning and teaching but also centers on a teacher and her actions in this situation.

Previously, the changes that the teacher made during the year of the classroom teaching experiment have been described in broad general strokes (Wood, Cobb, & Yackel, 1990, 1991). The purpose of this chapter is to depict in greater detail the transformation of that teacher's conception of what it meant to teach and to learn mathematics and its influence on her actions. My intention is to describe and interpret the events that reflected an emerging practice of teaching rather than to report the results of teacher change. In doing this, I show the reorganization that the teacher made in her thinking, revealed through her inner-directed actions as she attempted to alter her previous way of teaching mathematics to create a new form of practice. In choosing this approach, I contend that many of the characteristics that distinguish teaching as an interactive activity in which attempts are made by teachers to understand and negotiate meaning with their students are revealed in a different way of communicating. In the analysis, I describe the emerging practice from the perspective of a detached observer looking at the interaction, but I also attempt to take the perspective of a participant observer and consider the teacher's point of view by trying to infer her interpretation of the situation.

LEARNING IN CLASSROOMS

Over the past several years, research on children's cognitive development has created new understandings about the ways in which children make sense of mathematics. With this in mind, those interested in mathematics education were confronted with a central problem: How might they realize major assertions about the development of children's thinking in the collective setting of everyday school classrooms? In attempting to offer some insight into this process, several research and development projects have attempted to address the issue of children's learning of mathematics in school settings (e.g., Carpenter, Fennema, Peterson, Chiang, & Loef, 1989; Fennema, Carpenter, & Peterson, 1989; Hiebert & Wearne, 1992; Newman, Griffin, & Cole, 1989). However, what initially distinguished the classroom research project described in this book was a commitment to a radical constructivist view of learning mathematics and later an interactionist perspective (Wood, Cobb, Yackel, & Dillon, 1993).

Teaching

Any consideration of extending constructivist assumptions about learning to the setting of a classroom necessarily meant that teachers and their activity should be considered in the process. The activity of children's learning from a constructivist perspective has been well researched, but attempting to understand teaching from this perspective has been neglected. The self-reflective reports of Ball (1993) and Lampert (1988, 1992) provided some insight about the process as viewed by mathematics educators themselves attempting to teach in elementary school classrooms. However, this classroom teacher in the everyday setting of her own classroom found the constructivist view of mathematics and children's learning to be in strong opposition to her traditionally held views. From her perspective, learning mathematics was about following well-established procedures in order to arrive at right answers, and, as the teacher, she believed it was her responsibility to ensure that her students could do so.

It is not surprising that the emphasis on the individual, subjective construction of mathematical knowledge raises two questions that Noddings (1990) pointed out as critical for any teacher: What is important for the student to know in the sense of general mathematical knowledge, and how does a teacher judge when a student is constructing or has constructed meaning? These questions not only relate directly to the underlying epistemology of constructivism, but they also strongly influence the practicality of classroom life and create dilemmas that the teacher must attempt to resolve. Moreover, although one may consider learning to be an individual process, teaching is by definition an interactive collective activity in which it is the intention of the one teaching to influence the development of his or her students' thinking. This does not mean that the sole intention is to impose on children what to think or how to make sense of their experience. Instead, one must recognize that although teachers can create possibilities for learning by providing a focus for children's activity and support for their thinking, the act of learning itself is personal and subjective. Teaching, then, as viewed from this perspective, involves attempting to understand and negotiate meanings through communication. In this regard, teachers not only affect their students but, in turn, are affected by them in much the same way that parents influence and are likewise guided by their infants (Trevarthen, 1979).

However, a major difference that distinguishes schooling from learning that is accomplished in the home is the simple fact that a teacher's support for students' personal construction is offered while the child is participating as a member of a group. This creates for teachers the continual tension of trying to honor individual children's conceptions on the one hand and attempting to negotiate taken-as-shared (Streeck & Sandwich, 1979) understanding among the members of the class on the other.

Perspective Taken for Analysis

Although the intention of the research was to investigate children's mathematical learning in the classrooms from a constructivist perspective, it turned out to be a situation in which the teacher was learning as well. In this case, her learning involved not only a different meaning of mathematics and an increase in awareness of the ways in which children make sense of mathematics but, more important, an interpretation of the philosophical and theoretical perspective that guided the researchers' thinking. It was this construal that she could envision and on which she could rely to guide her actions. Moreover, it was her sincere attempts to accommodate our ideas of children's learning to her personal experiences in the classrooms that created the situations in which conflicts, contradictions, and surprises arose for her. These events ultimately led her to reflect on her activity and to create for herself a different way of teaching mathematics.

I take as evidence of the teacher's learning the changes that occurred in the patterns of interaction and the nature of the discourse that emerged between the teacher and her students over the course of the school year. These modifications in her practice did not progress in a linear sequence to a final end but, instead, reflected the "to-ing" and "fro-ing" that occurred as she shifted between an emerging practice and her previous traditional ways of teaching. These fluctuations reflected the continual, mutual orienting of the teacher and her students while they negotiated and renegotiated expectations and obligations for one another's behavior as they created an atmosphere of inquiry in the classroom. Major shifts and changes in the underlying social norms that form the patterns of interaction extended throughout the year, as the teacher and students found themselves continually redefining their roles in the classroom. As these patterns were jointly and interactively constituted, the teacher and her students established certain regularities in their interaction with one another. These regularities that the participants developed enabled them to maintain a continuity in their everyday life without the necessity of continually deliberating among various options for action.

Thus, the routines that were developed by the individual members of the class allowed many aspects of a situation to be taken for granted and enabled them to focus their attention on making sense of the unfamiliar or problematic aspects of a mathematical situation. More important, these routines provided the underlying foundation for the ways in which the members of the class interacted with one another and the nature of communication that occurred in the classroom. The variations that occurred in these routines and the concomitant dialogue created a source from which to investigate an emerging form of teaching. As a final comment, relating these changes as an observer I do not intend to act as an evaluator, but rather as a documenter of the shifts and vacillations that occurred. My purpose, then, is to describe and interpret the teacher's activity in light of her experiences and to consider the nature of the interaction and the relationship of this to an evolving practice of teaching mathematics.

Data Source

In this analysis I have selected, from a central data resource of lessons spanning a time from January to March (with two additional lessons from December and April), only those tapes that depict whole-class discussion. From this analysis I have chosen five lessons on which to conduct further microanalyses. The examples from the illustrative episodes are edited for the purpose of presenting those elements that are relevant to the analysis without reproducing substantial segments of transcribed data. Thus, the intention is to provide succinct characterization rather than a comprehensive description.

TRADITIONAL TEACHING AND EMERGING PRACTICE

In the traditional class, the "disparity in background knowledge between the teacher and students," to which Voigt (1992) referred, generally assumes that the negotiation of meaning that occurs between the teacher and students is one in which the intention is for the students to learn what the teacher already knows. The purpose of this didactical interaction is derived from the view that teaching is a matter of socializing students to the meanings held by society of which the teacher is already a member. In this role, the teacher is viewed as one who tells students information and then evaluates them to see how well they have acquired this knowledge (Sinclair & Coulthard, 1975; Weber, 1986). In this form of teaching, the teacher's purpose in asking questions is to elicit from children information previously presented and to evaluate immediately whether the students have understood what was intended. Because such teachers are only listening for correct responses from students, they do not engage in the kind of discourse in which attempts at genuine communication take place. In this regard, the nature of the talk in classrooms is very different from everyday conversations and reflects a situation in which the teacher is viewed as the one who knows, and students are viewed as those who do not know.

In the type of classroom under consideration, however, the students' experiences and thinking are an important aspect and central focus for instruction (Confrey, 1990; Labinowicz, 1985; Noddings, 1990; von Glasersfeld, 1987). A teacher who does not know about students' ways of thinking is at a distinct disadvantage in a situation in which children's sense making is the central focus. The necessity to engage in continuous genuine negotiation of meaning in order to participate becomes equally as important to the teacher as it is to the students. The teacher, in this case, must listen carefully to children's explanations and attempt to make sense of them in order to be effective in his or her role in the classroom.

At first, the teacher in this particular classroom was not aware of the ways in which the children's mathematical understanding developed, and in light of the tradition of school mathematics had not had sufficient opportunities to learn about them. It is not surprising then, that the "disparity in knowledge" to which Voigt (1992) referred took on a different meaning in this classroom. It was against this backdrop that I viewed the teacher as a participant in an evolving and emerging form of practice that involves a transformation in philosophy to consider a different vision of teaching and learning mathematics. In addition, in her view mathematics was a static subject, with set rules and procedures that needed to be carefully followed in order to arrive at correct answers. The changes that occurred in her thinking had a profound effect on her beliefs about the nature of mathematics. These shifts in her perspective resonated with the underlying view of the reform initiatives (National Council of Teachers of Mathematics, 1989, 1991). However, when the project began, the various reform documents for mathematics had not yet been released, and the conventional wisdom of what counts as school mathematics and "effective teaching" still prevailed (Rosenshine & Stevens, 1986).

Recreating the Classroom Atmosphere

The teacher's immediate concerns at the onset of the school year were to find ways to make it possible for children to work together productively as partners and to express their thinking to others during class discussions. Specifically, with the pairs she was confronted with the problem of being assured that the children would work cooperatively to solve the problems and complete the activities without her close supervision. If they could do so, then she would have an opportunity to move among the groups, listen to children as they solved problems together, and intervene to aid their learning as necessary. This meant that she did not want to spend her time monitoring their behavior as she circulated among the children.

Further, during class discussion she wanted the children to express their ideas and explain the methods they used for solving the instructional activities. She realized that it would not be possible to predetermine students' responses, because the goal of the instructional activities and the class discussions was to encourage individual solutions that would necessarily be unique. She also lacked prior knowledge of children's solutions, which might have allowed her to anticipate, at least to some extent, their ways of thinking. Finally, the nonevaluative stance that the researchers encouraged her to take with regard to students' responses in class and the lack of daily grading of papers added to her concerns. These changes in a long-established and very stable form of practice (Goodlad, 1984) created a great deal of uncertainty and unpredictability for her in a setting in which she was traditionally expected to be the authority and sole purveyor of knowledge. In addition, by her own admittance, she did not feel her understanding and knowledge of mathematics was strong, and in the past she lacked the enthusiasm and enjoyment for teaching mathematics that she held for language arts and reading.

Consequently, her main concern was to reestablish with her students the social norms that would form the underlying network of individual rights and responsibilities to others essential to working in pairs and in class discussion. She commented off-handedly after one of the project meetings before the school year began that "the rules would have to be different." Although the initial renegotiation of a different set of norms occurred within the first few weeks of school, those norms were continually being readjusted throughout the year. Even though the teacher realized the need to establish different social norms, the nature of these was not readily apparent to her. Thus, although the nature of the interaction that was established early remained relatively stable throughout the year, some renegotiation occurred as different routines were needed to accommodate new dilemmas that arose for the teacher and students.

At one level, the class discussion consisted of talk for which the main purpose was to constitute interactively the appropriate behavior necessary to create and maintain a problem-solving atmosphere. At the other level, the talk dealt with discussions about the students' mathematical thinking. These two levels have been previously described as "talking about talking mathematics," to refer to the situations in which the discussion was focused on expectations and obligations for ways of behaving, and "talking about mathematics," in which the talk was mathematical in nature (Cobb, Wood, & Yackel, 1993; Cobb, Yackel, & Wood, 1989).

At the beginning of the year, as one might suspect, the talk in class more frequently focused on talking about talking about mathematics than did the discourse that occurs later in the year. As the year progressed, the talk about mathematics often followed the established ways that are considered essential in conducting formats of argumentation (see Krummheuer, chap. 7, this volume). However, lest the reader misunderstand, I emphasize that the talk that concerned social or mathematical problems did not occur as separate entities. Quite to the contrary, it was not unusual for the discourse to shift from one to the other in the midst of an ongoing discussion, making it evident that the two levels of talk were intertwined.

Establishing a Different Pattern of Interaction

During the early weeks of school, the teacher and students mutually negotiated and established the individual routines that would form the enduring patterns of

6. AN EMERGING PRACTICE OF TEACHING

interaction and communication during class discussion. The following episode, which occurred in the first month of school, briefly illustrates many of the continuing characteristics of the whole-class discussions that occurred during the time frame of the analysis (January through April). In this example, the class was discussing the solutions to a word problem written on a card.

1	Teacher:	There are 6 tulips behind the rock. [4 tulips shown in the picture]. How many are there in all? Kathy?
2	Kathy:	10.
3	Teacher:	She got 10. Raise your hands if you got 10. (Several children raise their hands.) Raise your hands if you got a different number? What number did you get, Chris?
4	Chris:	11.
5	Joe:	I got 4.
6	Teacher:	How can we figure this out? Since I called on Kathy first, I'll let her describe how she got the answer.
7	Kathy:	I got 6 and then added 4 more.
8	Teacher:	Did anyone else get that answer or maybe do it a different way? Alex.
9	Alex:	11. Well—wait a minute. There would be 10 flowers.
10	Teacher:	How did you discover that?
11	Alex:	It is 4 flowers, one pair would be 6 flowers. 6 flowers in the front, 6 flowers in the back. That would equal up to 12. If [I] took 2 away [that] makes 4 in front and 6 in back would make 10.
12	Teacher:	Exactly! Did anybody else do it a different way?

The pattern of interaction that underlied the remainder of the class discussions for the rest of the year may be seen as emerging in this episode. Even though the teacher had previously told the children that she was interested in how they solved the problem, by reading the question given on the card she found herself in the situation of asking them for an answer. As a way to keep the conversation open, she asked the children to raise their hands if they agreed or disagreed with the answer Kathy gave. This act also served as a way for her to check their answers. When she discovered that different answers existed, she asked Kathy, who had given the correct answer, to give her method for solving the problem. Her following question, "Did anyone else get that answer or maybe do it in a different way?" illustrated the shifting that was occurring as she tried to adhere to the view that children's ways of solving problems are of central interest and to disregard the well-established tradition of school mathematics in which the correct answer is the only important consideration.

This shifting that occurred in this situation continued throughout the year and, as such, provided insight to the teacher's way of thinking about the immediate situation at hand and her struggles to create a different environment for learning mathematics in her classroom. From the observer's perspective, a tension was created for the teacher as she decided to accept children's incorrect answers while she still felt the obligation to ensure that they arrived at the correct answers. In this particular episode, she attempted to resolve her dilemma by returning to Kathy, who had given the correct answer, and asking her to give an explanation for her answer. Kathy's explanation, however, was merely a description of the procedure that she used to arrive at the correct answer, and in the tradition of school mathematics, would be considered an appropriate response. The teacher's following question can be seen as an attempt to continue the exchange and allow for further discussion. This reflected a deviation from the traditional form of questioning found in most classrooms and served to create an opportunity for Alex to reconsider his solution and resolve his incorrect answer. The unexpected novelty of his explanation was surprising to the teacher and, as such, was the beginning of her becoming aware that children have ways for solving problems that she, in many cases, had not even considered as possible solutions. In these discussions, she also began to experience occasions in which children revealed that they were perfectly capable of correcting their own thinking without her direct intervention.

The typical pattern of interaction that became established in these early lessons later became taken for granted in those situations in which the discussion focused on different methods children use for solving problems. The pattern involved a sequence:

- The teacher read the problem and called on a pair, although in most cases only one of the partners was to be the major spokesperson.
- The student gave an answer.
- The teacher repeated the answer and/or requested a solution method.
- The student gave his or her solution method.
- The teacher repeated the student's strategy, accepted the solution, and then asked for a different way of solving the problem, and called on another pair.
- The next student generally gave only a solution method.
- The teacher repeated the student's method, and then may or may not have asked for another different way.

During the months prior to Christmas, the teacher and students mutually established the routines and patterns of interaction for their working together. Initially, the teacher's central intention was for children to "cooperate" and "work together" to solve problems in small-group work. In these settings, she expected children to discuss and solve the problems as a joint activity and to agree on an answer. Although she began to recognize and even accept the constructivist assumption that children did have a variety of personally meaningful ways for solving problems, she could not possibly foresee the multitude of methods. She still viewed mathematics from her adult perspective, in which single procedures were used to find answers. Moreover, as a member of a long tradition of school mathematics, she was still concerned that the students arrived at the correct answer. Initially, she also did not fully anticipate that the individuals within a pair might in fact agree on the answer, but have different ways for solving the problem. It was not unusual, then, for the partners to find that arriving at a consensus about the way they solved the problem created a conflict (see also Cobb, chap. 3, this volume; Yackel, chap. 4, this volume). Frequently, the teachers' intervention was merely to help a pair resolve fundamental social problems they encountered when trying to work together.

In the class discussions, it was expected that students would not only present their answer and their ways of solving the instructional activities to others, but that they would also listen to the explanations of others. It was anticipated that differences in perspectives would arise during the discussion. In order for these differences to arise, it was expected that the teacher would not explicitly evaluate students' answers or direct them to the ways of solving problems that were presented in traditional elementary school mathematics instruction. It is not surprising, however, that when children gave what she considered to be an incorrect answer to a problem, she was still uncomfortable in leaving it. As a way to resolve the tension, she often devised ways to lead students to the correct answer. In addition, on more than one occasion in which the answers were valid, the students' methods did not fit expected conventions and were difficult for her to understand completely.

A look at an example of a whole-class discussion that occurred just prior to the months considered for analysis illustrates the ways she attempted to resolve her dilemma, and provides a basis for comparison of the later lessons. This example also exemplifies the to-ing and fro-ing that occurred between her traditional ways of teaching and an emerging practice. In the ensuing episode, the children were solving the problem (14 + 19), presented horizontally, for which three children give the answer as 23. In relating their solutions, each seemingly forgot to include one of the tens. Alex, on the other hand, had an answer of 33.

1	Teacher:	He says the answer's 33. (She writes 33 next to the 23.)
2	Alex:	Because there's a 10 in 19 and 14 and you take away the 4 and the 9 from the tens um . . .
3	Teacher:	(interrupts) Okay. So he's go . . .
4	Alex:	(interrupts) . . . it makes 20.
5	Teacher:	All right. Wait a minute. Stop [to Alex]. He's saying if I take this 4 off and this 9 off (gestures to problem). I know that 10 and 10 make 20 (writes 20). Okay. Let's go from there.
6	Alex:	And 4 plus 9 . . .
7	Teacher:	Uh uh. 4 and 9 . . . ? (Underlines the 4 + 9 in the sentence.)
8	Alex:	4 and 9, well, 7 plus 4 equals 11.
9	Teacher:	Uh uh.
10	Alex:	And 2 more equals 13.
11	Teacher:	Okay, 13.
12	Alex:	So 20 + 13. There's a 10 in 13, so I added 10 to 20 [to] get 30 and there are 3 left, so I get 33.
13	Teacher:	Ooh. Did you see that [to the class]?

Alex's solution indicated that his understanding of place value was fairly complete. He partitioned two-digit numbers into tens and ones and combined and recombined these to solve the problem. In addition, he did not seem to know 9 + 4 as a routine fact, so he used a known fact, 7 + 4, and the thinking strategy of add two to find the answer. The teacher's reaction suggested that his solution was also more complex than she had anticipated. Another child did not understand and commented, "I don't get it."

14	Teacher:	Okay. What he's saying is . . . maybe if I . . . I'll say it again. I'll try and explain it. He says in the 14 and 19 here is a 10 (points) and here is a 10 (points). . . . Then he took his 4 and his 9 . . . that's where he got this 13. Now he's going to put all those numbers together now. What is 20 plus 13?
15	Alex:	Well, actually, there's a 10 in 13 . . . and you add it to the 20 makes 30 . . .
16	Teacher:	(Interrupts) Okay, I see what you're saying.
17	Alex:	(Continues) Add 3 more from that 13 . . .
18	Teacher:	Okay. So you're saying 20 plus 3 actually. (She writes this as a number sentence on the board.) Okay. It's the same thing, isn't it?

In contrast, moments later, while still solving the same problem (14 + 19) Holly gives an answer of 23.

19	Holly:	First I had 20, then I added 3.
20	Teacher:	First you had 20. Did you turn this into 20? (Pointing at the 19.)
21	Holly:	(Shakes her head yes.)
22	Teacher:	Okay. She turned that into 20 by adding 1 to it. Then she did what? You took away one from this side? Is that what you did?
23	Holly:	(Shakes her head yes.)
24	Teacher:	To get your 3? Okay. So far I agree with you. That makes 23. But what about this poor lonesome critter here? (She points to the 1 in 14.)
25	Holly:	Hm . . .
26	Teacher:	This isn't just 4, is it? It is a . . . ? (voice rising)
27	Class:	14.
28	Teacher:	Which is a 10 and a 4, 4 ones. So what are we going to do now (to Holly)?
29	Holly:	(Says nothing.)
30	Teacher:	You're on the right track.
31	Holly:	So I add 3 more.
32	Teacher:	That's right, 23 and 10 more is a . . . ? (voice rising)
33	Holly:	33.
34	Teacher:	33.

In this exchange, the teacher assumed that she knew Holly's solution method and, thus, filled in much of the details of the explanation for her. Holly, for the most part, simply nodded her head or gave single-word responses in an attempt to respond to the teacher's leading questions. This shift from an inquiry routine to the traditional exchange with the highly predictable student responses to the observer seems like an anomaly.

But from the teacher's point of view, this is quite sensible. Realizing that even after the discussion of Alex's solution it was still possible for students to have an incorrect answer, as the teacher she felt it was her responsibility to ensure that students knew the correct answer. In order to do this, she directed Holly's attention to the point at which she thought Holly had made an error of forgetting to add the ten. It is not surprising, in these unanticipated situations, that because the teacher had not yet learned a more indirect way to carry out her perceived obligation, she relied on a well-known procedure that had been successful in the past.

In this form of interaction, described as the "funnel pattern" by Bauersfeld (1988) and Voigt (chap. 5, this volume), typically the teacher asks questions about information for which the answer is already known, in order to evaluate whether the student knows the same answer or not. In this interaction, if the student does not give the appropriate response, the teacher either tells the student or directs him or her in a step-by-step manner to the correct answer.

Teacher's Comments

In an interview in mid-December during a project meeting, the teacher described her new role as evolving from being an:

> authority person. Person to come to for the answers. This year I am not . . . and not giving the answer, it's forced them to take on the responsibility of figuring it out for themselves. . . . It's not frustrating, it's challenging. . . . They feel good about it. They have come to the point where they can learn, you know, themselves [pause] in spite of me.

JANUARY, FEBRUARY, MARCH

In December, a series of activities had been introduced to the students with the intention of providing experiences in which ten, although comprised of units of one, could be counted as a single unit (cf. Cobb, Yackel, & Wood, chap. 2, this volume). All this set the stage for the activities developed that formed the basis for helping children make sense of the major concept for second-grade mathematics—place value and methods for solving two-digit addition and subtraction problems with regrouping. Additionally, during this time activities that emphasized other multiple units were introduced and intertwined with addition and subtraction. Multiplication considered in this regard was an extension of students' natural, ongoing construction of units into groups other than ten. As children learned to move beyond counting by ones to using ten as a unit, they also learned to form other multiple units (e.g., fours, sixes).

Although the activities that were developed were intended to provide students with an opportunity to learn the concepts traditionally taught in second-grade mathematics, two things were quite different for the teacher. First, the activities were unique in their design and, hence, were unfamiliar to the teacher as well as to the students. In the project meetings we explained to her what students should do as far as the format, but the specific strategies children might use to solve them were rarely discussed, because they were part of our research agenda. This meant the teacher, like ourselves, was left to learn about these during the lesson as students talked about the ways in which they solved the problems. Therefore, the teacher found herself in situations in which it was necessary to listen to children in order to find out how they solved problems and then to relate what she learned to what she already knew as an adult about place value and the arithmetic operations.

The second major difference the teacher faced beyond the unique layouts on the student papers was that the "mathematics"—notions of tens and ones and the standard algorithms for solving addition and subtraction problem—did not resemble traditional school mathematics and were not directly presented in the instructional activities. Instead, numbers were thought of in terms of single units or multiple units and children were encouraged to generate their own algorithms for solving them. For the teacher, the well-established reproducible arithmetical content found in the mathematics textbooks on which she had always relied was no longer identifiable, and the essence of mathematics was being reinterpreted.

With this in mind, I began to examine the process of the teacher's evolving practice in detail during the months of January through March. I noted that she was still faced with the dilemma of trying to create situations in which her students could develop their personal meanings for mathematics. In order for this to happen, it was necessary to be aware of certain obligations she had to her students in her new role. This meant that during the interaction with her students it was her responsibility to probe, question, and offer suggestions, but not in a way that would hinder or interrupt the flow of student thought. In addition, she was not to help students make sense of their solutions by directly clarifying their ideas. Instead, she was expected to provide situations in which children would be allowed to clarify their own thinking. Moreover, she was not to validate using one solution over another (and this was the hardest), but instead to create situations in which children would take notice of others' ideas. From my perspective as a researcher this seemed quite reasonable, given the theoretical perspective, and, moreover, I was not responsible for maintaining the ongoing activity of the lesson or the collective attention of the class.

Consequently, because she knew very little about the nature of children's methods, she was unable to anticipate either the direction of their growth and understanding or the quality of their mathematical responses. Close listening to children's ideas was essential, and in fact became even more critical as the complexity of their thinking and their ways of coordinating units of different ranks

6. AN EMERGING PRACTICE OF TEACHING

developed over the school year. From her perspective, this created another tension: She was unsure of what the mathematics was that she was trying to teach. All this gave rise to a situation in which she found herself needing to be able to improvise and respond to the situation at hand rather than being able to rely on her previously well-established routines for teaching. This was in contrast to comments she had made earlier in the term as she described herself to us: "I am a creature of nature. I like routine. I like to know what is going to happen. I'm not terribly crazy about surprises."

January

In the following episode, the activity the children were working on is the familiar balance design with two-digit numbers (see Fig. 6.1). On this day, the teacher was sitting on a stool in the front of the class preparing to introduce the activity. This introductory section typifies the manner in which talking about talking about mathematics occurred during the second half of the year.

1 Teacher: Now remember, it's your responsibility if you understand and your partner doesn't see it, to try and do your best to have your partner understand through your explanation. Remember, we talked yesterday; we can't read each other's minds, so you are going to have to talk to your partner about it. Altogether I am going to put out pages M9, M10, M11, M12. So that's four. I don't think—I don't know if anybody will get through to all of

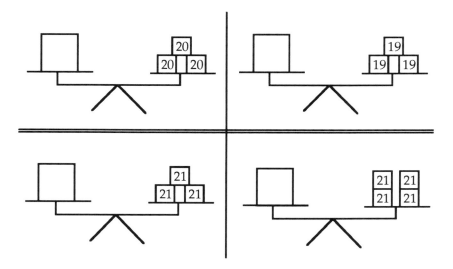

FIG. 6.1. Balance instructional activity.

		them or not, but like I said, it doesn't matter as long as what you are doing is working and . . . (voice rising)
2	Class:	Cooperating.
3	Teacher:	Trying to help your partner understand, cooperate, and learning yourself. Okay, any questions?

The emphasis of this introductory discourse was directed at a situation that often arose in which one member of a pair knew how to solve the problems and the other did not. In this brief opening, the teacher stressed that in that situation it was each child's responsibility to help his or her partner understand by explaining his or her solution. Unfortunately, it was not uncommon for one partner to dominate the activity by solving the problems, leaving the other with little opportunity to participate (cf. Cobb, chap. 3, this volume). Thus, when it was time for class discussion, the latter student would acknowledge that he or she did not know how to solve a problem. It was not the teacher's intention that children solve these problems in an independent manner, but, instead, she wanted them to talk with one another about their solutions. She believed children could enhance their understanding if they were given opportunities to explain their thinking; and if children would listen, they could learn from one another.

It became necessary during this time for the teacher to extend the existing social norms to include the expectation that the students would try to understand one another's solution methods. The teacher approached this in a very direct and explicit manner as evidenced during the introduction to this lesson, in which she told the children exactly what they were expected to do as they worked together. These extended norms, however, created an interesting dilemma for her students, as they were obligated to try "to help your partner understand, cooperate" and at the same time "learning yourself" (Yackel, Cobb, & Wood, 1991).

After the introduction, the children worked for 20 minutes in their small groups at solving the problems, which was followed by class discussion. The segment that follows provides an example.

1	Teacher:	Now on this page we have a balance and we are supposed to balance a 20, and a 20 and a 20. Now who . . . let's hear from Peter and Bonnie. What was your solution?
2	Peter:	60.
3	Teacher:	You think 60.
4	Bonnie:	Because we agreed that $2 + 2 + 2$ was 6, and 20 and 20 and 20 was 60.
5	Teacher:	All right. They said we knew that 2 and 2 and 2 made 6. They also know that this is really a 20, not just a 2 but a 20 in this case. 20, 20, and 20 must make . . . ? (voice rising)
6	Bonnie:	60.
7	Teacher:	(Simultaneously) 60. Who did it a different way? All right. Let's hear from Ryan and Carol [she means Katy]. Are you ready to help us out here? How did you and Ryan get this answer, Katy?

6. AN EMERGING PRACTICE OF TEACHING

7	Ryan:	We got 60.
8	Teacher:	All right. Now, did you do it differently than Bonnie and Peter? They said they knew that 2 and 2 and 2 made 6, and 20, 20, and 20 must make 60. Did you do it a different way? (Ryan shakes his head yes.)
9	Ryan:	I knew 20 and 20 was 40. And then, um . . .
10	Teacher:	Okay.
11	Ryan:	And I knew 2 more was 6. Two more added together all that made up to 60.
12	Teacher:	All right. Good. Anybody have a different way of doing it. Alex and Joy?
13	Joy:	Well, I knew that 20 plus 20 makes 40, and add 20 to that (she hesitates).
14	Teacher:	Add 20 to that? Okay, that's the way Ryan and Katy did it. Good. All right. Let's go on to the next one, then.

It is not surprising that this exchange reflected the initial pattern of interaction that was established early in the year, with two major modifications. First, the teacher now emphasized to the students that they should determine if their way was similar to or different from the solution that someone else had given before they volunteer. From the observer's perspective, it appeared that each student well understood that he or she was expected to find his or her own way to solve the problems. What was being negotiated at this point was the importance for each person to decide whether his or her way of solving was the same as or different from those already given before offering a solution to the class as a "different way." In this manner, the teacher tried to create an opportunity for each child to reflect further on his or her thinking as they compare their solutions to others'.

Second, although she continued to interact in the established manner, she often allowed the children to give a complete explanation before interrupting them. Previously, the students typically gave only part of their solution and then paused to allow the teacher to indicate whether she agreed or not. At that time, she often took the opportunity to expand or extend their comments, and then the children completed the remainder of their explanation. As the students became able to better express their thinking and to give their solution method as a continuous complete thought, such as Bonnie did in Line 4, she let them finish before commenting. Thus, the manner of breaking the solution into step-by-step procedures was gradually eliminated as the students learned to give their explanations succinctly.

During the lesson, however, the teacher continued to highlight those aspects, of either the problem or the solution that she thought were important. In this instance, it was the quantity the number represented when in the tens place (e.g., "2 and 2 and 2 made 6, and 20, 20 and 20 must make 60") and the relationships between the problems in which a thinking strategy could be used [e.g., " . . .if you already know, Joy, that 20 and 20 and 20 make 60, how can you figure this out?" for the problem (19, 19, 19)]. Added to this, she also encouraged two particular strategies that were given by the students. One was when they considered quantities as collected objects, collect the tens and add on the ones as single units (as in John's solution to the problem (21, 21, 21) "20 + 20 would be 40 and 2 [20] more would be 60 . . . 60, 61, 62, 63)." The other was to collect the tens and add on the ones as multiunits (e.g., "They say 20, 20, 20 makes 60," "this 4 and this 4 make 8," "8 and 4 more make . . . ?"). Although using the hundreds board to count surfaced several times during the lesson, she acknowledged it as if it were a well-known strategy and did not pursue it any further (e.g., "instead of counting with your fingers you used the hundreds board and counted with that. All right, that's another way to solve it").

The teacher was now well aware of the fact that in order for children to have opportunities to make sense of the problems and, thus, of mathematics, it was necessary for her to create situations in which she encouraged them to figure things out for themselves. Alternatively, she seemed to realize that it was her responsibility to try to make sense of the children's explanations, even though they might seem erroneous to her. These obligations created certain situations that were highly contradictory to her previous practice. She did want children to get the correct answers (and so did the researchers), but she also recognized that the focus of her teaching was on how children got their answers. What was important was their growth in understanding, not their ability to perform. Certain dilemmas still confronted her during the class discussion: How could she help children to reflect on their explanations as they were giving them so that their ideas could be understood by others? How could she get the other students who were listening to take notice of differences from and similarities to their solutions and those of others? The lesson that occurred on January 13 illustrated a pattern of interaction that was beginning to emerge as she attempted to resolve this predicament.

On this day, the students were having a great deal of difficulty interpreting the problems during the pair time such that they struggled to discuss their solutions during the class discussion that followed. This, consequently, created a difficult situation for the teacher, as she tried to facilitate the class discussion in her usual manner. For the observer, her attempts to conduct the exchange provided an opportunity to investigate the ways in which she was making sense of her activity and the strategies she was using as she tried to resolve her dilemmas. In this episode, the students solved the problem (35, 45/___) in the balance design which by now was familiar to all the students. Next, the problem was (36, 46/10, ___), which was difficult for the students to interpret because many were not able to determine whether to add or subtract the 10. Katy had arrived at the correct answer, but her explanation to the class was complex and, hence, unclear. Thus, the teacher continued:

1	Teacher:	In order to do this problem, what did you first have to figure out? Jamie?
2	Jamie:	Um. What was on the side that was all covered up.
3	Teacher:	Okay . . . first of all you had to figure out what 36 and 46 was. What was 36 and 46?
4	Jamie:	82.
5	Teacher:	It balanced out (writes 82 on overhead). They [researchers] could have stuck an 82 on that side and it would have been equal

> to the same as a 36 or 46. Okay. Together 36, 46, equals 82. They could have used an 82 instead, but they used two numbers. *So first you had to figure out what 36, 46 was. Then you had 10 on that side and you had to add something to it to balance it out. How did you figure that out, Katy? (Short pause.) We know there's 82 on this side, we know there's 10 on that side. Now, how did you figure out what to put in the box?*

The italicized sentences represent the way in which the teacher attempted to resolve the tension that existed between her inclination to step in and clarify what seemed to be confusing for the children and her wanting them to give the explanation for the problem themselves. Because it had been mutually constituted in these discussions, the children did not expect the teacher to direct them to a specific method for solving the problem, but rather that they find their own way of solving problems. Consequently, it would have been a violation of the established norms if the teacher had done so. Thus, the teacher found herself in a situation in which she was trying to find a way to provide them with assistance while still meeting her obligations to her students.

Focusing Pattern of Interaction. In this episode the sequence of questions, from the observer's perspective, were for the teacher a way of trying to build from students' existing understanding to focus the joint attention to a point of confusion. The teacher began with a series of questions to summarize the aspects of the problem that she thought were understood by most of the children. Then she drew attention to the aspect of the problem that was difficult for students to interpret, posed that as the problem to be solved, then turned the discussion back to the student to provide the explanation, "Now how did you figure that out, Katy?"

In many ways, this is similar to Ainley's (1988) structuring questions. She described these questions as those for which the purpose is to use student's prior knowledge to enable new connections to become clear. In this case, the difference is that the question the teacher asked was intended to have students reflect on the very point at which confusion existed. The act of focusing can be viewed as a possible intent a teacher may have for asking a particular question in a certain way, and can also be the effect, intended or otherwise, of asking a question. As such, the question serves to narrow the joint field of attention to one specific aspect, but still enables students to move forward in their thinking. This pattern, in many ways, appears to be reminiscent of the funnel pattern referred to earlier, in which a series of directing questions are asked. These could initially be interpreted by observers as leading the students to a specific end as predetermined by the teacher. However, the difference is that in the focusing pattern a sequence of questions are asked that serve to narrow the joint field of attention to the one specific aspect that is important, but then to allow the student an opportunity to provide the resolution of the problem. This, then, allows the student in question an opportunity to reflect on his or her reasoning as he or she reexplains his or her thinking and provides an opportunity for the others to make sense themselves of the unique aspect of the solution to the problem (Wood, 1994b).

February and March

During the months of February and March, the class discussions remained very much the same as in January. The teacher continued to follow her routine of taking two or three solutions and then moving to the next problem. The children, however, began to find ways to continue the discussion in order that they might offer more solutions. It seemed that as they developed deeper understandings of multiple units, including ten, and the relationships between numbers and operations, they had a wider variety of ideas about which they wanted to talk. Each single problem began to have the possibility of generating many different solutions or ideas about which the children wanted to talk. They soon found that by claiming a "discovery," which meant noticing something different about a problem or disagreeing with an answer, they would have an opportunity to continue discussing a problem. Thus, the children themselves began to interrupt and intervene in the normal flow of the discourse. An illustration of this is given in the following episode, in which the students were finished discussing their solution methods for (7×6) and (9×5), and the next problems were (11×5) and (11×10).

1	Teacher:	Good. All right. Now, these next two are kind of fun (referring to (11×5) and (11×10)).
2	Alex:	We made a discovery on the other one.
3	Teacher:	All right. Alex has another d . . .
4	Class:	(All are talking and waving hands hoping to be called on to solve the next problem, (11×5).)
5	Teacher:	Wait a second. Hang on. Alex's first. Because he said he made a dis . . .
6	Alex:	(Interrupts) 9 times 5. We got 45. And then [pause]. Well, plus 7 times was 35! So we figured 2 more 5s would be 45. 5 plus 5 is 10.
7	Teacher:	All right.
8	Alex:	And 10 added to 35 is 45.
9	Teacher:	Okay. Good. Let's hear from Andrea.

Alex, excited to share his thinking, appeared to know that not only should he give an answer, but he also had to be able to tell what he had discovered. He proceeded to explain the relationship he and his partner found between (7×5) and (9×5). This entailed using the previous solution to (7×5) and recognizing that (9×5) could be solved simply by adding 10 more to 35. As he gave his solution, he proceeded in a manner suggesting that the established routine was to provide reasons while giving an explanation. Unlike the previous month, the teacher simply indicated her agreement and did not attempt to restate or highlight his explanation. It appeared that at this point it had been interactively constituted in the class that

6. AN EMERGING PRACTICE OF TEACHING

in giving an explanation the children should provide a verification for their answer that included giving their rationale for the solution. By this time, it was clear that the teacher was willing to let the children take more responsibility for constituting an acceptable mathematical explanation. However, it can be noted that this exchange was still an interaction between the teacher and Alex, although the other members of the class simply listened.

In addition, the earlier renegotiated expectation for the way in which they were to work in pairs appeared to have become well established. In giving their explanation for (___ × 3 = 18), Betsy announced:

1 Betsy: First we had another answer [7]. We worked together and he told me it was wrong and he showed me why it was wrong.
2 Teacher: Did you hear that? I think that's kind of important. They had an answer, and Duane said, "Well it's wrong." He didn't stop just saying it's wrong, he said this is *why* it's wrong. And then he went on and explained how to get the right answer. So, it's one thing to just tell your partner, I disagree, but then you need to tell them why you do. That's important.

The emphasis the teacher placed on agreeing or disagreeing with a partner's answer and being able to support one's contention with reasons created an additional opportunity for students to learn that did not exist previously. Comparing their answers and determining if their solutions were similar or different provided them with a situation in which they could monitor their own thinking by reflecting on their previous activity.

April

As a final example, I have selected an episode that occurred late in April. The children were discussing the problem (39 + 33 =__) to which John had given the answer of 72.

1 Class: Agree, Agree.
2 Teacher: How did you think that?
3 John: I just knew that 30 and 30 make 60 and that [inaudible]
4 Teacher: All right. He said, I knew 30 plus 30 made 60, and then what?
5 John: I added the 9 and the 3.
6 Teacher: All right. That's exactly what I wanted you to say. He said, I added up the 9 and 3. And that's how he got 12. So he's saying 60 plus 12 made . . . gave him 72. That's great. Who did it a different way?
7 Ryan: . . . 3 and 3 is 6 and, um, 9 and 3 were 12, I mean—yeah, 12 but, um, I took 1 from the 3 [the ones of 33] (points to the 9 of 39) and then 10 and 2 make 12 and that would make 72.
8 Teacher: Okay. Who did it another way?

Two things of interest seemed to be occurring in the example. First, even though she knew that the students had a variety of methods, the teacher was still concerned about the traditional mathematics that she had been teaching over a number of years. One of the hallmarks of schooling for second-grade mathematics is the learning of the standard algorithms for addition and subtraction. This tradition is in direct contrast with the project goal that children construct their own efficient, albeit nonstandard, algorithms. In this instance, John's adding of the tens and then the ones was reminiscent of the partitioning of numbers that underlies the standard algorithm. The teacher, recognizing the similarity, responded by returning to a previous routine of interrupting, finishing the explanation, and highlighting what to her were the important aspects of the problem. Her comment, "That is just what I wanted you to say," which was offset by her next statement, "He said, I added up the 9 and 3," illustrated the shifting that occurred in her actions as she tried to balance the tension between the project goals of allowing individual children to construct meaningful personal methods, and the traditional goal of schooling, which is to ensure that students know the accepted practices held by society. In this regard, she often found herself inserting her intentions, catching herself, and then framing these intentions within the children's comments.

The second aspect to note is Ryan's solution to the problem. Typically, as we have seen before, he tended to add groups of ten, then make a ten from the ones and add that to the tens, and then add the extra ones. In this episode, he appeared to continue using this strategy with a slight alteration. As he added the one from 33 to the 9 in 39 giving him a 10, he said, "Nine and 3 were 12, I mean—yeah, 12, but I took, um, 1 from the 3 and then 10 and 2 . . ." From this example, we can infer that the norms that were extended to include the expectation that students would listen to others' solutions and decide if their ways were the same or different were guiding his response. In this situation, his explanation consisted of a combination of initially agreeing with John's way and identifying the point at which his method was different. In a sense, he had not considered adding 9 + 3 when he solved the problem, but he realized as he gave his explanation that he also had combined numbers in such a way that he got 12, just as had John. Ryan, in meeting his responsibility for participating in the class discussion, found it necessary to relate his ideas to the solution given previously. Although his explanation was still given as a description of his actions, the fact that he reflected on the thinking of another and used what he understood to inform his own thoughts is an important aspect in the development of mathematical reasoning. In order to take the perspective of others, children need to be able to make inferences about what others might mean and to use their understanding to determine if their ideas are similar and different.

Shortly after Ryan's explanation, Michael commented on the same problem (39 + 33 = ___).

1	Michael:	Well, there is also another way to do it. Just add up the 60 and count to 70 and add that 2 more on.
2	Teacher:	Okay. All right. Say that again, now, Michael?
3	Michael:	Well, that's adds [pause] up to 50, add on the 60, and on the 20.

6. AN EMERGING PRACTICE OF TEACHING 223

    4 Teacher: Oh!
    5 Michael: I took off the 2.

This exchange accurately illustrated the challenge and uncertainty teachers encounter when they engage their students in open-ended discussions to allow for a variety of solutions. In this particular case, the student previous to Michael gave a method that involved doubling 33 and adding on the remaining ones. Michael's first method, however, was based on a totally different conceptualization, and it was necessary for the teacher to switch from thinking about doubles. Not certain that she understood, she asked him to reexplain, which he did, only this time giving a solution that was different and difficult to understand. She knew that from Michael's perspective his explanation made sense to him. Yet, in the immediacy of the situation, she needed to decide whether to take the time to pursue his explanation further or simply to accept it in order to keep the discussion flowing for the other students.

    6 Teacher: [Pause.] All right, that would be another way to do it.
    7 Alex: What did he say?
    8 Teacher: I'm still not real sure. He said something about 50. Where did you get the 50, Michael?
    9 Michael: I just changed the 33 into a 34.
  10 Teacher: Oh. Okay [pause], and then what did you do?
  11 Michael: Well . . .
  12 Teacher: This 33 you changed to a 34. Is that what you say?
  13 Michael: Let me think about it [very long pause]. No! Now I remember. . . I changed [it] into a 32 instead.
  14 Teacher: All right. And then what did you do? Then what did you do with the leftovers over there?
  15 Michael: 3 plus 2 equals up to 5. Then 30 plus 20 equals up to 50. 50 plus 20 . . . equals up to 70, plus 2 equals up to 72.

In recapping, although she decided not to pursue Michael's method, another student was listening and trying to make sense of the solution given, and asked for further clarification. The teacher, quite comfortably now, admitted her own lack of understanding and changed her original decision in response to what she perceived was an interest in Michael's solution on the part of the class. It is also important to notice that, unlike the early discourse, the questions she asked created a situation in which the dialogue seemed to have some "genuine communicative purpose" (Stubbs, 1974). In each case, rather than paraphrasing and extending his comments, her questions were such that Michael was now the one who provided the explanation.

The mathematical meaning that evolved over the course of his explanation indicated a change in Michael's solution method. It appeared that initially he added the tens as $30 + 30 = 60$, then counted by ones to 70 making another ten, and added on the remaining two units. On the surface, his next comments seemed confusing, but in reality were possible changes he may have made in his thinking as he reflected on his mathematical activity. His final explanation involved taking 20 from 33 to make familiar combination (30 + 20 = 50), and then realizing that the remaining 10 plus another 10 (from 9 + 1) equaled 70, with 2 units remaining, making the answer 72. Notably, for Michael each refinement was an extension of his initial explanation, but in the immediacy of the situation for the teacher these were interpreted as different solutions.

As it was not always possible for her to understand the students' explanations, the teacher was often caught in the dilemma of deciding whether to unravel an individual solution or to just accept it in order to keep the discussion flowing for the other students. It was the awareness of the tension in these situations that enabled the observer to begin to appreciate the complexity of teaching. As a teacher, she had to be able to respond to the ongoing activity of the present situation in a way that created the best possible solution. Simply put, in such settings, opportunities to engage in any formal rational planning are nonexistent. It is also not possible for teachers to consider how their actions will affect others; their activity flows with the situation at hand. Although significant patterns of interaction can be identified by an outsider on reflection after the fact, this is not the understanding the teacher has to work with as the situation is developing. It is this tension that exists between ensuring that students have individual opportunities to learn and the necessity to keep the whole class involved in the activity and keep it proceeding in a productive direction that Heidegger (1962) referred to as *throwness*. Moreover, it is the ambiguity that is created in this situation that makes teaching in this manner a highly demanding form of practice. It is with this in mind that we return to the discussion.

16 Teacher: Okay. He has an idea here, though. He started something. You see what he's done here with taking this 33 here and making it into a 32? Does anybody see what else you might do with that?

Still not sure she understood Michael's solution, she tried to resolve her dilemma by responding with a question that directed the joint attention of the class to Michael's changing of 33 to 32. Although she appeared to have a specific solution in mind, rather than imposing her idea by explicitly telling or funneling she hinted at a possibility and asked the children for their ideas. Her attempt to focus their attention was seemingly ignored, because Peter commented:

17 Peter: I have something to add to this.
18 Teacher: Okay. What?
19 Peter: Well, it could be you add that 1 you took to 39, that equals 40 and add 32 to 40 equals 72.
20 Teacher: See what I'm saying to you? Michael sort of here (points to overhead) when he took one away from the 33 and it gave it over to 39. And turned that into a 32 and that into a 40 and then he was [pause] he converted that into a 72. So, that's another way.

CONCLUSION

It has been suggested that classrooms in which children talk about their personally constructed meanings have discussions that reveal higher levels of reasoning than do those discussions found in traditional classrooms (Prawat, 1991; Wood, 1994a). The process by which this develops is claimed by some to occur "naturally" (Palincsar & Brown, 1984). However, from my perspective, the teacher has a definite role in the process that reflects the complexity of teaching in this setting.

In this chapter, I have tried to describe the dynamic process by which the teacher's practice emerged during the months from January to April. To this end, I have described the "to-ing" and "fro-ing" that was reflected in the ways she interacted with her students. A tension emerged for the teacher as she attempted to cope with the variety in conceptual understandings held by the children and yet maintain established patterns of interaction. These interacting patterns relied on students' ability to contribute to the evolving taken-as-shared meanings. The dilemma became one of how to let children give their meanings on the one hand and, on the other, decide what she herself should tell them. Because her intention was to create a situation in which children were talking, it required a renegotiation of the expectations and obligations that underlie the patterns of interaction.

During the year, she watched and listened as children discovered and created mathematical meanings in the course of their own activity and she began to realize "how much children had within themselves—the power of their own thinking." Her role shifted from one of not giving information to one in which she created opportunities for children to think about mathematics. She struggled throughout the year to establish a way of teaching that shifted from an interaction involving questioning that was intended to check to an interaction in which negotiation of meaning was of central interest. This created for the students a different classroom setting, one in which children did not feel they were in a situation of constant evaluation. Instead, they found themselves in an atmosphere in which their ideas were listened to and in which their teacher attempted to understand their thoughts. She stated, "I've learned to let go of my classroom. Empowering students with responsibility gives them the feeling that they are needed and most important, that they have ownership in what they are learning" (Merkel, 1990, p. 2).

Further, she found that allowing children to talk required a great deal more cooperation and a more commonly held basis of understanding than she had used in her previous mathematics classes. The teacher's implicit conceptualizations of children's understanding were partly influenced by their novel solutions, their ability to give clear and concise explanations, and her prior experiences with traditional mathematics. The differences that emerged were more than a shift from focusing on the product of children's mathematical activity to an emphasis on their solution methods. Instead, the departure involved interpreting children's mathematical activity during the discussion and making qualitative distinctions in the solution attempts they gave. In this regard, her understanding of the situation developed in the ongoing activity with the children. As such, her actions were not based on detached reflective activity, but instead developed as she engaged in the interactive process of teaching and learning.

ACKNOWLEDGMENT

This chapter was written while the author was the Snodgrass Scholar in the School of Education at Purdue University.

REFERENCES

Ainley, J. (1988). Perceptions of teachers' questioning styles. *Proceedings of the Twelfth International Conference on the Psychology of Mathematics Education* (pp. 92–99). Veszprem, Hungary: Psychology of Mathematics Education.

Ball, D. L. (1993). With an eye on the mathematical horizon: Dilemmas of teaching elementary school mathematics. *Elementary School Journal, 93*(4), 373–397.

Bauersfeld, H. (1988). Interaction, construction, and knowledge: Alternative perspectives for mathematics education. In T. Cooney & D. Grouws (Eds.), *Effective mathematics teaching* (pp. 27–46). Reston, VA: National Council of Teachers of Mathematics and Lawrence Erlbaum Associates.

Carpenter, T. P., Fennema, E., Peterson, P., Chiang, C. P., & Loef, M. (1989). Using knowledge of children's mathematics thinking in classroom teaching: An experimental study. *American Educational Research Journal, 26*, 499–532.

Cobb, P., Wood, T., & Yackel, E. (1993). Discourse, mathematical thinking, and classroom practice. In E. Forman, N. Minick, & A. Stone (Eds.), *Contexts for learning: Social cultural dynamics in children's development* (pp. 91–119). Oxford, England: Oxford University Press.

Cobb, P., Yackel, E., & Wood, T. (1989). Young children's emotional acts while doing mathematical problem solving. In D. B. McLeod & V. M. Adams (Eds.), *Affect and mathematical problem solving: A new perspective* (pp. 117–148). New York: Springer-Verlag.

Confrey, J. (1990). What constructivism implies for teaching. In R. B. Davis, C. A. Maher, & N. Noddings (Eds.), *Constructivist views on the teaching and learning of mathematics* (Journal for Research in Mathematics Education Monograph No. 4, pp. 107–122). Reston, VA: National Council of Teachers of Mathematics.

Fennema, E., Carpenter, T. P., & Peterson, P. L. (1989). Learning mathematics with understanding: Cognitively guided instruction. In J. Brophy (Ed.), *Advances in research in teaching* (pp. 195–221). Greenwich, CT: JAI.

Goodlad, J. (1984). *A place called school*. New York: McGraw-Hill.

Heidegger, M. (1962). *Being and time*. (J. Macquarrie & E. Robinson, Trans.). New York: Harper & Row.

Hiebert, J., & Wearne, D. (1992). Links between teaching and learning place value with understanding in first grade. *Journal for Research in Mathematics Education, 23*(2), 98–122.

Labinowicz, E. (1985). *Learning from children: New beginnings for teaching numerical thinking*. Menlo Park, CA: Addison-Wesley.

Lampert, M. (1988). The teacher's role in reinventing the meaning of mathematical knowledge in the classroom. In M. Behr, C. LaCampagne, & M. M. Wheeler (Eds.), *Proceedings of the Tenth Annual Meeting of the Psychology of Mathematics Education—North America* (pp. 433–480). Dekalb, IL: Northern Illinois University.

Lampert, M. (1992). Practices and problems in teaching authentic mathematics. In F. K. Oser, A. Dick, & J. Patry (Eds.), *Effective and responsible teaching* (pp. 295–314). San Francisco, CA: Jossey-Bass.

Merkel, G. (1990). A metamorphosis. *School Mathematics and Science Center Newsletter, 4*(4), 1–2.

National Council of Teachers of Mathematics. (1989). *Curriculum and evaluation standards for school mathematics.* Reston, VA: Author.

National Council of Teachers of Mathematics. (1991). *Professional standards for teaching mathematics.* Reston, VA: Author.

Newman, D., Griffin, P., & Cole, M. (1989). *The construction zone: Working for cognitive change in school.* New York: Cambridge University Press.

Noddings, N. (1990). Constructivism in mathematics education. In R. B. Davis, C. A. Maher, & N. Noddings (Eds.), *Constructivist views on the teaching and learning of mathematics* (Journal for Research in Mathematics Education Monograph No. 4, pp. 7–18). Reston, VA: National Council of Teachers of Mathematics.

Palincsar, A., & Brown, A. (1984). Reciprocal teaching of comprehension—Fostering and comprehension—Monitoring activities. *Cognition and Instruction, 1*(2) 117–175.

Prawat, R. (1991). The value of ideas: The immersion approach to the development of thinking. *Educational Researcher, 20*(2), 3–10.

Rosenshine, B., & Stevens, R. (1986). Teaching functions. In M. G. Wittrock (Ed.), *The handbook of research on teaching* (3rd ed., pp. 376–391). New York: MacMillan.

Sinclair, J., & Coulthard, R. (1975). *Towards an analysis of discourse: The English used by teachers and pupils.* London: Oxford University Press.

Streeck, J., & Sandwich, L. (1979). Good for you.—Zur pragmatischen und konversationellen analyse von Bewertungen im institutionellen diskurs der schule. In J. Dittman (Ed.), *Arbeiten zur konversations analyse* (pp. 235–257). Tübingen, Germany: Niemeyer.

Stubbs, M. (1974). *Organizing classroom talk,* (Occasional Paper 19). Centre for Research in the Educational Sciences, University of Edinburgh.

Trevarthen, C. (1979). Instincts for human understanding and for cultural cooperation: Their development in infancy. In M. von Cronach, K. Foppa, W. Lepenies, & D. Ploog (Eds.), *Human ethology: Claims and limits of a new discipline* (pp. 530—571). Cambridge, England: Cambridge University Press.

Voigt, J. (1992, August). *Negotiation of mathematical meaning in classroom processes.* Paper presented at the International Congress of Mathematics Education, University of Laval, Québec, Canada.

von Glasersfeld, E. (1987). Learning as a constructive activity. In C. Janvier (Ed.), *Problems of representation in the teaching and learning of mathematics* (pp. 3–17). Hillsdale, NJ: Lawrence Erlbaum Associates.

Weber, R. (1986, April). *The constraints of questioning routines in reading instruction.* Paper presented at the annual meeting of American Educational Research Association, San Francisco.

Wood, T. (1994a). *The interrelationship between social norms, communicative interaction, and learning mathematics.* Submitted manuscript.

Wood, T. (1994b). Patterns of interaction and the culture of the mathematics classroom. In S. Lerman (Ed.), *Cultural perspectives on the mathematics classroom* (pp. 149–168). Dordrecht: Kluwer.

Wood, T., Cobb, P., & Yackel, E. (1990). The contextual nature of teaching: Mathematics and reading instruction in one second-grade classroom. *Elementary School Journal, 90*(5), 497–513.

Wood, T., Cobb, P., & Yackel, E. (1991). Change in teaching mathematics: A case study. *American Educational Research Journal, 28*(3), 587–616.

Wood, T., Cobb, P., Yackel, E., & Dillon, D. (Eds.). (1993). *Rethinking elementary school mathematics: Insights and issues* (Journal for Research in Mathematics Education Monograph No. 6). Reston, VA: National Council of Teachers of Mathematics.

Yackel, E., Cobb, P., & Wood, T. (1991). Small group interactions as a source of learning opportunities in second grade mathematics. *Journal for Research in Mathematics Education, 22*(5), 390–408.

7

The Ethnography of Argumentation

Götz Krummheuer
Pädagogische Hochschule Karlsruhe

In this chapter the main issue is the analysis of argumentation in the project class. An attempt is made to approach this both from a microethnographical description of classroom processes and by discussing several theories concerning argumentation.

The first overwhelming impression one had when observing the interaction in this classroom was the prevalence of situations of arguing and the obviously high regard these argumentations have for the teacher (see also Wood, chap. 6, this volume). This was presumably an effect of the underlying teaching experiment and, therefore, worthwhile to examine in greater detail (cf. Cobb, Yackel, & Wood, chap. 2, this volume). But having an argument in order to find appropriate mathematical solutions and definitions was not only bound to the teaching experiment in the project class: Discussing, explaining, justifying, illustrating, analogizing, and so on are all features of reasoning in many mathematical classrooms. In the project class this was systematically developed and extensively used. Therefore, focusing the analysis on the social genesis of argumentation is of theoretical interest and is commensurable with many situations of teaching mathematics, bound not only to the specificities of this class.

Argumentation is seen here primarily as a social phenomenon, when cooperating individuals tried to adjust their intentions and interpretations by verbally presenting the rationale of their actions.[1] Under the specific conditions of a classroom situation this aspect of negotiation seemed to have a strong impact on learning: Argumentatively based cooperation was obviously construed under the assumption that there existed a positive relationship between active participation

---

[1]Very often the word *rationale* or *rational* is used almost like a synonym for logic or logical. But *rationality* means first of all taking the best choice out of a set of options whereby what counts as the best is a matter of negotiation (see Follesdal, Walloe, & Elster, 1986). Thus, rationality is bound to many kinds of argumentation that necessarily cannot be subsumed under the criteria of formal-logical considerations.

in such social processes and the conceptual mathematical development of the students. This assumption about such a relationship is primarily a social construction, apparently widely accepted in the community of educators and learners in schools. It is seen here as a feature of a "folk psychology" (Bruner, 1990) of classroom learning.

This notion of folk psychology is intended to describe a stance that is somehow independent from the aims of the project. Certainly, the following ethnography of argumentation serves as a report of aspects of the interaction processes in the class, and one can draw some conclusions about the verification of project aims. But through the attempt of binding these argumentative processes in a framework about the everyday practices of mathematizing in the class, one establishes with this ethnography of argumentation a theoretical approach concerned with the conceptualization of the relationship of sociological and psychological aspects of learning.

In the following sections, several episodes from the project class are presented and analyzed. They serve in this chapter as illustrations of theoretical considerations. In order to find relevant episodes, an empirical analysis was done in advance and is not presented here in detail. The results show a certain circularity of social occasions in the sense that lessons are often structured along the following sequence: introductory phase—small-group activity—whole-class discussion.

In the introductory phase, the students and the teacher discussed the type of tasks included on the worksheets to be done next. Usually the teacher played an important role in this phase, especially when a relatively new kind of task was given. In the small-group activities generally two or three children worked together on several worksheets. Here, the children were to explain to each other what ideas and ways they had for solving the given problem. Additionally, they were to come to a commonly accepted solution (for a further discussion of the social norms in the project classroom see Cobb, Yackel, & Wood, chap. 2, this volume). Sometimes the teacher or a member of the project joined such peer group work. In such a case the teacher offered a variation of the idea of group work. The whole-class discussion finished this cycle by opening the floor for a wider discussion of different solutions and results. Here, the teacher was frequently acting in the foreground by organizing the turn taking or summarizing special solutions (cf. Cobb, chap. 3, this volume; Yackel, chap. 4, this volume; Wood, chap. 6, this volume).

This pattern had been called a "mini-cycle." As one can imagine, cycles did not appear in every lesson and sometimes they were carried into the following lesson. This organization may not have been intended by the project aims. In general, a cycle is a sequential order that can be identified time and again, and is meaningful for the analytical interest of this chapter.[2]

In the following section, the concept of argumentation is discussed and its connection to interactionistic and ethnomethodological approaches is outlined. The third section describes the contours of the relationship between the social and the individual with regard to learning mathematics in classroom interaction.

---

[2]The mini-cycles stem from January 21/22, January 26, January 27, and February 18. The first mini-cycle continued over 2 days.

THE CONCEPT OF ARGUMENTATION

Was ein triftiger Grund für etwas sei, entscheide nicht ich.[3]

In this section, the theoretical background of the concept of argumentation is outlined. The aim is to explore the rationality of the participants' actions in the project class, with specific attention to how they applied argumentative means to reach this goal.

The concept of argumentation is the main domain of rhetoric that can be traced back to Aristotle (Billig, 1989; Perelman & Olbrechts-Tyteca, 1969). Bound to this tradition, argumentation is seen mainly as a process that is accomplished by a single person confronted with an audience that is to be convinced. The interactive constituents of such activities are then only examined under the restrictive setting of a speaker and a group of listeners (Perelman & Olbrechts-Tyteca, 1969)—a setting that is uncommon in modern classroom communication.

Recent approaches to a theory of argumentation emphasize this communicative aspect more strongly, but usually they understand argumentation as a kind of isolated metacommunicative activity, and they underestimate the argumentative aspects in basal everyday activities. From an ethnomethodological stance, the conjunction of acting and showing the rationality of these actions is essential and should not be examined as exclusively separate topics. In a word, the notion of argumentation needs to be enhanced by aspects of interactionistic and ethnomethodological theories.

Theoretical Background of Argumentation

The notion of argumentation is examined in this section step by step. First, it is discussed with regard to approaches concerning a theory of argumentation. A provisional definition is given whereby specific features of argumentation processes in mathematics classroom life are taken into account. In a next step, the relation of this concept to basic assumptions of ethnomethodology is discussed and then reformulated within this theoretical framework. Then, functional aspects of an argumentation are presented that show that statements, which together are taken as an argumentation, play different roles in the establishment of an argument. Thus, not only the external relation of the concept of several theoretical approaches is presented, but also the internal structure of an argument. Finally, these considerations are summarized.

The Notion of Argumentation Within a Theory of Argumentation.
Empirically, the concept of argumentation will be bound to interactions in the observed classroom that have to do with the intentional explication of the reasoning of a solution during its development or after it. This could happen during the

---

[3]Wittgenstein (1974, no. 271): "What is a telling ground for something is not anything *I* decide."

small-group work and in the whole-class activities (introductory phase and closing whole-class discussion) as well. In general, argumentation is understood here as a specific feature of social interaction.

If one or several participants accomplish an assertion like "$4 \times 10 = 10 \times 4$" or "$31 + 19$ is the same as $32 + 18$,"[4] they do not only produce a sentence; rather, they make a declaration inasmuch as they claim such a statement to be valid. By proposing it they are not only indicating that they try to act rationally, but also that they could establish this claim in more detail, if desired. Usually, these techniques or methods of establishing the claim of a statement are called an *argumentation*. Thus a successful argumentation refurbishes such a challenged claim into a consensual or acceptable one for all participants (see Kopperschmidt, 1989).

With regard to classroom interaction, such accomplishments of argumentations generally do not appear in the form of a monologue but rather as direct face-to-face interaction. Because of the emergent nature of social interaction, argumentations are usually accomplished by several participants. Such a case is called a *collective argumentation* (Krummheuer, 1992; Miller, 1986, 1987). In addition, the development of a (collective) argumentation does not need to proceed in a harmonious way. Disputes in parts of an argumentation might arise that could lead to corrections, modifications, retractions, and replacements. Thus, the set or sequence of statements of the finally consensual argumentation is shaped step by step by surmounting controversy. The result of this process can be reconstructed and is called an *argument*. In order to sufficiently reconstruct an argument, the participants or observer must recall these interaction processes.

Very often, argumentation is understood as a metacommunicative activity that follows an ordinary action when its claimed validity is doubted or challenged. However, with regard to ethnomethodological approaches and to the data of the observed project class, it seems more appropriate to make use of the concept of argumentation when describing certain aspects of the ordinary action. For example, when children of the project class tried to solve a problem and accomplish for that aim a comprehensible method of reasoning, then this whole process already contained an argumentative feature. Very often this solution is already an argumentation and does not need to be secured by an additional and separate metacommunicative procedure.[5]

Two episodes might illustrate the kind of empirical data collected in the elementary mathematics classes concerning argumentation.

---

[4]These examples stem from the project class and were analyzed in more detail by Cobb (chap. 3) and Voigt (chap. 5) in this volume.

[5]This is especially the case when children apply a ratified calculation procedure, such as counting on the hundreds board. Here, the generation of a solution also leads, if acceptance is given, to the accomplishment of a complete argument. Obviously, especially in such calculational-related activities, the process of getting a conclusion and that of producing a supporting argument coincide: "It may well be, where a problem is a matter of calculation, that the stages in the argument we present in justification of our conclusion are the same as those we went through in getting the answer, but this will not in general be so" (Toulmin, 1969, p. 17). However, in primary mathematics education, calculations occur persistently and, hence, coincide with argumentation.

7. THE ETHNOGRAPHY OF ARGUMENTATION

Episode 1, January 26—Introductory Phase. The teacher introduced a worksheet that had to do with missing addend problems represented in a strips-and-squares form (for a more detailed description of the activities see Cobb, Yackel, & Wood, chap. 2, this volume). In the introductory phase the teacher discussed with the students the first problem of this page, which is shown in Fig. 7.1 (for a more developed analysis see Voigt, chap. 5, this volume).

Students develop several ideas to solve this problem. Two of these will be discussed. First is Carol's solution:

1 Carol: (Goes to board) Add 2 ten-bars take away 3 (points to top of problem).
2 Teacher: Add 2 ten-bars take away 3. Hm. OK. So she says (writes Carol's solution on overhead projector) add 2 tens and take away 3 ones. All right, that's one solution. I don't know if it is right or wrong, but that is what she says might work.

Neither Carol nor the teacher stated a numerical answer or drew the answer as ten-bars and squares at the screen. But they presented a method for how to find the result, which at the same time was also an argumentation. A little later Katy offered another explanation:

10 Katy: I add 17.
11 Katy: We're going to add 17.

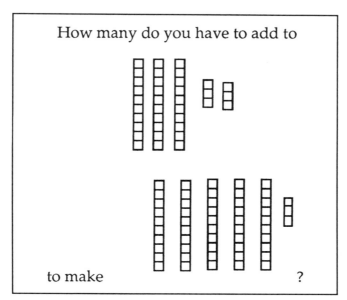

FIG. 7.1. A missing addend problem using strips and squares to represent "How many do you have to add to III::: (36) to get IIIII:. (53)?"

12	Katy:	Yea, to the top number.
13	Students:	Yea, yes.
14	Teacher:	OK. So she says add 17 . . .

Katy suggested "add 17" and got a positive reaction from several classmates (13) and also from the teacher (14). Katy stated the numerical standard name of the result, but she offered only this number without any further explanation.

Thus, comparing the two examples, one would say that Carol's demonstration contained features of an argumentation, although she was not asserting any concrete number as result. In contrast, Katy's statement was just an assertion that seemingly met with approval. But with regard to argumentation, she was not presenting any explanatory hint.[6] (For a further discussion of these solutions, see Cobb, chap. 3, and Yackel, chap. 4, this volume.)

Frequently, the concept of argument is used in the sense that it is displayed in order to support an assertion. In this case, a person tries to secure the assent of other participants by delivering an argument that he or she hopes will be convincing. In such situations, the concept of argument is used for a specific part within the process of argumentation (Kopperschmidt, 1989), whereas in the earlier definition an argument was seen as the product of an argumentation and not a specific part of it.

Both understandings are not mutually contradictory: A developed argument in the first sense will be applicable as a commonly accepted reason in an argumentation that is more complex in the regard that several separate arguments can be reconstructed. In this case, the single branches of the argumentation satisfy the definition given earlier. Often, one says that a single accomplished branch is a "good argument" with respect to the accomplishment of the more complex argumentation. Thus, one can speak on the one hand of the product of a minimally complex argumentation as argument. On the other hand, one can talk about the macrostructure of a complex argumentation that entails the accomplishment of several arguments.[7]

Episode 2, January 22—Small-Group Work. Two children, Andrea and Andy, were working on the problem shown in Fig. 7.2.

1	Andrea:	Okay, 7 and 7. 7 and 7 is 13 . . . 14 . . .
2	Andy:	Huh? Oh . . . let's check it . . .
3	Andrea:	1, 2, 3, 4, 5, 6 ,7, 8, 9, 10, 11, 12, 13, 14 (counting fingers). Yeah?
4	Andy:	Right.
5	Andrea:	And plus . . . add 7 more . . . 15, 16, 17, 18, 19, 20. (writes answer).
6	Andy:	I thought it was 22. That's why I said "huh."

---

[6]This does not mean however that Katy was not acting rationally or "accountably" (see the section entitled "Rationality in Ethnomethodological Approaches").

[7]Some authors try to keep the differentiation between these two kinds of understanding of argument clear by naming the arguments as parts of an argumentation as "'argument" (Paschen & Wigger, 1992; see also Kopperschmidt, 1986).

7. THE ETHNOGRAPHY OF ARGUMENTATION

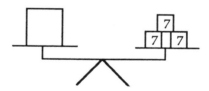

FIG. 7.2. Balance problem represented in text as _/7, 7, 7.

"Okay, 7 and 7. 7 and 7 is 13 . . . 14. Yeah, 14." One can interpret this utterance in the following way: First, Andrea tried to solve the problem linearly (find sum of the first two numbers—add third number to it), and second, she tried to refer to 7 + 7 as a known number fact but did not remember exactly. She mentioned (in Line 1) the numbers 13 and then 14. Andy was not quite sure about this partial sum and asked for a check (Line 2). The check, then, was a counting from 1 to 14 (in Line 3), which was agreed on by both children (see Lines 3 and 4). In Line 5 Andrea continued counting up 7 starting from 15. She got 20 and wrote this down as a result. Andy seemed to agree indirectly (in Line 6).

In this case, the argumentation contained a product of two arguments. First, the children demonstrated that 7 + 7 = 14. Second, they demonstrated—mathematically incorrectly—that 14 and 7 is 20. In both cases they counted by their fingers, which obviously could have been unified to a single counting process from 1 up to 21. But the children kept these two statements clearly separate, so that two distinct arguments were accomplished, which together build a total one.

For the purpose of this chapter both aspects—argument as outcome of an argumentation and argument as part of a complex argumentation—are of importance, although the foremost interest in analyzing the interactional structuring of argumentations favors the first conceptualization.

The Differentiation Between "Analytic" and "Substantial" Argumentation. If one uses the concept of argumentation in the field of mathematics, one might tend to bind it closely to that of proof. The analysis of argumentation in a classroom, then, could be misleadingly understood as a treatise on proof. Therefore, one should notice that both the concept of argument and that of argumentation need not be exclusively connected with formal logic as we know it from such proofs or as the subject matter of logic. There are more human activities and human efforts that are argumentative, but not in a strictly logical sense. As Toulmin pointed out, if these formally logical conclusions would be the only legitimate form of argumentation at all, then the domain of rational communication would be extremely restricted and argumentation as a possible way of a communication, which is based on rationality, would be rather irrelevant (Toulmin, 1969).

A logically correct deduction, for example, contains in its conclusion nothing that is not already a potential part of the premises. It explicates aspects of the meaning of the premises by means of deduction. These kinds of argumentations are "analytic" (Toulmin, 1969, p. 113). In contrast, "substantial" arguments (p. 113) expand the meaning of such propositions insofar as they soundly relate a specific case to them by actualization, modification, and/or application. Thus, substantial argumentations are informative in the sense that the meaning of the premises increases or changes by the application of a new case to it, whereas analytic ones are tautological, that is, a latent aspect of the premises is elaborated visibly.[8] Usually, these kinds of substantial argumentations do not have the logical stringency of formal deductions, which is not to be taken as a weakness, but rather as a sign that fields of problems exist that are not accessible to formal logic. According to Toulmin (1969), "The only arguments we can fairly judge by 'deductive' standards are those held out as and intended to be analytic, necessary and formally valid" (p. 154).

As Toulmin strongly emphasized, a substantial argumentation should not be subordinated or related to an analytic one in the sense that the latter is the ideal type of arguing and that one can always identify in substantial arguments the logical gulf in comparison to an analytic one. Substantial argumentation has a right by itself. By substantial argumentation a statement or decision is gradually supported. This support is not conducted by a formal, logically necessary conclusion, nor by an arbitrary edict such as declared self-evidence, but is motivated by the accomplishment of a convincing presentation of backgrounds, relations, explanations, justification, qualifiers, and so on:

> The very nature of deliberation and argumentation is opposed to necessity and self-evidence, since no one deliberates where the solution is necessary or argues against what is self-evident. The domain of argumentation is that of the credible, the plausible, the probable, to the degree that the latter eludes the certainty of calculation. (Perelman & Olbrechts-Tyteca, 1969, p. 1)

This distinction helps to clarify the conceptual framework chosen for the analysis of argumentation in mathematics classroom situations in primary education. It is the substantial argumentation that is seen here as more adequate for the reconstruction. Again, this does not imply that such argumentations have to be judged as poor or weak; rather, it is claimed that the impact of argumentations will be reconstructed in a more appropriate manner with regard to folk psychological assumptions of learning when the analysis is based on the concept of substantial argument. At this point, at least two reasons are given to support this decision:

- Children generally do not act on an axiomatic mathematical system: The mathematical knowledge of children at the primary school level is rather at an empirical–theoretical status (Struve, 1990) and their mathematical statements carry the significance of acting on experientially real mathematical objects (Cobb & Bauersfeld, chap. 1, this volume). This may be counting on fingers, or concluding from an existing concrete embodiment, or taking evaluations of an authoritative person for granted.
- In addition, children at the primary level usually do not exclusively draw conclusions of the analytic kind. Studies about the ontogenesis of the ability to

---

[8]See also Wittgenstein (1963): "The propositions of logic are tautologies. Therefore the propositions of logic say nothing. (They are the analytic proposition)" (6.1, 6.11).

7. THE ETHNOGRAPHY OF ARGUMENTATION

argue a claim and make deductive conclusions, like part–whole inclusions, in their fully developed verbalized form appear relatively late, from a developmental point of view. Specifically, Piaget's theory about the development of causality shows that deductive conclusions are based on a process of several prior stages and even at the latest stadium of "representative causality" reminders of previous stages still exist.[9]

For Perelman and Olbrechts-Tyteca (1969), this substantial kind of argumentation is the basis for a theory of argumentation in general. For them, argumentation has something to do with the art of convincing and not with the necessity of logical conclusions: "The object of the theory of argumentation is the study of the discourse techniques allowing us *to induce or to increase the mind's adherence to the theses presented for its assent*" (p. 4).

Rationality in Ethnomethodological Approaches. The previous discussion indicates that the rationality of everyday activities, such as those in the project class, should not be measured by the standards of a scientific model of rationality, which in mathematics is oriented toward deductive logic. Specifically, ethnomethodology emphasizes that the participants in the affairs of everyday life constitute their model of rationality interactively, which one might call the "informal logic" (Klein, 1980), emerging in these situations. One of the aims of ethnomethodology is to explore such interactive constitutions of rational acting (Garfinkel, 1972). The application of a formal-logical or scientific model of rationality for this purpose is vehemently rejected:

> In a word, the [scientific] model furnishes a way of stating the ways in which a person would act were he conceived to be acting as an ideal scientist. The question then follows: What accounts for the fact that actual persons do not match up, in fact rarely match up, even as scientists? In sum, the model of this rational man as a standard is used to furnish the basis of ironic comparison; and from this one gets the familiar distinctions between rational, nonrational, irrational, and arational conduct. (Garfinkel, 1967, p. 280)

Thus, from the stance of ethnomethodology, the rationality of everyday actions is interactively generated while acting in a social setting; it is not an invariant part of such actions. On the contrary, the participants of everyday encounters are continuously concerned with showing and clarifying the rationality of their actions for themselves and for the others as well (see Lehmann, 1988).

The participants use so-called accounting practices, which are techniques and methods that help to demonstrate the rationality of the action while acting. In the process of accomplishing an action the participant is already trying to make his or her actions accountable. The concept of "account" is a key one in the conceptual system of Garfinkel's work, which, unfortunately, he does not define precisely. Attewell's (1974) characterization is helpful: "Garfinkel uses the nuances of

---

[9]See Inhelder and Piaget (1985), Keil (1979, 1983), Piaget (1928, 1930, 1954), Siegal (1991), Völzing (1981); a poignant summary about Piaget's approach is in Siegal (cf. pp. 39–45).

English to express this equivalence between making sense of something and explaining that sense. The word 'account' carries this equivalence; to account for something is both to make it understandable and to express that understanding" (p. 183).

The application of this concept is broader than that of argumentation. It envelopes all kinds of rational processing of experience. The unifying character of all accounts is that they make the action to which they refer understandable, and the proposed claims intersubjectively acceptable (for a further discussion of intersubjectivity, see Voigt, chap. 5, this volume). In order to do so, one endows it with the "status of an intersubjective object" (Leiter, 1980, p. 162). One important class of such accounts includes those demonstrations bound to languaging (Bauersfeld, chap. 8, this volume), to which argumentation belongs. In many ethnomethodological works these language-bound activities seem to be the regular practices in demonstrating the accountability of the ongoing action. "To do interaction, is to tell interaction" (Attewell, 1974, p. 183). Garfinkel very often replaced the word *accountable* with words like *telling, storyable,* or *reportable* (see Lehmann, 1988).

Argumentation is, therefore, primarily a discourse technique and not a feature of the solely reasoning subject. Thus Klein (1980) could characterize it as the attempt to "transfer the collectively doubted into the collectively accepted by collectively shared means" (p. 19; translation by Krummheuer). With regard to the concept of substantial argumentation, one should add here that the collectively accepted basis as well as the collectively shared means for this transfer might also change through the interactive establishment of such an argumentation.

The Reflexivity of Interactional Activities. For the ethnomethodological approach, it is essential that the accomplishment of an action and the demonstration of its accountability are not necessarily separate activities. Garfinkel understood it as the central issue of his studies, that "the activities whereby members produce and manage settings of organized everyday affairs are identical with members' procedures for making those settings 'account-able'" (Garfinkel, 1967, p. 1).

This theorem, which asserts that in the process of social interaction the participants try to make their actions understandable and accountable as well, is often termed *reflexivity* (Lehmann, 1988; Mehan & Wood, 1975; Voigt, chap. 5, this volume; Yackel, chap. 4, this volume). Garfinkel's spelling of *account-able* already pointed out that this reflexivity does not necessarily mean that the rationality of one's actions is explicated entirely or sufficiently while accomplishing these acts. Usually only hints or rough outlines are given, from which the participants have to produce the account by themselves while interacting (e.g., when a child declares that he or she has just used the hundreds board). Sometimes, however, participants offer their reasoning voluntarily, as shown in the following example.

Episode 3, January 22—Small-Group Work. Andrea and Andy were working on the balance problem __ /6, 6, 6. Both children had already agreed on the answer 18. Andrea offered her argumentation:

1	Andrea:	Okay. Okay, 12, 13, 14, 15, 16, 17 . . . 18.
2	Andy:	18.
3	Andrea:	You know how I got that?
4	Andy:	Mmm hmm.
5	Andrea:	6 and 6 . . . 6 and 6 is 12 . . . 6 more: 18.

Andrea elaborated a solution by starting with an obviously well-known number fact of $6 + 6 = 12$ and then continuing counting on 6 more steps (in Line 1). This has been already reconstructed as their way to solve this kind of problems (see Episode 2). She ended up with 18, which seemed to meet Andy's approval (Line 2). Nevertheless, Andrea offered to explicate the reasoning of the just-presented argumentation. In Line 5 she rephrased her argumentation, whereby the previously obviously problematic addition of the third 6 did not appears to be a problem again.

Habermas (1985) described the interactional development from the accomplishment of activities to the explicit thematization of their accountability as the change from "communicative acting" to a "rational discourse." The interactional function of this subsequently rational discourse can then be seen as the production of acceptance for the continuation of the previous activities. However, from an ethnomethodological point of view and also with regard to the observations in the project class, it should be emphasized that the interaction process is not to be divided into a phase of communicative acting and another phase of rational discourse during which an argumentation is accomplished. Reflexivity is the incessant and more or less expressively designed conjecture of doing something and intimating its accountability. Argumentation merely describes the aspect of any interaction that is concerned with the expression of the accountability.

The Functional Aspect of Argumentation

This chapter has thus far presented a discussion suggesting that argumentation is the method or technique to (re)establish the challenged claim of an assertion. This section revolves around the questions regarding how it happens that a certain sequence of interactive moves serve this way. It is essential here to take into account that an argumentation usually contains a sequence of statements, each of which plays a different role in the emerging argument. The following discussion examines which kind of statements should necessarily entail an argument and how they work in relation to each other for the accomplishment of an argumentation. This is called the "functional" aspect of an argumentation (Kopperschmidt, 1989).

Toulmin (1969) proposed for this purpose a "layout" of an argument—a scheme that represents the ideal model of a substantial argumentation. He developed it by analyzing the numerous occasions of "rational enterprise" in the multiple fields[10] at rational communication. The layout describes a common basis for these attempts at argumentation. Usually in a concrete situation it appears only partially and/or in

---

[10] For Toulmin's concept of "field" see the section entitled, "Toulmin's Concept of 'Field' and the Concept of 'Framing,'" later in this chapter.

a varied version. It helps to reconstruct the informal logic of an argumentation and the kind of accountability developed for the resolution of a quarrel. This scheme is not to be misunderstood as a method for identifying the different components of that model in concrete interaction—this needs to be done by a related analysis of interaction. The scheme merely points out the different roles that utterances play in an interaction when reconstructed from the perspective of the emergence of a substantial argument.

Globally, an argumentation functions in a way that the challenged assertion is to be secured by presenting it as a conclusion from undoubted facts. This implies an inferential step from these facts to the conclusion, which also needs to be taken as undoubted by all participants. In the following section the creation of such inferences is first studied in more detail. As it is shown, beyond the identification of a certain core of an argument, which are the data a conclusion is based on, and the warrant of the entailed inference, there exists an additional means to secure the whole argument by specifically backing the applied and warranted inference.

Episode 4, February 18—Small-Group Work. First, a short transcript from the project class is presented to help exemplify the theoretical considerations that follow. At this point it is not necessary to display the hermeneutic process of coming to a relatively valid and dense interpretation of the transcribed strip of reality. The transcript will only be used for illustrative purposes.

The following transcript deals with the multiplication task $4 \times 4 = \_\_$. Multiplication had been introduced in this lesson. Previously in the small-group work the children, Jack and Jamie, solved the problem $2 \times 4 = 8$ just before this (for analyses of the social and cognitive aspects of this dyad, see Cobb, chap. 3, this volume; and Yackel, chap. 4, this volume).

1 Jack: What's 8 plus 8?
2 Jamie: 16. It's 4 sets of fours, 8 (pause) 16.
3 Researcher: Why did you say what's 8 and 8 Jack?
4 Jamie: 'Cause 4 sets, um 4, 2 sets make 8.
5 Researcher: Yes.
6 Jack: (Holds up fingers on one hand) You have 2 more sets. Like it's 2 and 2 make 4.
7 Researcher: OK, OK, very good.

In the following discussion, this example is interpreted line by line. Toulmin's layout for this accomplished argument appears at the end of section.

The Creation of an Inference. As seen previously, the basic idea of the functionality of an argumentation is that one tries to support a questioned assertion by inferring it from another statement. The argumentative force, which lies among others in the assumption, is that the inference is widely accepted. Further, the validity of the added statement, on which the inference is based, is unquestioned. If we call the primary assertion C (for conclusion) and the supporting statement D

7. THE ETHNOGRAPHY OF ARGUMENTATION

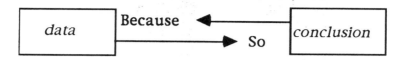

FIG. 7.3. Symbolic representation of the argument "D, so C."

(for data), then we can symbolize this part of the argument as "C, because D" or "D, so C" (Toulmin, 1969, p. 107).[11] In a pictorial scheme one can represent this relation as shown in Fig. 7.3 (see Toulmin, 1969).

Data. As we can see in the example, the two boys seem to refer to a (number) fact $8 + 8 = 16$. They appeal to it and try to present it as a foundation on which their assertion $4 \times 4 = 16$ is supposed to be based. Any argumentation needs such a kind of factual initiation on which the conclusion can be grounded. Toulmin (1969) called these facts data and stated, "Data of some kind must be produced, if there is to be an argument there at all: A bare conclusion, without any data produced in its support, is no argument" (p. 106).

Utterances, which serve in the functional role of data in an argumentation, can be doubted, of course. In the example there might be some questions about the correctness of $8 + 8 = 16$.[12] The way in which Jack and Jamie established this number fact points to this issue: Jack asked for the result (Line 1), possibly indicating that he did not know the solution with certainty. Jamie gave the answer (in Line 2) and after that the statement was not doubted. By this it was given the status of a datum.

For the example one could show this relationship between data and conclusion in the scheme shown in Fig. 7.4.

If there is no agreement about the validity of such an utterance supposedly functioning as data, another argumentation would be needed to give acceptable evidence to it. This would be a recursive application of the general scheme of argumentation and would not lead to a new functional category of its own.[13]

Warrants of Inferences. Besides the doubting of the factuality of some data, the questioning of the conclusion based on these data can also arise. It can happen that one agrees with the data but does not find them to support the assertion. In our example a child could totally agree with "$8 + 8 = 16$," but could problematize what this had to do with the assertion "$4 \times 4 = 16$."

Taking the data, the function of an argumentation is to be committed to a certain step toward the conclusion. This step might not be found appropriate or legitimate.

---

[11] Both expressions are equivalent. For easier reading in the following, only the second form—"C, so D"—will be used.

[12] The question here is not whether this equation is mathematically correct; rather, the focus is on the acceptance by the participants. At this time in the project class this equation does not belong to the set of undoubted number facts for all children (see Cobb's analysis, chap. 3, this volume).

[13] The arising problem of an infinite regress is discussed later in the chapter.

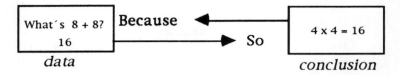

FIG. 7.4. Symbolic representation of argument presented in example from project class.

Such criticism can, according to Toulmin (1969), only be rejected if one presents propositions that are of a functionally different nature than the data, namely "general, hypothetical statement," that act as "bridges" (p. 98), over which one can lead the particular inferential step. They function like "inference-licenses" (p. 98) and authorize the step that is built in the concrete argument. Toulmin called them "warrants" (p. 98) and stated: "its task being simply to register explicitly the legitimacy of the step involved and refer it back to a larger class of steps whose legitimacy is being presupposed" (p. 100).

In our example, one can interpret the second part of Jamie's utterance in Line 2 as part of such a warrant. After the accomplished agreement about the data $8 + 8 = 16$, the legitimate question can be asked, what did this addition have to do with the given multiplication problem 4×4? Referring back to the definition of multiplication, which had been given in this lesson before, one could easily reinterpret $8 + 8$ as 2×8, but not as 4×4. The number 8 does not even visibly appear in 4×4.

Thus, if there is a legitimate step from the data $8 + 8 = 16$ to the conclusion $4 \times 4 = 16$, then one has to justify how the eights in the data appear in this inference. Jamie said "It's 4 sets of fours, 8 (pause) 16." Formally speaking, he stated a connection between the interpretation of 4×4 as 4 sets of fours and 8. This still seems rather opaque. But it contains a more general approach than the datum $8 + 8 = 16$. It refers to a general definition of multiplication in terms of sets, as it has been introduced in this class. Just this slightly inscrutable utterance functions like a bridge builder: It poses a hypothetical connection between the addition of 2 eights and the problem 4×4 by reformulating the multiplication task as a statement in terms of sets.

In Line 4 this becomes clearer, "'cause 4 sets, um 4, 2 sets make 8." Here Jamie explicitly showed what the 8 in the data had to do with the multiplication concerning sets of fours. It was the result of 2 sets of fours. By this he added another datum: 2 sets of fours equal 8. This number fact had been found just before, as the two boys started working on the problem sheet. Further, Jamie declared in Line 6 that one has "2 more sets." This utterance worked again as a warrant in the sense that it authorized the application of the number fact $8 + 8 = 16$.

Thus, the argumentation up to this point can be reformulated as: It is clear that $8 + 8 = 16$ and that 2 sets of fours equal 8. So, because 4×4 can be interpreted as 4 sets of fours, and because one has 2 more sets of fours and one agrees with the datum that "2 sets of fours equal 8," it seems acceptable that $4 \times 4 = 16$. It remains unstated how it is possible to divide the 4 sets of fours into 2×2 sets of fours.

7. THE ETHNOGRAPHY OF ARGUMENTATION

We can expand the schematic pattern of an argument by a box for the warrant, as shown in Fig. 7.5. This skeleton of an argumentation containing data, warrants, and conclusion will henceforth be called the "core" of an argument. Its formal shape is "D, because W so C." Formally, this is the minimal form of an argumentation. Toulmin (1969) wrote, "Provided that the correct warrant is employed, any argument can be expressed in the form 'data; warrant; so conclusion' and so become formally valid" (p. 119).

On the surface of spoken language, it is generally not possible to differentiate between data and warrants. Also, as the example shows, in the interaction the participants do not necessarily structure their contributions according to the functional categories of the scheme. Additionally, the distinction is often difficult to make because data and warrants relate to each other. The data cited depend on the warrant being chosen. One should, therefore, keep in mind that data is supposed to "strengthen the ground" (Toulmin, 1969, p. 98) on which the specific argument is supposed to be constructed. Warrants, in contrast, are rather general, "certifying the soundness of *all* arguments of the appropriate type" (p. 100).

Backing the Warrant. In a particular case the soundness of the step from data to conclusion can be ratified by warrants. But one still can ask "why *in general* this warrant should be accepted as having authority" (Toulmin, 1969, p. 103). Toulmin called the functional category securing a warrant its *backing*. "Standing behind our warrants . . . there will normally be other assurances, without which the warrants themselves would possess neither authority nor currency—these other things we many refer to as *backing* (B) of the warrants" (p. 103).

This new concept is clarified first by the example of this chapter. The argumentation of Jack and Jamie had developed to the point where there was agreement about the data $8 + 8 = 16$ and "2 sets of fours equal 8." The warrants are that 4×4 can be interpreted as "4 sets of fours" and that "there are 2 more sets of fours than in the previous task $2 \times 4 = 8$. So, $4 \times 4 = 16$ can be taken as a conclusion of the data.

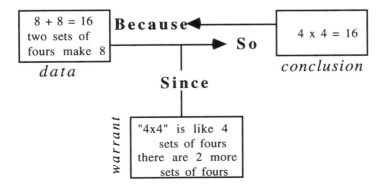

FIG. 7.5. Expanded symbolic representation of argument including warrant.

Possibly urged by the questions of the interviewer (Line 3), Jack gave an explanation (Line 6) by holding up fingers on one of his hands "you have 2 more sets. Like it's 2 and 2 make 4." The middle part, "you have 2 more sets," had already been identified as a warrant. The nonverbal action of holding up fingers of one of his hands and the utterance "like it's 2 and 2 make 4" can be interpreted here as a backing of this warrant.

The last warrant states that in 4×4 are 2 more sets of fours than in 2×4. Jack's backing of this is that he referred to the fundamental insight that one can additively decompose a number. In the particular case, this means that 4×4 can be decomposed in 2×4 and again 2×4. He did not say this in a generalized way; he used an analogy, remarking with the expression "like it's 2 and 2 make 4." Additionally, he demonstrated this fundamental approach of decomposing a number by using his fingers. Among second graders, counting by fingers or symbolizing numbers and/or one of their additive decompositions is one of the most undoubted and most ratified strategies of proving an arithmetical statement. Thus, this can be taken here as a backing.[14] The diagram can now be completed (see Fig. 7.6).

Backing refers to global convictions and primary strategies that can be expressed in the form of "categorical statements" (Toulmin, 1969, p. 105) like understanding addition as a counting procedure that can basically be demonstrated by fingers. Backing explicitly binds the core of an argument to such collectively accepted basic assumptions. Toulmin's layout of an argument is outlined here so far[15] and in the following summary.

Two additional remarks may be allowed here: (a) In the specific case of the example, different backings can be offered to the given warrants, some of which might even be assessed as more relevant from a mathematical point of view. For example, one could have tried to express as a backing why it is possible to split a set of 4 sets of fours in 2 × 2 sets of fours. One could have demonstrated this iconically by referring back to the dot representation of sets during the introductory phase, as shown in Fig. 7.7.

In this case, the whole information of the conclusion is already included in the backing. In such cases, Toulmin (1969) called an argument "analytic": "An argument from D to C will be called analytic if and only if the backing for the warrant authorizing it includes, explicitly or implicitly, the information conveyed in the conclusion itself" (p. 125). Thus, the concrete argumentation of Jack and Jamie could be reformulated by applying another backing, which would turn the originally substantial argumentation into an analytic one. This might be one of the reasons why the researcher reacted in such an enthusiastic way in Line 7: By replacing the backing by one like the mentioned one,[16] the accomplished argument would appear as one that tentatively adhered to mathematical standards.

---

[14]Mathematically, this can be justified using the law of distribution: $4 \times 4 = (2 + 2) \times 4 = 2 \times 4 + 2 \times 4$. In the example it is not meant that Jack was arguing in this way. His backing was rather an analogy and not a deduction from an axiomatic law. Obviously, in this case it is possible to invent different backings. One should not mix them up (see related discussion later in this chapter).

[15]It should be mentioned here that Toulmin's layout contains two more categories, so-called "qualifiers" and "rebuttals" (see p. 102), which are not outlined here.

[16]See also footnote 14.

7. THE ETHNOGRAPHY OF ARGUMENTATION

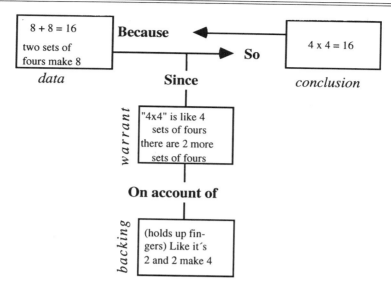

FIG. 7.6. Completed schematic representation of argument from class example.

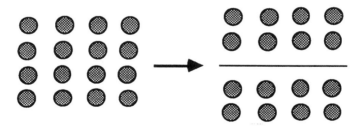

FIG. 7.7. Dot representation illustrating how to split a set of 4 sets of 4 into 2 × 2 sets of 4.

(b) As one can easily see, the accomplished argumentation in this episode was a collective one. Here, it was mainly Jamie who gathered all data and made the conclusion exclusively by himself. The warrants were added by both children, and the backing was solely produced by Jack. Figure 7.8, which contains the insertion of the speakers' names in the scheme, might elucidate this.

As one might see by this graphic presentation, both boys made essential contributions by their statements to the development of this argumentation. Although in the interpretation given here the main contributions of Jack and Jamie relied on different components of the argument, one could gainsay that this argumentation was only based on one of the children's mathematical abilities.[17]

---

[17] For the relationship between these two boys and its effect on their learning, see Cobb's case studies, chap. 3, this volume.

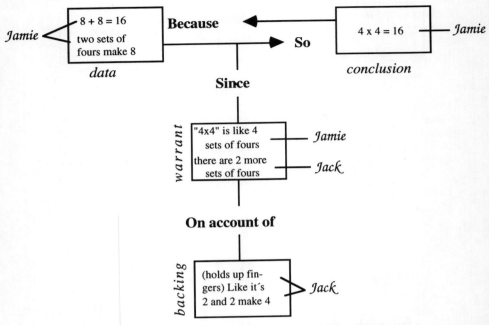

FIG. 7.8. Graphical representation of class example with addition of speaker names.

Summary: Argumentation

One main assumption of microsociological theories is the reflexivity of social interaction:[18] The participants do not only coordinate their *individual* goals and intentions by adjusting their actions and by negotiating their definitions of the situation, but they also try to demonstrate the seriousness and accountability of their participation through the same actions.[19]

With regard to the complexity of everyday interaction, one should avoid the idea that this ethnomethodological concept of rationality can be sufficiently subordinated under categories of formal logic, or that it should be described by the violations or deficiencies of formal-logical reasoning. The emerging rationality in everyday interaction has its own dynamics and rights reflecting the complexity and variety of social affairs and the ever–present human endeavor of giving structure to their lives together in a reasonable way.[20]

One common and obviously also more frequently successful way of giving structure is by using language; that is by choosing verbal means in order to express the sense made in this moment and the reason for this specific sense-making. These

---

[18]See the earlier section entitled "The Reflexivity of Interactional Activities."

[19]See the earlier section entitled "Rationality in Ethnomethodological Approaches."

[20]See the earlier section entitled "The Differentiation Between 'Analytic' and 'Substantial' Argumentation."

7. THE ETHNOGRAPHY OF ARGUMENTATION 247

language-bound methods or techniques of expressing the reflexive claim of acting rationally will be called *argumentation*. Its aim is to convince oneself as well as the other participants of the property of one's own reasoning and to win over the other participants to this special kind of "rational enterprise." The final sequence of statements accepted by all participants, which are more or less completely reconstructable by the participants or by an observer as well, will be called an *argument*.[21]

Ideally an argumentation in everyday affairs contains several statements that are related to each other in a specific way and that by this take over certain functions for their interactional effectiveness. This functional aspect is described here in accordance with Toulmin (1969). The fundamental approach of this functional analysis is that the claimed validity of an assertion or statement is established by an argumentation in a way that the questioned assertion appears as the conclusion of other assumed undoubtedly valid statements. This system contains, if sufficiently elaborated:

- A conclusion (C), the validity of which was doubted.
- Data (D), on which the conclusion is grounded.
- Warrants (W), which give reason for the legitimacy of the applied inference from D to C.[22]
- Backings (B), which support the warrants by giving categorical statements about principal convictions leading the thinking of the arguing individuals.[23]

Schematically one can represent this system as shown in Fig. 7.9.

It should be mentioned that the different categories in Toulmin's layout cannot, in general, be recognized on the surface of spoken formulations; they must be identified by an appropriate analysis of interaction. Additionally, not all of the positions of the scheme shown in Fig. 7.9 have to be filled in a concrete argumentation. This scheme represents a general layout of an argumentation, which might vary in concrete interaction. It helps to reconstruct the emerging rationality, that is, (informal) logic of everyday affairs.[24]

CONTOURING THE RELATIONSHIP BETWEEN ARGUMENTIZING AND LEARNING

After developing the concept of argumentation and its functional structure, its connection with learning in mathematics classes can now be outlined. To this end, three main issues are discussed:

---

[21] See the earlier section entitled "The Notion of Argumentation Within a Theory of Argumentation."
[22] See the earlier section entitled "The Creation of an Inference."
[23] See the earlier section entitled "Backing the Warrant."
[24] See the earlier section entitled "The Functional Aspect of Argumentation."

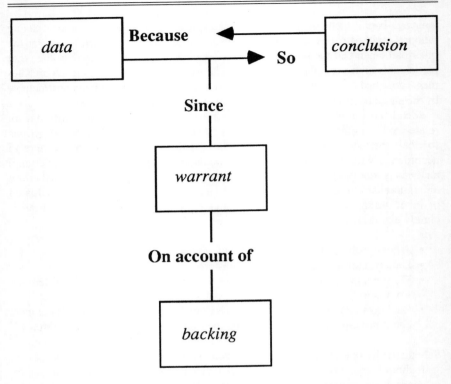

FIG. 7.9. Schematic representation of an argumentation system.

- The individual.
- The interaction.
- The reflexive relationship between the two.

The first section shows the way in which argumentation relates to the individual process of the definition of the situation. Those individual processes of sense making are usually prestructured and socially founded. The concept of framing is developed for the purpose of describing this issue. Ultimately, argumentations are based on those framings and can, therefore, be characterized as framing dependent. But there are parts of an argument—namely the components of its core—that are not directly bound to a specific framing. This allows one to more precisely describe what argumentatively accomplished agreements look like. Additionally, it is possible to clarify the already mentioned problem of infinite regress of arguments in the case that an agreement cannot be found regarding presented data.

Based on the clarification of the kind of agreement generated by argumentation, the second section describes the related interaction in more detail with regard to its impact on learning. Here, the nature of classroom interaction is taken into account. For this purpose, the concept of format of argumentation is introduced to describe a type of patterned interaction that is concerned with argumentation in mathematics classes.

Finally in the third section, this new concept makes it possible to describe the relationship between the social and the individual as a reflexive one. On the one hand, the child's avail for the accomplishment of a collective argumentation depends on his or her reliance on certain generalized forms of arguments that he or she knows and to which one can refer. These are called, with reference to the terminology of rhetoric, *topoi;* that is, the student's conceptualization of what is argumentatively commonplace. On the other hand, the student's experience of formatted argumentations qualifies him or her to recollect specific arguments as similarly structured, and to cognitively construct a related topos.

The Framing-Dependency of Argumentations

Toulmin's Concept of "Field" and the Concept of "Framing." As the final remark to the example in the second section has already shown, in specific cases the core of an argument can be ensured by different backings, which might change the character from a substantial to an analytic one. Beyond that, a major topic arises that Toulmin described as the "field-dependency" of arguments. These fields can be identified as certain societal institutions, such as legal courts, scientific congresses, seminars at a university, medical consultations, and so on. In all of these social events certain domain-specific standards of reasoning exist that cannot be reduced to one specific logical type.[25]

Referring to the example of Jack and Jamie, the field in which the two boys' argument could be located is that of the commonly shared experience of concrete counting activities. This seems to be a collective basis for the arithmetical reasoning of primary school children (see, e.g., Saxe, Guberman, & Gearhart, 1987; Steffe, von Glasersfeld, Richards, & Cobb, 1983). The field in which a mathematician could more likely try to address the alternative backing is that of the axioms of set theory for positive integers.

Toulmin's distinctions of fields are very global and he himself suggested studying this field-dependency in smaller units: "The distinctions [between different fields] we have made so far are very broad ones, and a closer examination could certainly bring to light further more detailed distinctions, which would improve our understanding of the ways arguments in different fields are related" (Toulmin, 1969, p. 42).

In pursuing this proposal, it seems appropriate to embed Goffman's (1974) concept of framing. Framing is defined in general as the schematized interpretation of a social event, which emerges by processes of cognitive routinization and interactive standardization (see also Krummheuer, 1992). The concepts of frame, framing, script, and so on are widely used in cognitive science. They refer in general to the cognitive constitution of meaning, and claim that an individual tries to create

---

[25]See Toulmin (1969, p. 13). Initially he even defined "field-dependency" by reference to "logical types."

a meaning for a situation according to prior experiences that are recalled and considered as similar.

> Framing provides a means of "constructing" a world, of characterizing its flow, of segmenting events within this world, and so on. If we were not able to do such framing, we would be lost in a murk of chaotic experience and probably would not have survived as a species in any case. (Bruner, 1990, p. 56; see also Müller, 1984, chap. 2.2)

By introducing Goffman's frame concept in order to explore more detailed distinctions in the concept of the field of an argument, two essential new facets take shape:

- The perspective has been turned from the rather macrosociological approach of Toulmin to the microsociological approaches of ethnomethodology and interactionism.
- The social phenomenon of fields related to institutions has been converted into a rather individual-related concept, that of framing.

Whereas the first point is somehow comprehensible as a goal of doing a more detailed analysis, the second issue, in contrast, demands more explication.

From the interactionistic point of view this individual construction of one's own world (see Bruner, 1990) happens in consonance[26] with the constructing activities of other individuals by the social means of negotiation. Thus, framing is not only an individual routinization of sense making, but also a social standardization of these individuals' processes by which a taken-as-commonly-shared reality emerges. This is also more or less emphasized in more psychological approaches.[27]

In Goffman's conceptualization of framing, the social aspects of this process are broadly developed (cf. Krummheuer, 1992) and, therefore, appropriately applicable for the purpose of differentiating Toulmin's field concept. Framing within this sociological context refers to Schütz's ideas of the "natural attitude" (*die natürliche Einstellung*) and the "unquestioned given" (*das fraglos Gegebene;* Schütz & Luckmann, 1979). After becoming accustomed to a certain kind of framing, the strip of reality interpreted accordingly appears for the individual as natural, evident, somehow logical. The production and the impact of an argument, as well, have to be judged in the light of framing processes, which are the predominant and embracing processes of sense making. This means arguing is framing-dependent.

Episode 5, January 26—Introductory Phase. In Episode 1 two small solutions were presented to illustrate proposals for the solution of the problem, "How many do you have to add to III::: [36] to make IIII::. [53]?" Carol offered, "Add

---

[26]The word *consonance* is often used synonymously with *agreement.* Here, the rather music-related connotation is meant: "a pleasing combination of sounds simultaneously produced" (see Webster's Dictionary).

[27]See, for example, Schank (1976): "It is impossible to make sense of new inputs without some sense of the place of these inputs in the world" (p. 167). Also interesting here is Bruner's reception of Bartlett's scheme concept (see Bruner, 1990).

7. THE ETHNOGRAPHY OF ARGUMENTATION 251

2 ten-bars and take away 3." Katy stated, "I add 17." After this, Jamie presented the following argumentation:

50 Jamie: That there was 36.
51 Teacher: OK, that's very important (writes 36 down next to strips and squares). To turn that picture into a number. OK. So we have 36.
52 Jamie: And I agree with all of them.
53 Teacher: You agree with all of them, uh? OK. What's this number down here if we had to write that in numbers (writes 53 next to strips and squares).
54 Jamie: And just like I put 20 take away 3 more is 17.
55 Teacher: Say that again. Did you hear what he said? (Looks at class) Say that again, Jamie.
56 Jamie: 20 just like we had these 3 and these 2 together. That would be 20 altogether. Take away 3 would be 17.

The interpretation of this solution affords special attention. Jamie started with "that there was 36." The teacher agreed with that very strongly and wrote the number 36 under the strips and squares picture III::: [36]. This can be seen as a confirmation of the datum, that there is a picture, which is to be conceived of as the number 36. Here one can see that the way a picture is to be taken as data for an argument can be disputed: Does one have to pretend that the picture is a numeral, such as when the teacher said, "Turn the picture into a number" (Line 51), or is it also possible to take the picture purely as an image?

Jamie's approach, as understood here, was as follows: He first compared the images of the ten-bars and recognized a difference of 2 ten-bars between III::: and IIIII:. ("These 3 and these 2 together"). Then he compared the single squares, where he realized 3 more in III::: (see his statement in Line 54: "I put 20 take away 3 more is 17"). This leads him to the conclusion that 17 is the result.

His solution was still image dependent, although it contained numbers: He deduced the number 20 from the comparison of the ten-bars of the images, but at the same time he offered a numerical subtraction "20 take away 3 is 17." In this sense, his solution was a combination of Carol's image-dependent and Katy's image-independent argumentations. This can be seen as a switch between different framings: One part of the argument, which dealt with the tens or the ten-bars, seemed to be iconically or image-dependently framed, whereas the other part, which was concerned with the ones or the single squares, seemed to be numerically or image-independently framed.[28]

In general, the field-dependency of an argument is, as already mentioned, primarily connected with the backing of a warrant. The core of an argument can be validly presented despite different possible framings.[29] The warrant is just the

---

[28] In this way he restricted the numerical facts of addition and subtraction to the number range from 0 to 20, which usually can be seen as very fundamental knowledge for second graders.

[29] Basically, this led Jamie to his framing-switching argument in this episode.

explication of appropriately applied rules of inferences. Basically, this is a framing-independent action. As Toulmin (1969) said, warranting the step "D, so C" leads generally to a "formal" validation (p. 119) in the sense that the performing of the concrete inference is legitimized by the outlined principles of the warrant. Conversely, the backing of a warrant refers to categorical statements, which now can be described more precisely as framing-related convictions and certainties: "An argument expressed in the form 'Datum; warrant; so conclusion' can be set out in a formally valid manner, regardless of the field to which it belongs; but this could never be done, . . . for arguments of the form 'Datum; backing for warrant; so conclusion'" (p. 123).

The Structuring of Argumentatively Accomplished Agreements. As already outlined, during an argumentation process the framings of participating individuals do not necessarily have to be the same or nearly the same. From an interactional point of view, they only have to fit insofar as the running framing processes of each individual do not lead to framing-immanent contradictions, emotional rejections, cognitive oversophistications, or cognitive oversimplifications for the individual.

Interactionally, this fit is the essential characteristic of a consensus emerging through interaction. With reference to Goffman (1959), this is called a *working consensus:* "We have then a kind of interactional *modus vivendi*. Together the participants contribute to a single over-all definition of the situation which involves not so much a real agreement as to what exists but rather a real agreement as to whose claims concerning what issues will be temporarily honored" (p. 9).

As the elaboration of the example in the last section already showed, the core of an argument can appear to be acceptable under different framings. The different framings, especially between the teacher and the students, seems to be a typical occurrence in teaching–learning situations (see Krummheuer, 1982, 1983, 1992). A working consensus under the condition of a framing difference in mathematics classrooms can be called a *working interim.* Obviously, they are very fragile and transient, but their persistent accomplishment seems to be a crucial part in the process of learning (see Voigt, chap. 5, this volume).

The Finite Regress of Arguing. By their specific way of framing, participants of a social group constitute a central element of their culture (Goffman, 1974). They are part of the *Lebenswelt* with its—as unquestionably given and naturally seen—plausibilities. Argumentations are founded in these certainties. In particular, the backing is bound directly to them.

Now recourse can be taken to the earlier mentioned problem of an infinite regress of securing doubted data by running a new argumentation. Pragmatically, this regress stops when data can be sensibly framed by all participants. This happens when they frame the situation in a consonant way or when they refer back to the embracing plausibilities of their common *Lebenswelt.* Framings function as resources of sense-making processes and have an implemented horizon of self-evidences. *Lebenswelt,* as the ultimate resource of sense making, cannot as such be

7. THE ETHNOGRAPHY OF ARGUMENTATION

challenged. As an unthematic and unproblematic horizon, it always sets the claims in which a question can arise. As Wittgenstein (1974) phrased it: *"Wer an allem zweifeln wollte, der würde auch nicht zum Zweifeln kommen. Das Spiel des Zweifelns selbst setzt schon Gewißheit voraus"* (Nr. 115).[30]

The Format of Argumentation

Definition of the Concept. Coming back to the functional role of argumentations within an interaction, it is essential to see the crucial role they play for the accomplishment of a working consensus when this process is assumed to be rational. Argumentation had been introduced as the communicative techniques or methods used to come to such an agreement. The way in which such an agreement can be interactively accomplished must now be stated more precisely. From the newly developed perspective, this can only mean that the emerging, working consensus is based on the construction of an argument that is acceptable to all participants.

The minimal (i.e., no more reducible) request is the conjoint acceptance of the core of the argument. This leads, as outlined earlier, to a formally valid argument.[31] The unspecified backings, then, make sense in accordance with all actualized framings. This condition of the acceptance of the core of an argument actually appears to be the maximal one, when the consensus is of a working interim type. Due to the underlying framing differences, the possible attachable backings do not make sense to all actualized framings but only to a single one. The status of the concept of format of argumentation, which will unfold in the following, is concerned with this problematic situation of accomplishing a conjointly acceptable argumentation under the condition of a framing difference.

This concept is tailored to the analysis of interaction processes in primary mathematics classes. Its full poignancy is, therefore, only recognizable if some, as typically conceived, empirical peculiarities of the interaction processes are taken into account. We have to consider that classroom interaction is:

- Mostly organized as collective argumentation.
- Typically accomplishing a working interim.
- Generally proceeding along specific patterns of interaction.

The concept, now, is supposed to take this interactional atmosphere into account when describing the accomplishment of commonly to-be-shared argumentations. A format of argumentation is a patterned interaction in which the core of an argument emerges.

One can easily see how the three conditions are involved in this definition. Because this concept:

---

[30]"If you tried to doubt everything you would not get as far as doubting anything. The game of doubting itself presupposes certainty."

[31]This is a sequence of utterances functionally containing the structure "D, since W so C."

- Describes an interaction, the performed argument is based on a collective argumentation.
- Refers only to the core of an argument that is compatible with the argumentative potential of a working interim.
- Is bound to patterned interaction and hence, by definition, the third aspect is also taken into account.

Implications for Learning. This section is concerned with the theoretical implications for learning. The concept of format, which is referred to in an earlier definition, was introduced by Bruner (1983) in his later studies of language acquisition. His approach is bound to the fundamental insight that learning a native language is inextricably interwoven with learning its situations of application. A child, for example, does not learn the words in order to request a cookie; he or she primarily learns the social event of a request, where these words fit in. According to Bruner, "in order for the young child to be clued into language, he must first enter into social relationships of a kind that function in the manner consonant with the uses of language in discourse" (Bruner, 1985, p. 39). For Bruner, this coupling of learning something and interacting in situations that are suitable to the learning object is not restricted to the acquisition of one's native language.

Bruner characterized the social relationship as a format. Formally this is a patterned interaction, usually containing an adult in the role of a teacher and children in the role of learners, that functions as a principal vehicle of the Language Acquisition Support System.[32] But the importance of this concept for theoretical implications of learning arises rather by extending this formal definition and hereby making it not only applicable for the acquisition of language. A format is:

- Contingent in the sense that the interactional moves of the participants depend on each other.
- An instance of a plot in the sense that the goals of the participants need not be the same.
- Developing or moving toward coordination of the goals, not only in the sense of agreement but also in the sense of division of labor.
- Turning after a process of routinization into something seen as having exteriority and constraints.
- Asymmetrical with respect of the consciousness of the members about the issues going on.[33]

The attempt to specify these extensions to the concept of format of argumentation makes some of them become more lucid and important, whereas requiring others to be modified.

---

[32]See Bruner (1982, 1983, 1985). He stated: "A format is a standardized, initially microcosmic interaction pattern between an adult and an infant that contains demarcated roles that eventually become reversible" (1983, p. 120).

[33]For further elaboration, see Bruner (1985, p. 8).

7. THE ETHNOGRAPHY OF ARGUMENTATION

The matter of contingency becomes a very crucial point, because several functional aspects have to be considered for the development of an argumentation. Formatted implementations of an argumentation are used to order the utterances in a way that the relevant three topics—data, warrants, and conclusion—appear in a fairly comprehensible way.

A formatted argumentation appears like a plot, because the participants do not necessarily have to understand each move congruently. Due to the ascribed framing difference, the participants have different understandings of the argument. Cooperation based on a working interim depends on trust in the seriousness of the participants' contributions and the confidence that there is a prospect of getting along fairly. This is the main idea of describing a format as a plot.

The contributions of the participants who are replaying a given format may change, as was evident in the project classroom. The more the participants get used to them, the more they take over moves that originally had been performed by others. This seems to be the impact of the individuals' thinking or reflection about those formatted experiences. Through this they cognitively construct more general types of arguments that are at hand for them as if they had a kind of an objective status. The formatted arguments are an experiential basis for the individual's cognitive construction of a topos, a central idea that will be formulated in the following section.

Often, formatted argumentations are generated by participants of qualitatively different mathematical "capacities," such as when the teacher is involved. In this case the teacher very often proposes a guideline for the production of appropriate arguments. This generally conforms to his or her framing and, thus, one can usually describe him or her as more competent (cf. Wood, chap. 6, this volume). But this asymmetrical pattern is not necessary for the accomplishment of a formatted argumentation. Additionally, when students work together on several problems that they solve in a similar way, a format of argumentation can emerge with the described impact. It is a matter of empirical reconstruction to clarify the importance of symmetry or asymmetry in conjunction with the concept of format. Bruner's empirical data is gathered in the mother–child interaction as a basis for language acquisition in early childhood. Here the asymmetry is obvious. With regard to the numerous occasions of small-group work in the project class, the issue of symmetry becomes more noteworthy.

Summarizing, the concept of format of argumentation within a theory of learning mathematics is concerned with the conceptualization of the suitability of socially accomplished regularities for cognitive development. In addition to the notion that social processes accompanying the individual cognitive processes are patterned and hence somehow recognizable and easily memorizable for the individual, the more important issue is that the argumentatively accomplishable agreements can only be reached in a formal way. The format of argumentation is related to the core of an argument, and this means that the crucial differences in the underlying framing remain tacit. The development and restructuring of a framing, which are seen here as the impact of a successful cognitive learning process, are still the purely individual achievements of sense making. (The kind of relationship between a format of argumentation and the individual's learning is discussed after the following example.)

Episode 6, January 21—Whole-Class Discussion. In order to exemplify the concept of format of argumentation, a larger part from a whole-class discussion is presented and analyzed. The worksheet shown in Fig. 7.10 is on the floor.

For the first problem, Katy proposed the solution:

1 Katy: I knew 4 and 4 is 8 and 2 more fours and all that together makes 16.
2 Teacher: Exactly. She says I knew 4 and 4 is 8, 4 and 4 is 8 and 8 and 8 are 16. (Points to screen.) Is there anybody that did it a different way? Totally different than the two we just talked about. Michael and Holly, did you come up with something?
3 Michael: Well 8 and 8, well [sound out] we added those 2 fours together and that made 8 and 8 equals 16.
4 Teacher: Okay, so that's the same way Katy had done.

Clearly one can see here how Katy accomplished an argumentation in order to secure the assertion that the answer to the first problem was 16. First she referred to the data that "4 and 4 is 8" and that there were 2 more fours. Then she concluded that, "all that together makes 16" (Line 1). The teacher rephrased her argumentation by demonstrating in more detail why this inference was legitimate: She seemed to show more clearly where the two different sums of 4 + 4 came from and that one had to add them together. Insofar, her rephrasing contained auxiliary utterances that functioned as warrants. Thus in sum, the core of an argument had been collectively outlined.

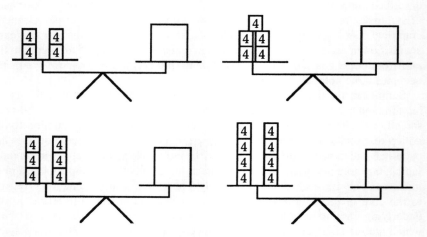

FIG. 7.10. Balance worksheet for whole-class discussion.

7. THE ETHNOGRAPHY OF ARGUMENTATION 257

For the second problem, 4, 4, 4, 4, 4/__, she asked for a special kind of argumentation: As we see, she hereby referred to a specific format of argumentation:

10 Teacher: How many of you used the first one to get the second one?

Peter developed an explanation that seemed to include a more global view than the teacher was asking for, but by this it elucidated the relevant format very well:

20 Peter: Well, all the problems you just add more fours. They just put more fours than the first problem.
21 Teacher: All right.
22 Peter: So it must be 4 more.
23 Teacher: What would that be?
24 Peter: 16 plus 4 equals 20.
25 Teacher: All right . . .

Peter's utterances are interpreted here in the way that he saw that all the given problems varied from the first one by more blocks of fours. In Lines 22 and 24 he explained how this insight could be used to solve the second problem: "So it must be 4 more . . . 16 plus 4 equals 20." The rather sophisticated use of mathematical phrases is striking at the first glance. He began with a formulation that denoted an inference, "So it must be" and he continued with a mathematically exact sentence "16 plus 4 equals 20."

As the processing of the following problems shows, this mathematical enunciation did not become standard, but the entailed format of argumentation, which laid out the reference to a prior problem, was established fairly well.

Holly and Katy presented an argumentation for the third task, which is on the bottom left of the worksheet.

30 Holly: We aah . . . we just aah, could be added all like we started on the first one and then we added the 4 and then we go on and added the 4 and then we go over here and added the 4.
31 Teacher: All right, what did you get for a solution to this one?
32 Katy: Oh . . . we got 30, no, 24.
33 Teacher: OK. 24! Any problems with this? Everyone agrees? (Looks around at class.) OK!

Here, one sees how Holly and Katy together presented an argument according to the outlined format of argumentation. Obviously this format was well established in this class. After this, Linda presented a solution for the problem 4, 4, 4, 4, 4, 4, 4, 4/__ on the screen of the overhead projector in front of the classroom:

40 Linda: We knew that this 4 here was 16.
41 Teacher: . . . now listen to what Linda has to say. This is how they came up with the solution.

42	Linda:	We used this for the 16 and ... (points to last problem on the screen).
43	Teacher:	OK.
44	Linda:	... and this was 16 and we knew that 16 and 16 is 32.
45	Teacher:	All right. They saw this as one set (draws a ring around fours on overhead projector) and then this set of fours has another 4 fours and they said that's 16 and that's 16 and they said 16 and 16 is 32. All right! That was a good solution! Who has a different way of doing it?

Linda identified in the balance picture two blocks of four boxes of fours (Line 40). The result of one of these blocks was already known to be 16, and the number fact $16 + 16 = 32$ was also well known (Line 44). The teacher rephrased this solution by pointing out on the screen how the 8 boxes of fours were divided in 2 sets of 4 boxes. After this, she declared the presented solution to be a good one (Line 45). Here, again, the same format of argumentation had been accomplished. The given image of boxes of numbers was separated in subsets of already known and/or previously solved problems, the results of which then had to be added (cf. Voigt's discussion of thematic patterns of interaction, chap. 5, this volume).

Topos and Format

This section deals with the relationship of the individual's reliance on an argumentation and his or her avail of the interactional accomplishment of this argumentation. Because of the framing dependency, the conjointly accomplished argument might be assessed differently by the contributing individuals. The argument, or more precisely the core of an argument, does not only mean something different to each participant with regard to each of their actualized framings; it is also, therefore, differentially important and convincing for each of them.[34] Thus, the way the individual attempts to influence the emergence of an argument is, at the same time, the way he or she lays out his or her confidence in this argument. This reformulation of the notion of reflexivity[35] is utilized here with regard to the clarification of the relation between the sociological and psychological dimensions of learning.

The development of an argumentation is intended to ensure the claimed validity of a statement. One can perform this by trying to adopt an argumentation that was already successful in prior cases. This means that one wants a concrete argumentation in which the participants individually recognize structures of more general argumentative accounting practices. Cognitively, that means that each individual constructs a structural similarity among several argumentations experienced in different situations. In rhetoric, this abstract pattern or general form of similarly

---

[34] It should be recalled here that the issue is substantial argumentation, as described in the section entitled "The Differentiation Between 'Analytic' and 'Substantial' Argumentation."

[35] For the notion of reflexivity, see the earlier section entitled "The Reflexivity of Interactional Activities."

7. THE ETHNOGRAPHY OF ARGUMENTATION 259

structured arguments is called a *topos* or *locus*.[36] "Loci are headings under which arguments can be classified. They are associated with a concern to help a speaker's inventive efforts and involve the grouping of relevant material, so that it can be easily found again when required" (Perelman & Olbrechts-Tyteca, 1969, p. 83).

This concept is used here as one that is bound to the individual. It is the individual's inventive efforts and his or her achievement to group relevant material that, in the terminology of this chapter, is the reconstruction of several argumentations that are experienced as similarly structured ones. Here, it is assumed that the emergence of several similarly formatted argumentations will stimulate this cognitive process.

Now one can describe the relationship between the contribution of an individual to the accomplishment of an argumentation and his or her cognitive development as a reflexive one between topos and format: On the one hand, the child tries to contribute argumentatively in a social interaction by offering utterances that are sensible for him or her with regard to a specific available topos and may possibly lead to the collective emergence of a formatted argumentation. On the other hand, it is this participation in a sample of similarly formatted argumentations that leads him or her to construct cognitively a topos.

Two episodes might help clarify this reflexive relationship. The first example illustrates how two children know a specific format of argumentation but cannot cognitively make use of it. This means that for this format no individual topos is available. In the second example, a student recognizes two different argumentations as similarly formatted and obviously seems to be able to apply it as a topos.

Episode 7, January 22—Small-Group Work. An adult spoke to Andrea and Andy just in passing. The two children were working on the problem __/7, 7, 7. Prior to this episode, they had solved the following problems:

__/11, 11
__/11, 11, 11
__/11, 11, 11, 11
__/11, 11, 11, 11, 11
__/6, 6, 6, and now
__/7, 7, 7.[37]

7 Researcher: That's one of interest. Now you've got this one: 6, 6, and 6, is 18. Can this help you do this next one?
8 Andrea: Yeah . . .
9 Researcher: How?
10 Andrea: Well, 6 is 1 lower than 7, and 7 is 1 higher than 6.

---

[36]In this chapter, the word *topos* is preferred. The alternative preference for *locus* in some of the following quotations is not changed.

[37]This episode is a direct continuation of Episode 3 in the section entitled "The Reflexivity of Interactional Activities."

11	Researcher:	So, how would you . . . how would you use this to help you figure this one out?
12	Andrea:	Nope.
13	Researcher:	Nope? Did you do it, Andy? Or is that too tough?
14	Andy:	It could help, but I don't know . . .
15	Andrea:	Not that much.
16	Andy:	. . . how. It could. I know it can help.
17	Researcher:	OK. Let's try the next one.

The researcher appeared and asked whether the solution of the problem before could help solve this one (Line 7). Andrea answered spontaneously "Yeah" (Line 8) and explained in more detail (Line 10) "Well, 6 is 1 lower than 7, and 7 is 1 higher than 6." In response to the researcher's additional question asking whether this could help to figure out the problem (Line 11), Andrea answered, "Nope" (Line 12). Additionally, Andy could not give any constructive hint as to how this could help, although he also seemed convinced that this would be a possible way of solving (Lines 14 and 16).

It might be that Andy knew about these kinds of argumentation. Possibly, he learned about them in prior whole-class discussions. Also, Andrea's spontaneous answer pointed in the same direction. It seemed that they knew about this kind of argumentation but that they were cognitively not ready to apply a topos, which could have helped them to develop a corresponding argumentation. The children's recognition and belief that there existed a useful and forceful way of argumentizing by referring to previous problems signified experiences with argumentations formatted in this way. Cognitively, it appeared that neither of them had made much sense out of this mode of argumentation. Presumably, they had yet to develop a topos that could be seen as based on this experience.

Episode 8, February 18—Small-Group Work. Jack and Jamie worked on the multiplication task 8×4. Before they had solved $9 \times 4 = 36$ and $10 \times 4 = 40$.

1	Jack:	Look! Look! Just take 4 from that (points to previous problem 9×4). Just take 4 away from that (points to previous problem 9×4) to get that (points to problem 8×4). See! 10 add 4, then add 4 to that.
2	Jamie:	Just take 2 from there (points to previous problem 10×4). Take 8 away from there (points to problem 10×4).
3	Jack:	No. Take away, um, 4 from there (points to 9×4).
4	Jamie:	8 from there (points to 10×4). That makes 32.
5	Researcher:	How did you get 32?
6	Jack:	[Almost inaudible] Take away 4.
7	Jamie:	Just take away 8 from that (points to 10×4).
8	Researcher:	From which one?
9	Jamie:	This one (points to 10×4), 10 plus, take away, 10.
10	Jack:	Times.

11	Jamie:	Times 4.
12	Researcher:	And how did you do it, Jack? Did you do it the same way?
13	Jack:	Yeah, same way.
14	Researcher:	OK.
15	Jack:	But I used that and (points to 9 × 4). Take away 4. It makes 32.

Although the two boys at the beginning strongly disagreed with each other's solution proposal, at the end Jack called his approach the "same way" (Line 13). Jack proposed (Line 1) to refer to the previous problem (9 × 4) and to subtract 4 from its result. Jamie suggested to go back to another previous problem (namely 10 × 4), and to subtract 8 from that (Line 2). In Lines 3 and 4 both children insisted on their idea, and Jamie attached the result 32. It seemed at this moment as if the two boys assumed they were working on different kinds of argumentations.

At this point, the adult researcher intervened and asked for their different approaches (Line 10). Jamie reports to him (Lines 11–16). The researcher then asked Jack whether he thought that he proposed an argument in the "same way" (Line 13). Jack agreed but added that he referred to another previous task (Line 15). Cognitively, he seemed to recognize a similarity between these two approaches. This meant that he had cognitively constructed or applied a topos with regard to this similarly formatted argumentation (see also Krummheuer & Yackel, 1990).

Learning and Argumentizing in the Mathematics Classroom

One assumption of folk psychology suggests that the argumentative presentation of mathematics allows children to learn more easily and more reflectively. Starting with this assumption, the previously stated theoretical elaborations assert that, especially under the condition of a framing difference, which is seen as constitutive for teaching–learning situations, the argumentatively generated collaboration in mathematics classes is based on an agreement about the core of an argument rather than on the framing-dependent backing. Certainly conceptual mathematical learning also has to do with the changing or restructuring of the actualized framing, but the functional category of an argument, which directly is related to the individual's framing, does not seem to be the interactive focus of collective argumentation.

At first glance, this position might be seen as a claim that learning mathematics happens by chance. A closer look suggests that one has to give up the idea that learning mathematics can take place in a "direct way." As Bauersfeld (chap. 8, this volume) outlines, learning happens in an indirect way by participation in an appropriately developed classroom culture. The less direct access an individual has to the socially constituted reality, the less one has direct access to the cognitive structure of its participants through social interaction. This is merely the consequence of the reflexive relation between the individual and the social perspectives.

Although argumentation motivates students to reconstruct their framings, there is usually no way to force cognitive reconstruction by argumentative necessity. This again elucidates the need for substantial arguments. One argument is not as convincing as another one for an individual, and one can assume that cognitive development would be more stimulated when the accomplished and similarly formatted argumentations have a strong, convincing force for the student. In such a case the obvious motive for the individual is to cognitively construct new topoi. A clarification of the concept of the force of an argument might be helpful at this point.

The force of an argument refers to Toulmin's functional categories: The core of an argument must be soundly developed if it is to be strongly convincing. The participants agree with the concrete inference according to the outlined warrant taken as presented. In other words, the warrant explains the soundness of the inferential step. This aspect will be called "explanatory relevance" (see Krummheuer, 1992; Miller, 1986). It is, as already mentioned, framing-independent. For the involved individual, cognitive capacity and/or emotional eagerness determine whether this inference is comprehensible to him or her or not. But, additionally, cognitively understanding the explanatory relevance of an argumentation does not imply that the individual finds this argument totally convincing. Its internal soundness cannot be doubted, but it does not count for a single individual if it does not sufficiently clarify the relevance according to his or her emerged framing.[38] This "framing specific relevance" refers to the attached framing-dependent backing of the core of an argument.

Obviously, in mathematics classes the accomplishment of acceptance of an argumentation with regard to the explanatory relevance already seems to be a major topic of their regular interaction. The demonstration of the soundness of an argumentation, especially in arithmetic, makes great cognitive demands on the students with regard to both its convincing presentation as well as the individual comprehension. Moreover, under the condition of an emerging working interim, this seems very often to be the maximum that is attainable.[39]

The assessment of the argument according to its framing-specific relevance generally differs between the participants. Their articulation is not impossible, but the ambition to expand the agreement toward this aspect of an argument endangers the accomplished working interim. It seems that what can be done in mathematics classes, typically when establishing a working interim by argumentative means, is to secure the explanatory relevance of an argumentation. Thus, under these classroom conditions the agreement about a concrete argumentation is an agreement about its formal validity.

In a certain sense the emphasis on formally valid agreements reflects on the level of classroom interaction that, from a broader viewpoint, could be seen as the basis for human cooperation in postmodern times in general. Collective argumentation can be seen as a paradigm for the search for truth. Further, rational negotiation

---

[38] Here, Wittgenstein's quote from the beginning of this chapter comes to the fore. With regard to the child's framing it might be difficult, sometimes even too difficult, for him or her to comprehend the system of statements forming the core of an argument. But in case this core is interactionally ratified as formally valid, then "what is a telling ground for something is not anything I decide" (Wittgenstein, 1974, no. 271).

[39] Episode 6 in the section entitled "An Example" might be taken as an example of this.

appears as the locus of truth. In times in which ultimate certainties and authorities are lost, the concept of truth is bound to the argumentatively ratified consensus between interacting subjects, which necessarily can only be a formally validated one and, thus, only a procedural one (Habermas, 1985; Toulmin, 1972).[40]

The interactional restriction of collective argumentation in mathematics classes turns one's view toward the fact that in such teaching–learning situations communicative problems exist that cannot easily be coped with by argumentative means. The discussion about the legitimacy or appropriateness of actualized framings is beyond the potentiality of argumentative resolution. Of course, formatted argumentations do not exclude the individual framings, but they do not emerge thematically in the argument.[41]

The skillful presentation of different solutions by students in the observed project class was one of the most overwhelming impressions. It gave much food for thought and, among others, resulted in this chapter about argumentation. Pragmatically, the variety of arguments generated for one task seems to enable the class to cope with the tension of different framings. On the one hand, the option for arguing under different framings was given by the request for developing different solutions; on the other hand, the competition for different solutions involves, at least implicitly, the comparison of different framings.

The difference in underlying framings could have been pointed out more frequently and more poignantly in the interaction of the project class, although doing this would have endangered the working interim that emerged. In such situations, the participants and, here especially the teacher, should try to push the communication as close as possible toward this point of breakdown, thus enabling a change in the individuals' framings while making the variance of framing explicit. As pointed out by Bauersfeld (chap. 8, this volume), languaging relies on everyday language description, metaphorizing, and modeling rather than on the schematized use of particles of mathematical terminology. The attempt to foster the generation of differently formatted and competing argumentations as well as their commenting and comparison by means of everyday languaging is necessary for the support of conceptual mathematical learning.

Summary: The Relationship Between Argumentizing and Learning

This chapter can be understood as an attempt to accomplish an argument about the relationship between argumentizing and learning in mathematics classes. The

---

[40] See also Bauersfeld's elaboration about the "reality illusion," chap. 8, this volume.

[41] The agreement about the core of an argument was said to be the formal validation (see the sections entitled, "The Creation of an Inference" and "Implications for Learning"). The emphasis on formal validation might also be one reason why students' mathematical knowledge is often rather focused on calculational and algorithmic aspects and less focused on conceptual reflection. As my own studies on secondary level show, generally students' activities are based on framings with such an "algorithmic-mechanical" character (see Bauersfeld, Krummheuer, & Voigt, 1985; Krummheuer, 1983, 1989, 1991, 1992; Krummheuer & Yackel, 1990 ; and also Voigt's concept of thematic pattern of direct mathematization, chap. 5, this volume).

notion of argumentation appears, then, on two levels: First, it has been used for the description of observed collective activities in the project class (argumentizing); second, this concept can be used as a means for characterizing the relationship between these collective activities and learning mathematics. In the next section the results of this exploration at this second level are outlined. Toulmin's layout, which was introduced for the analysis of the functional aspect of argumentation according to the first level, is reapplied, but this time for the presentation of these results. In conclusion, some limitations of this approach as well as some implications for the teaching and learning of mathematics are discussed.

Arguing and Learning. According to Toulmin's functional analysis, the accomplished argument is relatively complex: The whole argumentation can be divided into one main argument (which is based on two parts of data) and two subordinate arguments (which support the appropriateness of these data). For elucidation, the main argument is outlined here first, followed by the presentation of the two subordinate arguments. The whole argumentation is also presented pictorially, according to Toulmin's scheme. It might be helpful to refer to the layout shown in Fig. 7.11 while reading these summarizing comments.

The main statement is that successful, conceptual, mathematical learning is reflexively based on the active participation in the formatted accomplishment of the core of an argument that is acceptable for all participants.[42] This relationship is secured by presenting it as the conclusion of two basic insights (data) about the character of argumentation in mathematics classes.

1. Generally, in the interaction of mathematics classes, the main focus of an argumentation is on the accomplishment of an acceptable core of an argument.[43]
2. The students contribute to the accomplishment of a collective argumentation according to their available topoi of arguments and vice versa: The participation in such collective argumentation fosters the students to develop novel topoi.[44]

The appropriateness of these two issues is clarified in more detail later. First, it is taken as data of the main argument.

The inference from these data to the main assertion is legitimized by the postulate that through interaction there is no direct access that influences the individual framing-dependent aspects of an argument like the backing. However, the development of a new topos is embedded in the individual's construction of a new framing. This strict emphasis on the individual cognitive construction of his or her own reality is a basic assumption of any constructivist theoretical approach.[45]

Taken together, these aspects represent the main argument. Its pictorial layout (Fig. 7.11) shows the functional relation of the different statements once again. Still

---

[42] See the section entitled "Definition of the Concept."
[43] See the section entitled "The Structuring of Argumentatively Accomplished Agreements."
[44] See the section entitled "Topos and Format."
[45] See the sections entitled "Implications for Learning" and "Learning and Argumentation."

7. THE ETHNOGRAPHY OF ARGUMENTATION

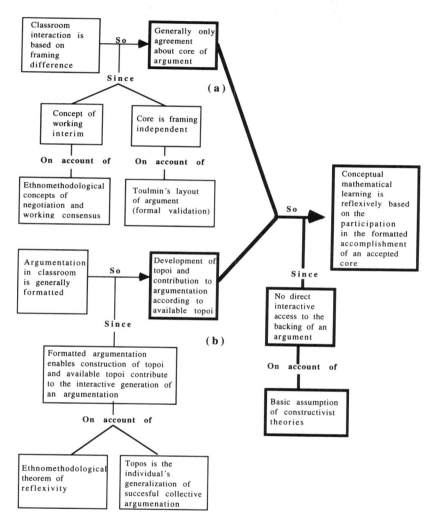

FIG. 7.11. Schematic representation of an argument.

open is the certification of the appropriateness of the two Data (a) and (b). This is clarified in the following.

The argumentative support for Datum (a) is that generally in mathematics classes an agreement is only accomplishable with regard to the core of an argument.[46] This has to do with the constitutive fact that mathematics class interaction is characterized by a qualitative framing difference, especially between the framing

---

[46]See the section entitled "Warrants of Inferences."

of the teacher and those of the students.[47] The inferential step from this empirical fact to Datum (a), which functions in this subordinate argument as a conclusion, is supported by two different conditions:

- It is possible to interact on the basis of framing differences. The kind of emerging agreement can be described as a working interim. This stresses the point that such agreements tend to be fragile and transient. This consideration is based on fundamental concepts of ethnomethodology and interactionism such as working consensus and negotiation of meaning.[48]
- The functional analysis of argument by Toulmin reveals a certain part of an argumentation, which is framing-independent, the so-called core. This leads to a formal validation of an argument and keeps the framing-specific relevance of an argument apart from the emerging agreement.[49]

The datum of the main argument can be seen as the conclusion of the fact that interaction in mathematics classes is often patterned in a way that can be described according to Bruner's concept of format. This kind of interaction helps the students to recognize similarly structured argumentations and to develop a topos for them. Topos is seen here as the individual's cognitive construction of a generalizing category of similarly structured and interactively successful arguments.[50] But, on the other hand, the availability of certain topoi enables these students at the same time to contribute to the accomplishment of an argumentation. Further, the mutually dependent relationship between the construction of a novel topos and the contribution according to already available topoi refers to the fundamental ethnomethodological position of reflexivity.[51]

Finally, looking at the whole argument one can identify two initiatory presuppositions (see the left margin of Fig. 7.11), namely that classroom interaction is based on framing differences and that argumentation processes in mathematics classes usually can be reconstructed as formatted ones. These statements function here as data of the two subordinate arguments. Additionally, one can distinguish several theoretical basis positions (bottom line of Fig. 7.11) on which the whole argument is founded. These are the:

- Ethnomethodological notion of reflexivity.
- The notions of argumentation and topos.
- The constructivist notion of the individual's construction of reality.

Altogether, then, it is apparent that access to the individual's cognition through interaction is first indirect and second reflexively bound to the contribution of these

---

[47]See the section entitled "Toulmin's Concept of 'Field' and the Concept of 'Framing.'"
[48]See the section entitled "The Structuring of Argumentatively Accomplished Agreements."
[49]See the section entitled "Toulmin's Concept of 'Field' and the Concept of 'Framing.'"
[50]See the sections entitled "The Format of Argumentation" and "Topos and Format."
[51]See the section entitled "The Reflexivity of Interactional Activities."

7. THE ETHNOGRAPHY OF ARGUMENTATION

cognizing individuals to the emergence of the related interaction.[52] The next section elaborates on this topic in more detail.

Limitations and Implications. It should be noted that the previous elaborated argumentation both relativizes and concretizes the assumption of folk psychology maintaining that argumentizing helps conceptual mathematical learning. Beyond the notion that this relation is a reflexive one, it also had been pointed out that under the framing condition of mathematics class interaction, argumentation is usually focused on the accomplishment of the framing-independent core of an argument. Essentially, this means that argumentatively emerging working interims are a kind of formal validation, checking whether the accomplished inference is acceptable according to the actualized (but not necessarily entirely explicated) and different framings of the participants.

Conceptual mathematical learning, however, has to be seen as the individual's construction of a new framing. There is no direct access by (collective) argumentation or any other kind of interaction to this individual cognitive process. It is the active participation in a more or less evolved culture of argumentizing that can be taken by the student as an experiential basis for his or her cognitive processes of sense making. Thus, learning is indirectly influenced by interaction.

This kind of a culture of argumentation can be characterized on the one hand by regularities, commitments, and obligations for the participants, in order to contribute positively in the reaccomplishment of such regularities in similarly defined situations. Such formats of argumentation help students to recognize different problems as similarly solvable and thus possibly enable a reorganization of their cognitive structure accordingly. On the other hand, this culture should be open to the development and presentation of alternative argumentations that are acknowledged as legitimate, but different formats for collective argumentation (see Voigt, chap. 5, this volume).

This second implication is to be seen as a practicable but also vulnerable way of coping with the interactional problem of collectively arguing without having a commonly shared basis of framings. It is not as much the explicit discussion of the difference of the framings, but rather the implicit and continuous intonation of these differences with the option and/or necessity of metaphorizing and languaging by everyday descriptions (see Bauersfeld, chap. 8, this volume). Thus, conceptual mathematical learning has as much to do with the establishment of different formally valid mathematical argumentations as it is related to a developed, everyday-language-bound platform for indicating the framing-dependent convictions and certainties in each individual's interpretation of the accomplished argument. The danger exists that, in favor of the formally valid demonstration of arguments, the latter aspect is easily neglected. As a consequence, it leads to a relatively mechanical and calculational oriented understanding of mathematics[53]—a consequence that has always been criticized in mathematics education.

---

[52]See the section entitled "Learning and Argumentizing in the Mathematics Classroom."
[53]See footnote 52.

REFERENCES

Attewell, P. (1974). Ethnomethodology since Garfinkel. *Theory and Society, 1,* 179–210.
Bauersfeld, H., Krummheuer, G., & Voigt, J. (1985). Interactional theory of learning and teaching mathematics and related microethnographical studies. In H. G. Steiner & H. Vermandel (Eds.), *Foundations and methodology of the discipline mathematics education (didactics of mathematics)* (pp. 174–188). Antwerp: University of Antwerp.
Billig, M. (1989). *Arguing and thinking. A rhetorical approach to social psychology.* Cambridge, England: Cambridge University Press.
Bruner, J. (1982). The formats of language acquisition. *American Journal of Semiotics, 1,* 1–16.
Bruner, J. (1983). *Child's talk. Learning to use language.* Oxford, England: Oxford University Press.
Bruner, J. (1985). The role of interaction formats in language acquisition. In J. P. Forgas (Ed.), *Language and social situations.* New York: Springer.
Bruner, J. (1990). *Acts of meaning.* Cambridge, MA: Harvard University Press.
Follesdal, D., Walloe, L., & Elster, J. (1986). *Rationale Argumentation. Ein Grundkurs in Argumentations—und Wissenschaftstheorie* [Rational argumantation. An introduction to the theory of argumentation and science]. Berlin: de Gruyter. (Reprinted from *Argumentasjonstoeri og vitenskapsfilosofi,* 1977, Oslo).
Garfinkel, H. (1967). *Studies in ethnomethodology.* Englewood Cliffs, NJ: Prentice-Hall.
Garfinkel, H. (1972). Remarks on ethnomethodology. In J. J. Gumperz & D. Hymes (Eds.), *Directions in sociolinguistics.* New York: Holt, Rinehart & Winston.
Goffman, E. (1959). *The presentation of self in everyday.* New York: Doubleday.
Goffman, E. (1974). *Frame analysis. A chapter on the organization of experience.* Cambridge, MA: Harvard University Press.
Habermas, J. (1985). *Theorie des kommunikativen Handelns* [Theory of communicative acting]. Frankfurt: Suhrkamp, 2 Bde., 3. Aufl.
Inhelder, B., & Piaget, J. (1985). *Growth of logical thinking from childhood to adolescence.* New York: Basic Books.
Keil, F. C. (1979). *Semantic and conceptual development: An ontological perspective.* Cambridge, MA: Harvard University Press.
Keil, F. C. (1983). On the emergence of semantic and conceptual distinctions. *Journal of Experimental Psychology: General, 112,* 357–385.
Klein, W. (1980). Argumentation und Argument [Argumentation and argument]. *Zeitschrift für Literaturwissenschaft und Linguistik, 38/39,* 9–57.
Kopperschmidt, J. (1989). *Methodik der Argumentationsanalyse* [Methodology of the analysis of argumentation]. Stuttgart: Frommann-Holzboog.
Krummheuer, G. (1982). Rahmenanalyse zum Unterricht einer achten Klasse über "Termumformungen" [Frame analysis of an 8th-grade class about "transformation of terms"]. In H. Bauersfeld, H. W. Heymann, G. Krummheuer, J. H. Lorenz, & V. Reiss (Eds.), *Analysen zum Unterrichtshandeln.* Cologne: Aulis.
Krummheuer, G. (1983). Das Arbeitsinterim im Mathematikunterricht [The working interim in mathematics classes]. In H. Bauersfeld, H. Bussmann, G. Krummheuer, J. H. Lorenz, & J. Voigt (Eds.), *Lernen und Lehren von Mathematik. Analysen zum Unterrichtshandeln II.* Cologne: Aulis.
Krummheuer, G. (1989). *Die menschliche Seite am Computer* [The human side on the computer]. Weinheim, Germany: Deutscher Studien Verlag.
Krummheuer, G. (1991). The analysis of social interaction in an "interactive" computer environment. In F. Furinghetti (Ed.), *Proceedings of the Fifteenth PME Conference, II* (pp. 262–268). Assisi, Italy: Program Committee of the 15th PME Conference.
Krummheuer, G. (1992). *Lernen mit "Format." Elemente einer interaktionistischen Lerntheorie. Diskutiert an Beispielen mathematischen Unterrichts* [Learning with "format": Elements of an

interactionist theory of learning. Discussed with examples of mathematical classes]. Weinheim, Germany: Deutscher Studienverlag.

Krummheuer, G., & Yackel, E. (1990). The emergence of mathematical argumentation in the small group interaction of second graders. In G. Booker, P. Cobb, & T. N. de Mendicuti (Eds.), *14th Conference of the International Group for the Psychology of Mathematics Education, III* (pp. 113–120). Oaxtepec, Mexico.

Lehmann, B. E. (1988). *Rationalität im Alltag? Zur Konstitution sinnhaften Handelns in der Perspektive interpretativer Soziologie* [Rationality in everyday affairs? The constitution of sensible acting under the perspective of interspective sociology]. Münster, Germany: Waxmann.

Leiter, K. (1980). *A primer on ethnomethodology.* Oxford, England: Oxford University Press.

Mehan, H., & Wood, H. (1975). *The reality of ethnomethodology.* New York: Wiley.

Miller, M. (1986). *Kollektive Lernprozesse* [Collective processes of learning]. Frankfurt: Suhrkamp.

Miller, M. (1987): Argumentation and cognition. In M. Hickmann (Ed.), *Social and functional approaches to language and thought.* San Diego, CA: Academic.

Müller, K. (1984). *Rahmenanalyse des Dialogs. Aspekte des Sprachverstehens in Alltagssituationen* [Frame analysis of dialogue: Aspects of language comprehension in everyday language]. Tübingen, Germany: Narr.

Paschen, H., & Wigger, L. (1992). *Zur Analyse pädagogischer Argumentationen* [The analysis of pedagogical argumentation]. Weinheim, Germany: Deutscher Studienverlag.

Perelman, C., & Olbrechts-Tyteca, L. (1969). *The new rhetoric. A treatise on argumentation.* Notre Dame, IN: University of Notre Dame Press.

Piaget, J. (1928). *Judgment and reasoning in the child.* London: Routledge & Kegan Paul.

Piaget, J. (1930). *The child's conception of physical causality.* London: Routledge & Kegan Paul.

Piaget, J. (1954). *The construction of reality in the child.* New York: Basic Books.

Saxe, G. B., Guberman, S. R., & Gearhart, M. (1987). *Social processes in early number development. Monographs of the Society for Research in Child Development* (Serial No. 216 Vol. 52, No. 2).

Schank, R. C. (1976). The role of memory in language processing. In C. N. Cofer (Ed.), *The structure of human memory.* San Francisco: Freeman & Company.

Schütz, A., & Luckmann, T. (1979). *Strukturen der Lebenswelt* [Structures of the everyday word]. Frankfurt: Suhrkamp, Bd. 1.

Siegal, M. (1991). *Knowing children. Experiments in conservation and cognition.* Hillsdale, NJ: Lawrence Erlbaum Associates.

Steffe, L. P., von Glasersfeld, E., Richards, J., & Cobb, P. (1983). *Children's counting types: Philosophy, theory, and application.* New York: Praeger Scientific.

Struve, H. (1990). *Grundlagen einer Geometriedidaktik* [Foundations of a didactics of geometry]. Mannheim, Germany: BI Wissenschaftsverlag.

Toulmin, S. (1969). *The uses of argument.* Cambridge, England: Cambridge University Press.

Toulmin, S. (1972). *Human understanding, Volume I, General introduction and Part I: The collective use and evolution of concepts.* Princeton, NJ: Princeton University Press.

Völzing, P. L. (1981). *Kinder argumentieren* [The argumentation of children]. Paderborn: Ferdinand Schönigh.

Wittgenstein, L. (1963). *Tractatus Logico-Philosophicus.* (D. F. Pears & B. F. McGuiners, Trans.) London: Routledge & Kegan Paul.

Wittgenstein, L. (1974). *Über Gewißheit* [On certainty]. Oxford, England: Basil Blackwell.

8

"Language Games" in the Mathematics Classroom: Their Function and Their Effects

Heinrich Bauersfeld
University of Bielefeld

> *Mathematics education begins and proceeds in language, it advances and stumbles because of language, and its outcomes are often assessed in language*
>
> —Durkin and Shire (1991, p. 3)

ABOUT THE THEORETICAL PERSPECTIVE

Communication in classrooms has been investigated from many different perspectives (see overviews in Cazden, 1986, and Erickson, 1986). However, pragmatic studies of the specificity to the discipline are rare, and these few are devoted to the material aspects (Durkin & Shire, 1991; Pimm, 1987) rather than to the functional aspects of language or to patterns of speech production (Krummheuer, 1992; Maier & Voigt, 1991). Bruner (1990) found too much concern in research with what people say compared with what they actually do and he found it "curious that there are so few studies that go in the other direction: How does what one *does* reveal what one thinks or feels or believes?" (p. 17). Admittedly, it is difficult to conduct investigations in the shadow of a discipline undergoing fundamental change like linguistics: "How can one have a scientific study of an ill-defined entity?" (Davis & Taylor, 1990, p. 5).

In particular, we know much from microsociological analyses about the quick-action switches and about the current, instantaneous structures of communication in classrooms from the famous triad of moves "elicitation-reaction-evaluation" (Bellack, Kliebard, Hyman, & Smith, 1966; Mehan & Wood, 1975), to the functions of *indexing, reciprocity,* defining the situation by *framing* (Goffman, 1974; Krummheuer, 1992), *patterns of interaction* (Bauersfeld, 1980; Voigt, 1984), and so on. Little is known about long-term effects: How do the actually functioning structures of the mathematical discourse in the classrooms—as they are identifiable from an observer's perspective—*emerge* or *become constituted* across time (Garfinkel, 1967; Heritage, 1984; Mehan & Wood, 1975), forming the specific history of this classroom with this teacher and these students? What are the effects on the participating individuals' (both students' and teacher's) biographies in general and on their individual mathematizing in particular?

The following reflections are to serve as a critical theoretical orientation for analyses of languaging in the mathematics classroom. They have developed more or less across our discussions in the joint project, but draw as well on our (the German group's) previous analyses and theoretical discussions. A brief summary is given of core convictions from philosophy of the sciences, language, linguistics, sociology, ethnomethodology, and related fields of reflection on languaging. The interest here is with the processes of communicating and their impact on personal development, rather than with language as an object or as a given medium that can be used. We prefer to speak of "languaging" instead of "speech" or "speaking," notions that evoke the connotation of language use. With this orientation, we follow the late Wittgenstein (1974).

Further, we hold the conviction that for different analytical purposes different perspectives and theoretical backgrounds can be useful (see Hörmann, 1977, p. 3, also p. 92 about Carnap). Related problems of compatibility are discussed when necessary. However, the positions of social interactionism and ethnomethodology on one side and the radical constructivist principle on the other side seem to be compatible. The primacy debate about the individual versus the social can be regarded as surpassed in recent linguistic debates. Both perspectives are accepted, necessary,[1] and useful, although they are not simply to integrate. This chapter neither pretends to produce a covering theoretical approach (for constructivists an impossibility anyway), nor does it pretend to aim at an integrated theory.

THE "REALITY ILLUSION" (VON GLASERSFELD): THE "OBJECT GIVEN" VERSUS PERSONALLY CONSTRUCTED AND "TAKEN-AS-SHARED" ISSUES

> Positivist history of culture . . . sees language as gradually shaping itself around the contours of the physical world. (Rorty, 1989, p. 19)

---

[1] Although Lerman (1992) recently stated: "I suggest that the answer may be found in the primacy of the social over the individual" (p. 2-46).

8. "LANGUAGE GAMES" 273

Since Plato, the philosophy of language has followed the conviction that names/words fit in with objects: "Names are natural, not conventional" (Platon, "Kratylos"). Consequently, the signifying or denoting function—names are taken as labels for objects—has dominated over the referring function. Like Plato, the later Aristotle was also convinced of the congruence between reality and human cognition. Not long before Kant, the philosopher Christian Wolff took an early constructivist view, particularly of language: *"Objectum Physici sunt res naturales, quia Physici actiones, quas tanquam Physicus suscipit, in iisdem terminantur, nec ultra eas progrediuntur."* (The physicist's objects are natural things, because the physicist's actions, which he takes as a physicist, are terminated in these actions and do not reach beyond; 1728, § 949, p. 684; author's translation). In other words: Things become "objects" only through the activities of a subject; what else they might be, aside from the subject's activities, we cannot say. In recent words: "The objectivity of objects is based upon the subject who relates his/her activities to them rather than upon the objects themselves" (Picht, 1987, p. 125).

The usual descriptions of "employing a perspective on an object" as well as "using language for describing something" carry two underlying presuppositions:

- Object is taken as something neutrally given or existing, as an entity one can look at "objectively" and one can describe adequately.
- Language is taken as a similar entity like the objects, existing objectively as a means for arbitrary descriptive purposes, which by themselves do not change the object (e.g., adding paint or a wrapper to an object).

From a radical constructivist's position, these presuppositions do not hold. Von Glasersfeld named this core concern of radical constructivist's critique the "reality illusion."

It was Vygotsky who first discussed the developing power of cooperating in a subculture, and the relation between language and thinking. In the years 1931 and 1932 he investigated the influence of collectivism on peasants' thought, together with Luria and colleagues. The final report could not be published because it fell victim to Soviet censorship. Bruner described this:

> The principal finding of their study—suppressed for years, and finally appearing not in the form in which Vygotsky wrote it but only in Luria's book (1979) of many years later—was that participation in an agricultural collective had the effect of promoting growth in the thinking of the peasants involved, which took them from childlike, primitive forms of thinking to adult forms of thought. Collective activity, in a word, led peasants along the way to adult thinking or—better—to socialist adult thinking, which in Vygotsky's account was a more rational, more "scientific" form of thinking. Such was the dogmatic romanticism among the leading figures in the official Russian establishment of the time that they took Vygotsky's conclusion as a criticism of peasants for their pre-scientific thinking—and banned the book. (Bruner, 1984, p. 94)

This in itself is a demonstration of the constructivist principle: The reality is generated in the head and is not an objectively given ontological reality.

Augustine's correspondence theory of language has influenced writing for more than 16 centuries, including the younger Wittgenstein's "Tractatus" (1922/1972). In that, he had claimed an aprioristic isomorphy between the logic of language and the logic of realities, which philosophical analysis can discover:

2.1. We picture facts to ourselves
2.2. A picture has a logicopictorial form in common with what it depicts.
4.01. A proposition is a picture of reality (concerning sentences).

Learning words, therefore, is giving names to objects.

For our purposes, the source of Wittgenstein's fundamental change is of interest. Wittgenstein had passed regular teacher education and taught in different rural schools in Austria from 1920 through 1926. His biographer Fann (1971) supposed: "Without exaggeration we can say that Wittgenstein's early theory of language, as developed in the ivory tower, has been brought back to the firm grounds through his school students" (Fann, p. 50, author's translation). The fundamental revision of his thinking, printed in his "Philosophical Investigations" (1945/1974), has influenced the following generations of philosophers, linguists, and other scientists: "Experiencing a word, we also speak of 'the meaning' and of 'meaning it.' . . . Call it a dream" (Wittgenstein, 1974, p. 216). "For a large class of cases—though not for all—in which we employ the word 'meaning' it can be defined thus: the meaning of a word is its use in the language" (Wittgenstein, 1974, § 43). And even "A sound is an expression only if it occurs in a particular language-game." (§ 261). "The term 'language-*game*' is meant to bring into prominence the fact that *speaking* of language is part of an activity, or of a form of life" (§ 23).

From an interactionist–constructivist observer's perspective, a speaker's utterances can function for the listener like "pointing at something" only, or better as directing the focus of attention, whereas the construction of what might be meant, the construction of references, is with the listener. The speaker's utterances and intentions have no direct access into the listener's system. What the listener's senses receive undergoes spontaneous interpretation in several areas of the brain (Roth, 1992). These interpretations have emerged across many social interactions, encounters through which the person has tried to adapt to the culture by developing viable reactions and trying to act successfully. What we call *understanding* appears from this perspective as the active construction of such meanings and references supported by the social interaction within the culture. Russel's famous "set of three"—a prime minister, a head of cabbage, and a donkey—can serve for quite different objectivations according to the perceiving subject: as one object (a model set for three), as three sufficiently separable objects, as a satirical pointing at certain revealing similarities, and many more. Additionally, these ascriptions of meaning are necessarily specific to the perceived situation. Bruner, who in his last books called himself a constructivist, finally adopted the same view: "Language is . . . our principal means of *referring*" (Bruner & Haste, 1987, p. 87).

8. "LANGUAGE GAMES"

In terms of Luhmann's (1984) systems theory, accordingly, recognizing "something" is making a difference and, therefore, recognizing is an "objectivating practice." In their theory of autopoietic systems, Maturana and Varela (1980) limited the function of language to the referring function. They identified "linguistic behavior" as "an historical process of continuous orientation" (Maturana & Varela, 1980, p. 34; see also Fischer, 1991). There is no transportation of information by language means between systems. Due to the closed nature of the systems there is active (internal) construction of meaning only, according to the present conditions of the perceiving/constructing system. The key function of languaging is orienting—helping the listener or system with an orientation within his or her internal language system.

The focus of attention, our "looking at," cuts something out of the perceived diffuse continuum and makes an object of it. This process does not necessarily result in the same consequences with all persons, because it is a historical, situated, and individual process. Only across social interaction and permanent negotiations of meaning can "consensual domains" emerge, so that these "learned orienting interactions" can lead an observer to the illusion of some transport of meaning or information among the members of the social group. What we call *natural language,* then, is understood as an evolutionary "system of cooperative consensual interactions between organisms" (Maturana & Varela, 1980, p. 31); a definition that, by the way, allows us to identify language with animals too. If we speak of "describing something for somebody" we only metaphorize the process outlined here.

Our accustomed and nonreflected ways of speaking about things, as we thematize them through languaging, facilitate the transfer of the reality illusion onto all kinds of mental constructions, ideas, systems, structures, and hierarchies, not only onto "physical objects." However, in every case it is our personal activity that defines what "it" is for us. There is no direct access to the "it," and, particularly, there is no discovery and no direct reading either of a text or of a diagram, chart, or picture, and there is no direct taking up of information. Even more radically, Varela (1990) claimed that there is no information at all, there is only subjective knowing.[2]

Consequences for Classroom Analyses

In the realities of mathematics education in schools and universities, the Aristotelian perspective still dominates not only in the treatment of embodiments ("Just look at it!" and "You can *see* it!"), but more general in the organization of classroom settings and the classroom communication as analyzed in several microsociological investigations. The failure of the classical methods of teaching can be interpreted as a manifestation of this classical principle's failure: The teacher knows and

---

[2]"The break with the traditional analysis of perception can also be traced back to Wittgenstein (1953), who probed the difference between *seeing* something and seeing it *as* something" (Bechtel & Abrahamsen, 1991, p. 159).

teaches the truth, using language as a representing object and means. Because there is no simple transmission of meanings through language, the students all too often learn to say by routine what they are expected to say in certain defined situations. Mostly they are left alone with their constructive acts of interpreting, understanding, reflecting, and integrating. The key issues, that is, what is beyond the techniques—the "structuring of the personal system's structures" (extending a word from Mehan)—remain unsupported and undiscussed; they are emerging in obscurity.

The reproduction of the reality illusion, the belief in verbal transport of knowledge and direct learning and teaching, still is a very common feature in the early school years—with devastating consequences. Under the pressures from limited time, prescribed curricula, and external control, and driven by their own engagement, ambitions, and anxieties, the teachers are inclined to replace the necessary dealing with the students' subjectively constructed meanings with the conventionalized exchange of symbols with poor taken-as-shared meanings and definitions. "There is abundant anecdotal evidence that much of what happens in school is driven by a need to maintain bureaucratic and institutional norms rather than scholarly norms" (Pintrich, Marx, & Boyle, 1993, p. 193). In particular, the teachers' internalized obligation to maintain a smooth, productive, and steady flow of the classroom processes functions toward a curtailment of the negotiations and toward their replacement by routines for the direct production of wanted actions and related verbal descriptions. Too many teachers fall victim to the reality illusion: "recitation" (Hoetker & Ahlbrand, 1969) and the "funnel pattern" (Bauersfeld, 1978, 1988), in which instruction dominates over interaction. The effects are well known; mathematics is not only the most hated school subject for many students, it is the most ineffective one as well.

THE REPRESENTATION ILLUSION: LANGUAGE AS A MEDIUM BETWEEN PERSON AND WORLD VERSUS SOCIAL CONVENTIONS

> Whereas the positivist sees Galileo as making discovery—finally coming up with the words which were needed to fit the world properly ... the Davidsonian sees him as having hit upon a tool which happened to work better for certain purposes than any previous tool. (Rorty, 1989, p. 19)

Traditionally, classical linguists have treated language as an object, as a kind of an objectively existing medium between person and reality. The interpretation of language as a given object carries ideas of correspondence[3] and representation. If

---

[3]Davidson reacted specifically to Rorty's critique (see the motto) of his "coherence theory" and the related notion of "correspondence" (see Davidson, 1990, especially the added paragraph, "Afterthoughts," pp. 134–137).

there would be correspondence between language and reality, then, surely, one could arrive at true verbal statements about the world. Descriptions (and teaching), then, would become a case only of an adequate selecting and of providing for sufficient precision of the verbal means (denotations), as well as an adequate fit of these means with the object. These demands, moreover, would appear to be achievable through observation, experimenting, and inferences.

This traditional or Aristotelian view—taking language as the third (objective) issue between the two givens (the subject and object)—again constitutes the subject–object dualism. The view raises questions such as: "Does the medium between the self and the reality get them together or keep them apart?" and "Should we see the medium primarily as a medium of expression—of articulating what lies deep within the self? Or should we see it as primarily a medium of representation—showing the self what lies outside it?" Hence, "the see-saw battles between . . . idealism and realism, will continue, more or less endlessly" (Rorty, 1989, p. 11). Saussure's separation of the signifier (words) and the signified (object) is problematic for the same reasons, but more so because his analyses are based on written language: Language neither duplicates nor maps nor represents the world. "There is no literal language in the traditional sense of a language that simply 'fits the world' perfectly and without remainder; nor is there metaphoric language that we can define once and for all, in Aristotelian fashion, as language that talks about something as if it were something else" (Gergen, cited in Leary, 1990, p. 270).

Rorty (1989), Davidson (1990), and others (Bruner, 1990; Roth, 1992) built their alternative view on the late Wittgenstein (1974). We make realities treatable in our mind mainly through languaging, and we have and communicate about realities through language games. Particularly, the reflection on language itself is possible only within specific language games, played namely in linguistics. However, "The logic or grammar of the signifying language game is not the logic of the signified per se. The logic of signification constitutes the logic of the signified: The pretended logic of the world is the shadow of the grammar thrown onto the world" (Fischer, 1991, p. 85, author's translation).

Related to the development of languaging in social groups is the notion that it is through conventions only that a "consensual domain" (Maturana & Varela, 1980) emerges across social interaction. The consensual domain enables communication and, by the way, supports the participants' illusions of dealing with neutral or universal objects, as each being the same object for all members in the social group and to be called up by a name or by adequate description. Thus, the alternative interactionist–constructivist perspective traces the sameness of an object back to a social convention, as brought forth through the members' taken-as-shared activities. To give an adequate description for such jointly generated[4] objects means that each member of the social group can produce an acceptable interpretation of or act acceptably with the object. Finally, there is no metalanguage behind the many language games and languages that could serve as the adequate description of realities and as a basis for translations.

---

[4]See footnote 2.

The discussion refers as well to technical languages and their limited domains. Technical languages have been construed as artificial means for use in the limited and reduced situations of technical worlds and other human-made settings. The price paid for their unequivocal denotations and logical relations are their high closed nature and exclusiveness and, especially, the exorcising of metaphors and figurative use of words. The sharper these criteria are made, the more difficult access to the introduction into the technical language becomes. Mathematics can serve as a paradigmatic case. It took a development of more than 2,000 years until Goedel could point at some fundamental difficulties and before a fundamental crisis shook the mathematical community (see Davis & Hersh, 1980): The closed descriptive systems are not self-sufficient; they inevitably rest on the richer natural language behind. There is no understanding and no access without the flexibility and metaphors of everyday languaging.

The consequences from this perspective are far-reaching: "Truth is a property of sentences, since sentences are dependent for their existence upon vocabularies, and since vocabularies are made by human beings, so are truths" (Rorty, 1989, p. 21). This includes mathematics and its truths as well, they appear as human-made conventions. Compared with the traditional view, this alternative perspective clearly carries a dramatic shift, which Rorty characterized as: "To drop the idea of languages as representations, and to be thoroughly Wittgensteinian in our approach to language would be 'to de-divinize the world'" (p. 21).

Consequences for Classroom Analyses

If there is no chance for any direct derivation from either what you see to how to say it, or from what is said to what you are expected to see, then teaching will be in severe trouble. There is no simple pointing at something and naming it adequately, because what is seen and recognized can be reasonably different with each of the students. Teachers often forget these differences by requesting "Just look at it!" as if an array of beads or sticks or any other didactical means such as signs, models, or embodiments were self-speaking issues. There is no direct-developing insight from any kind of organized reality experiences in the classroom, because all of such new encounters are shaped and interpreted by the students' subjective domains of experience and related special language games (Bauersfeld, 1988, 1993). At any stage the child is an informed child as the teacher is an informed expert. However, the individual "informedness" of the child can lead to contrafunctional or misleading constructions from the perspective of the teacher's aims.

Because the individual focus of attention puts a certain order over the continuous and chaotic flow of the sensations, our interpretation of the sensual intake—which is a drastic selection already—functions like throwing an organizing net over it and generates objects in the flow, supported and guided by learned language games. This process is a historical, situated, and personal process, and by far it does not lead all people to the same reactions. Both students and teacher, therefore, have difficulty mutually understanding the other's utterances in the sense the speaker means it. To the extent that the teacher and students jointly develop and deepen a specific and differentiated language game, based on taken-as-shared experiences, activities, and objects, there will be better chances for a sufficient mutual understanding and an effective interaction.

Assuming that there is a fit between language and world consequently leads to the possibility of discovery: Once you have the language means, you can discover something (in the sense of unveiling a hidden object) and can describe the properties adequately. There is no discovery unless the discoverer has well-developed perspectives, in terms of both language and experienced imaginations of what he or she is going to hunt for, or better—of what he or she is going to make an object. Discovering in Piagetian terms means assimilating the unknown to available interpretive structures, and, if necessary, accommodating these across the developing new experiences, thus generating new objects[5] (for the person).

THE ILLUSION ABOUT LEARNING RULES: DISCOVERY LEARNING AND METACOGNITION VERSUS NEGOTIATION OF MEANING

> The term "language-*game*" is meant to bring into prominence the fact that the *speaking* of language is part of an activity, or of a form of life. (Wittgenstein, 1974, § 23)

The classical understanding of school learning is as an acquiring of knowledge, supported by teaching, as a transmission of (cultural) knowledge. In its most sophisticated forms, teaching followed Plato's midwife model: to help what is already in the student's mind be born [see Struve & Voigt's (1988) microsociological analysis of Plato's "Menon" and their related critique]. In more recent descriptions, it appears as the organization of discovery learning (for a critique, see Bauersfeld, 1992b). However, knowing a rule does not imply the knowledge about when to use it; the same applies to a word and more general to knowledge itself. The growing awareness of the problems with teaching knowledge, then, has produced typical repair strategies in cognitive psychology: If knowledge rests on preknowledge, then a teacher will have to know what the child already knows. Consequently, this had been tried, arriving at repair approaches or, in the case of arranging chances for the child to demonstrate his or her preknowledge, arriving at discovery learning issues. Further, when learning something appeared to not cover the needs, the recommendation for "learning of learning" as an adequate

---

[5]This is, perhaps, the most misunderstood statement regarding radical constructivism (Johnson, 1991). The "observer/describer" does not generate any objective entity in the external world (as many critical commentators have sarcastically remarked). He or she indeed creates what the "something outside" means for him or her, and that is, by principle, the only access the describer has to the object at all, and more generally to the world.

teaching had been given. Additionally, arriving at the insight that the adequate use of knowledge requires "knowledge about knowledge" has led to teaching approaches with procedural knowledge (Hofstadter, 1980; see also Anderson, 1983), conceptual strategies, metaknowledge (Campione, Brown, & Connell, 1990; Weinert & Kluwe, 1986), and so on.

Such a categorical framework seems to suffer from a typical lack of flexibility: The treatment of issues as given objects, as knowledge, that is, the making an object of it (*Verdinglichung*), obscures the nature of the related processes as currently performed and as highly situation-specific accomplishments (Griffin & Mehan, 1981; Mehan & Wood, 1975). The key difference is with the concept of knowledge as an object and the alternative interpretation of knowing as a current performing. This is not merely falling back into the process–product dichotomy; it is the difference between the use of a product in a process versus the flexible fixation of meanings in the current flow of social interaction. We, therefore, avoid the notion of knowledge (the concept is as weak as its relative information), and prefer to speak of knowing or ways of knowing.

From our perspective, the failure of the cognitive models and their repair is predictable, because the attempts for repair do remain in the range and limits of the same background philosophy. It was only recently that cognitive science repaired the learning model by accepting the specificity of contextual conditions and speaking of *shared knowledge, situated learning,* or *situated action,* thus beginning to integrate the social dimension (e.g., Norman, 1993; Resnick, Levine, & Teasley, 1991; see also Bauersfeld, 1993).

Growing up in isolation, the actively constructing and creating human being neither develops personality nor arrives at language as we know it. Further, enculturation, if viewed as the ruling force for the development of personality (see Leont'ev's "activity theory" in Bauersfeld, 1992a), does not produce sufficient explanations for individual creativity and the permanent change of societies (and their languaging). With this statement we are in accordance with French poststructuralist psychologists like Walkerdine (1988), who regarded "the individual–social dualism . . . as a false dichotomy" and with Ernest's related recent critique (Ernest, 1993), but also with recent issues from pragmatic linguistics (e.g., Davis & Taylor, 1990) and with Mead's social interactionism.

Considering the intimate interrelation between the individual and the social leads to the concept of culture, when interpreted as an observer's description of the processual structure of a social system. Two characteristics of the broad (and as for the rest vague) concept are of relevance:

1. The indirectness of learning through participation (becoming a member).
2. The permanent constitution and change of the culture through the activities of the members, including the newcomers.

For the following reasons the two characteristics are of importance:

1. What does it mean to become an active member of a social group, and, especially, to participate in their discursive practice acceptably? As it is with cultures, their members learn through participation, through living in and with it. This kind of learning appears as the subject's permanent although mostly covert constructive effort toward viability, in order to become accepted and to do things as they have to be done. Bruner (1991) identified his concept of culture "as implicit and only semi-connected knowledge of the world, from which, through negotiation, people arrive at satisfactory ways of acting in given contexts," and this he contrasted with the "structuralist perspective" on culture as "a set of interconnected rules, from which people derive particular behaviors to fit particular situations" (p. 90). The often-used notions of culture as implicit knowledge (we would prefer to say *implicit knowing,* pointing at the subject's momentary flexible production) or as a stock of themes (Luhmann, 1984; for the concept of themes see Voigt, chap. 5, this volume) point at these mostly indirect learning processes, which accompany all everyday activities, including languaging.

2. A combined interactionist–constructivist view, on the other side, has to take into account reflexively the fundamental interrelation as well: both change mutually, the members and their culture. The everyday practices of the members reproduce or constitute the culture; without members there is no cultural process. However, members not only live their culture—through their creativity and deviant inventions they actively contribute to develop and change it. Thus, cultures—and particularly the subculture of a specific mathematical classroom—have a history like members have their specific biographies.

The experiencing of resistance and constraints from outside—the radical constructivist notions for describing the subject's encounters with the world (von Glasersfeld, 1991)—necessitates more detailed analyses in order to understand the social nature of learning–teaching processes, particularly the development of language games and their specificity to different mathematics classrooms. "The mastery of knowledge and skill requires newcomers to move toward full participation in the sociocultural practice of a community" (Lave & Wenger, 1991, p. 29). The overestimation—if not absoluteness—of the social, on the other side, necessitates correction as well, as recent further developments of activity theory demonstrate (e.g. Lektorsky, 1993, and many more articles in that periodical).

With this specialized (or reduced) understanding of the concept of culture we can speak of the *culture of a mathematical classroom.* Again, how do teacher and students arrive at taken-as-shared issues, such as activities, objects, language games, and social regularities? With an answer we could then tackle questions like: What characteristics distinguish classroom cultures in which mathematical learning and personality development have served our aims, and what long-term effects with regard to students' mathematical learning will emerge from different classroom cultures?

From quite different investigations, answers to the first question point at:

- Starting with open tasks, that is, tasks that open the chance for students to employ and develop their own interpretations when engaging in the task (see Cobb, chap. 3, this volume).

- Whole-class discussions, in which the teacher serves as a peer in languaging and problem solving or when intervening in small-group activities (see Wood, chap. 6, this volume).
- Ample chances and space for students to demonstrate their own way of thinking, such as when in small-group activities, when working in pairs under the obligation to mutually explain each others' ideas and solutions, or in teacher-guided whole-class discussions (see Yackel, chap. 4, this volume). "People can help each other learn when they use their differences in a joint activity. Much schooling in America has a contrasting focus, namely, finding reasons why each others' contributions are inadequate" (Bredo & McDermott, 1992, p. 35).
- An early introduction to the practice of argumenting, not only as a preparation for later mathematical proving, but also as a more general developing of rational discourse in the classroom. This would also serve as an adequate treatment of evident reasons, ambiguities, and contradictions (see Krummheuer, chap. 7, this volume).
- Encouraging a flexible and multidimensional languaging, through casually making an object of the languaging itself and particularly through furthering figurative speech, such as metaphors and metonymies (see Bauersfeld & Zawadowski, 1981; Pollio, Smith, & Pollio, 1990).[6]

Issues common in these findings and analyses are certain descriptions (thus drawing attention to meanwhile shared theoretical convictions among the authors) of: negotiation of meaning as a fundamental characteristic of the classroom discourse, from which taken-as-shared issues are constituted or emerge in terms of languaging and objects; arriving at mathematical practices such as social conventions; and taking into account the necessary asymmetry of classroom discourse with the power on the teacher's side as a matter of principle. From an interactionist–constructivist perspective, negotiation of meaning in social interaction functions as a starting point for the development of an effective language game, and is, from a long-term view, the only way for the adaptation and relative congruence of the students' subjective constructions toward reliable shared elements in a consensual domain.[7]

Consequences for Classroom Analyses

From the alternative perspective, experimenting with more sophisticated ways of teacher–student interactions appears to be necessary in order to provide better chances for the emergence of more sensibility and differentiated reflections of the

---

[6]Pollio et al. (1990) pointed at the neglect, "that cognitive psychology has dealt with figurative language as less important than literal language largely because of an implicit bias towards rationalistic philosophy and because of an unwillingness to deal with issues of ambiguity, novelty, beauty, and context" (p. 141).

[7]In order to avoid misunderstandings, it might be necessary to state here that the previous listing should not be taken as complete or representative or as *the* condition for success, nor do we claim exclusiveness or absolute truth for the offered theoretical background. The interactionist–constructivist position can be only one among other competing perspectives.

own experiences, and especially to provide for a related enrichment of the classroom culture. No wonder, therefore, that with the spreading awareness of the object illusion there is a growing awareness of the necessity to complement instruction by discussing issues and even negotiating meaning in the classroom discourse (see Greeno, 1991; Resnick, 1989; Schoenfeld, 1987[8]). Putnam, Lampert, and Peterson (1990) stressed "the importance of talking about mathematics," demanding that "classrooms should provide ample opportunities for students to verbalize their thinking and to converse about mathematical ideas and procedures" (p. 138). The practice of mathematics education is changing accordingly, albeit slowly. At least there is much more experimenting and investigating than before. However, because most of these recent recommendations still have their backup in theoretical models from cognitive science—now extending their bases into sociopsychological dimensions—they are made for fostering understanding, improving the use of embodiments, teaching problem-solving strategies (metaknowledge), and so on. "Verbal discourse constitutes an important means of revealing individual knowledge" (Putnam, Lampert, & Peterson, 1990, p. 138). Indeed, but speaking of "more discussion" or of "negotiation of meaning" can stand for quite different ideas and realizations in the classroom, according to the founding philosophy of the speaker or writer. What about the teacher's role?

When language is (for the subject) "an active *molder* of experience" rather than "a passive mirror of reality" (Leary, 1990, p. 357) or an adequate representation of it, then the qualities of the culture a person lives in become critical. The totality of everyday experiences forms the basis for the personal development, interactively, and in all dimensions of the senses. This is true not only with the eye-catching specialties of different cultures, but also with that what is *not* thematized or treated within that culture. From an observer's perspective, such missing qualities have no chance of becoming taken-as-shared knowledge or everyday use, unless an outsider invents the "new" issue. This is particularly true with languaging. As Barthes (1980) said, "je ne puis jamais parler qu'en ramassant ce qui traîne dans la langue [Always I can speak only by picking up from what is lying around in language]" (pp. 20–21). Consequently, the function of the teacher becomes crucial. For too long education has neglected the functioning of imitative learning, which indeed happens in every classroom as the most common form of learning in a culture (Tomasello, Kruger, & Ratner, 1993). As an agent of the embedding culture, the teacher functions as a peer with a special mission and power in the classroom culture. The teacher, therefore, has to take special care of the richness of the classroom culture—rich in offers, challenges, alternatives, and models, including languaging.

Coulmas (1992) discussed the following economic principle: "Speakers adapt their language to the realized needs" (p. 325); "They do not invest more energy than necessary in order to serve the actual communicative purposes" (p. 324).

---

[8]Commenting on "solutions to mixed sets of algebraic equations," Schoenfeld (1987) gave the general recommendation: "Discussing your solution methods with your students—discussing *why* you choose the approaches to mixed problems that you do, and comparing the methods you chose with the methods they chose—may be tremendously useful to them" (p. 29).

Consequently, what is not a functional part in a culture cannot contribute to the developing of linguistic means with the members. If you cannot recognize what you do not expect to see (no descriptive means available—no disposition for recognition), then you also do not see what is not in the focus of your attention (no eye catching, no curiosity, no peculiarity, and no deviant experiencing). This is the complement to the critique of discovery illusions: Poverty and emptiness are not the origin of diversity and differential refinement.

What emerges from the everyday practices in the mathematics classroom is not very different in its nature from the impact of everyday life in our culture, and what Schütz (1975) described as the epoché of natural attitude. It is just done, and it is usually taken for granted without reflection. Both insights found a strong argument for demanding: What is taken as natural at school as well as the nonreflected effects of the everyday practices—the missing as well as the enacted parts—require the most careful analyses for their functionality in relation to the goals of education.

ASPECTS OF CHILDREN'S LANGUAGING IN THE MATHEMATICAL CLASSROOM

To be is to communicate. (M. Bakhtin, quoted in Lektorsky & Engeström, 1990, p. 11)

If we had to identify common ground among the theories of Vygotsky, Piaget, and Bruner regarding the function and development of language, the answer would be the fundamental influence of social interaction. Even Chomsky's neglect of the development of language has—through other researchers' critical pursuits and their follow-up studies—given prominence to the role of context and to the social conditions. Because the interest here is with consequences from converging and powerful theoretical bases rather than with differences[9] among their foundations or distinguishing peculiarities, we add a few remarks only in complementing the preceding chapter.

Children enter mathematics education at school with more or less well-developed mother tongue languaging. What is new, then? Mathematics opens a new perspective on the world. The interest is no longer with properties like type of material, use, color, importance, like and dislike, and so on. What counts is the answer to "how many?" Further, dealing with the different "how many" issues (numbers) develops into an exceptional situation (the math lesson) with special materials (representations), very special manipulations and signs (operations and number sentences), and a related special language game. It is a very strict game

---

[9]Vygotsky's position has been analyzed much more carefully over the last few years (see Kozulin, 1990; Moll, 1991; van der Veer & Valsiner, 1991). Also, in his late years Bruner cleared his perspective (see Bruner, 1984; 1985; 1990; Bruner & Haste, 1987). For recent accounts of Piaget's work, see Garton, 1992, and Wood, 1988.

with strange rules about what is allowed and what is not, and the child's only access into this new subculture is active participation.

Linguists will agree that there is no language development without social interaction or personal construction of meaning. Hence, how does a child arrive at saying "Five!" and writing "3 + 2 = 5" when exposed to a picture of three birds sitting on a fence and two more birds approaching in the air? Because language is not simply "the storehouse of human knowledge" (Luria, 1979, p. 44), the students have to build up a whole network of viable activities before something can be called an *object* or *seven* or *minus*.

"In communicative encounters involving reference, however ritualized, one element of the 'ceremony' is not fixed. The unknown is the referent to which a joint focus of attention is to be achieved by the two participants" (Bruner, 1983, p. 122). This is especially difficult for beginners in mathematics lessons, because no one can see "five." There are only (already conventionalized) objects like birds, sticks, or beads, as well as related practices. Across sets of changing objects the child then learns to point at one object after the other and say successively "one, two, three. . . ." The imitation of both sounds and acting leads to the generation of the counting practice, later elicited by: "Find out how many." Thus, a possible answer is the emergence of regulations across the interaction on both levels: as conventions on the social level, and as a routines and habitual interpretations on the individual level. A one-to-one correspondence is not given by nature.

For many children, this is not an easy achievement. During a research project in a German school, a boy late in first grade seemed to produce number sentences quite haphazardly, such as "3 + 2 = 4," "4 + 1 = 3," and so on. Sitting aside with him for some rehearsal, it became clear that he just imitated the finger tapping and recited the first number words, both fairly fast. He had not realized that he needed to keep a one-to-one correspondence between finger and number word.

Preschoolers do produce a lot of different interpretations when asked to comment on a picture, a certain array of rods or blocks, or other representations (Bauersfeld, 1995: Radatz, 1990). It is the teacher's continuous insisting on certain standard descriptions that relatively soon forces the child into a narrowing adaptation of these originally open varieties to the expected standard utterances and taken-as-shared interpretations. The repetition of the desired words by the teacher and whole class (often like a Tibetan prayer mill), and the related negative sanctioning even of minor aberrations in talking or manipulating, have a strong selecting power over the children's productions.

Certain problems arise with the structuring of the internal processes: How does a child learn to correct an inadequate habit of constructing meaning? Teachers can easily correct the products, but there is no direct access to the individual (internal) processes of constructing. Thus, on the surface of the official classroom communication, everything can be said and presented acceptably, but the hidden strategies of constructing may lead the child astray in other situations or in the face of even minor variations.

Relatively speaking, the only way out is to maintain a classroom culture in which students get as many chances as possible to present their ideas and inferences, not only by verbally commenting (for many children this would be very difficult), but also by acting them out in actions and demonstrations. The opening for such multidimensional ways of negotiating of meaning, as favored by interactionists, demands a peculiar awareness and attention from the teacher for two reasons:

1. Most of what a child learns is done so indirectly. The effects emerge across the child's active participation in the culture. Simple cause–effect analyses have no chance of making these processes sufficiently transparent. Although there is no doubt that the teacher has an important peer function in these processes, it is not only through his or her verbal production.

2. The children's pre- and nonverbal development is much richer, as recent research studies with very young babies have shown (Bauersfeld, 1993). "Certain communicative functions or intentions are well in place before the child has mastered the formal language for expressing them linguistically" (Bruner, 1990, p. 71). With good reason, Siegal (1991) warned against overestimating the effects of official classroom talk and against interpretations of children's verbally presented knowledge that appear too easy: "Children's development is better characterized by development towards a conscious accessibility of implicit knowledge rather than a simple lack of conceptual knowledge or coherence at different stages" (p. 133).

Related to figurative means in languaging Pollio et al. (1990) criticized Piaget, who held that an utterance is figurative only if "it represents a deliberate and purposeful deviation from literal usage based on a deep understanding and its ramifications." They developed a more functional view, accepting a figure as valid if it "serves a communicative purpose, either reflectedly or unreflectedly" (p. 157). This is in accordance with our findings. Children combine verbal means from different language games as soon as the related domains of subjective experiences are in an activated state simultaneously. Thus, their current language production combines what is at hand in the present state of mind in order to serve the perceived demands of the situation. In other words: "The most significant mechanism for metaphoric construction seems to rest in our attempt to make the abstract world understandable" (Pollio et al., 1990, p. 161).

Consequences for Classroom Analyses

Following educational principles like "learning by doing" and "discovery learning," teachers try not to give away too much support and help. This tendency can limit their peer function in languaging to a minimum. (The psychoanalytical practices of C. R. Rogers are famous for not giving away the slightest hint, through the techniques of simply rephrasing the client's utterance into another question.) By questioning and waiting, they expect the students to find their way by themselves and to arrive at the wanted descriptions. In a given reality it would suffice to give space and freedom for activities in order to elicit students' exploring and

8. "LANGUAGE GAMES" 287

discovering of nature or the prepared structures in so-called "structured material" for didactical purposes. The engaged active learner would then arrive at the truth, according to Newton's classical view of nature by answering the researcher's questions. But, "The world does not speak. Only we do" (Rorty, 1989, p. 6).

Teaching and learning are intimately related issues; Steinbring (1991) spoke of "co-evolutionary processes" (p. 87). How will languaging in the classroom develop and differentiate if there is no peer and no model in function? Barthes' (1980) statement, "Always I can speak only by picking up from what is lying around in language" (pp. 20–21), points at a crucial issue: Children pick up elements of languaging, imitate and try them when communicating in similar situations, change the ascription of meaning according to the experienced reactions, and thus in time learn to fill these imitated elements with certain taken-as-shared meanings specific to the situation.

To function as a model does not mean that the teacher has to say everything by him- or herself. For very young children, Bruner (1983) described the "scaffolding" approach as a means for developing languaging: Mothers introduce little answer–response games that the child enjoys and easily adopts. Soon, the child takes over the leading role and now plays the game with mother, but with changed roles. In the elementary classroom, the teacher may introduce a special character—a puppet, a colored stone, something special that can be moved—with a special function: Whenever necessary, the teacher or the children can ask the "Jesbut" whether a current utterance can be said "clearer," "in a more understandable way," "a bit wittier" or more humorous, using "something similar" (metaphors), and so on. Jesbut's answers should somehow appear stereotyped as in, "Yes, but we can also say. . . . " Once the children begin to take over the Jesbut's role, the teacher can learn a lot about the children's capacity for playing with language. In time, children can learn that an utterance is not "easier" or "more understandable" for everyone. In this way, a new field for negotiation can be opened. The means have the key function of turning the students' attention toward languaging itself, that is, toward how things can be said. It should serve for enrichment and differentiation, rather than for simple corrections in a know-all manner.

The consequences from this complicate the teacher's professional task:

> The child's acquisition of language requires far more assistance from and interaction with caregivers than Chomsky (and many others) had suspected. Language is acquired not in the role of spectator but through use. Being "exposed" to a flow of language is not nearly as important as using it in the midst of "doing." (Bruner, 1990, p. 70)

Further, there is no help from mathematics itself, that is, through rational thinking or logical constraints, as teachers often assume. Mathematics does not have self-explaining power, nor does it have compelling inferences; for the learner there are only conventions. As Wittgenstein (1974) put it: "Am I less certain that this man is in pain than that twice two is four? . . . 'Mathematical certainty' is not a psychological concept. The kind of certainty is the kind of language-game" (p. 224). This means that certainty arises with the social conventions of use. Thus, we have come back to social interaction and the culture of the classroom: "Look on the language-game as the *primary* thing." (Wittgenstein, 1974, § 656).

REFERENCES

Anderson, J. R. (1983). *The architecture of cognition*. Cambridge, MA: Harvard University Press.

Barthes, R. (1980). *Lecon/Lektion-Antrittsvorlesung am Collége de France am 7. Januar 1977* [Lesson-Inaugural lecture at the College of France on January 7, 1977]. Frankfurt: Suhrkamp.

Bauersfeld, H. (1978). Kommunikationsmuster im Mathematikunterricht—Eine Analyse am Beispiel der Handlungsverengung durch Antworterwartung [Patterns of communication in the mathematics classroom—an analysis of the narrowing effect of the teacher's expectations for children's answers onto the teacher's options for action]. In H. Bauersfeld (Ed.), *Fallstudien und Analysen zum Mathematikunterricht* (pp. 158–170). Hannover: H. Schroedel Verlag.

Bauersfeld, H. (1980). Hidden dimensions in the so-called reality of a mathematics classroom. *Educational Studies in Mathematics, 11,* 23–29.

Bauersfeld, H. (1988). Interaction, construction, and knowledge—Alternative perspectives for mathematics education. In D. A. Grouws & T. J. Cooney (Eds.), *Perspectives on research on effective mathematics teaching: Research agenda for mathematics education* (Vol. 1, pp. 27–46). Reston, VA: National Council of Teachers of Mathematics and Lawrence Erlbaum Associates.

Bauersfeld, H. (1992a). Activity theory and radical constructivism—What do they have in common and how do they differ? *Cybernetics and Human Knowing, 1*(2/3), 15–25.

Bauersfeld, H. (1992b). Integrating theories for mathematics education. *For the Learning of Mathematics, 12*(2), 19–28.

Bauersfeld, H. (1993). Theoretical perspectives on interaction in the mathematical classroom. In R. Biehler, R. W. Scholz, R. Sträßer, & B. Winkelmann (Eds.), *Didactics of mathematics as a scientific discipline* (pp. 133–146). Dordrecht, Netherlands: Kluwer.

Bauersfeld, H. (1995). Structuring the structures. In L. P. Steffe & J. Gale (Eds.), *Constructivism and education* (pp. 137–158). Hillsdale, NJ: Lawrence Erlbaum Associates.

Bauersfeld, H., & Zawadowski, W. (1981). *Metaphors and metonymies in the teaching of mathematics* (Occasional Paper 11). Bielefeld, Germany: University of Bielefeld, Institut für Didaktik der Mathematik.

Bechtel, W., & Abrahamsen, A. (1991). *Connectionism and the mind*. Oxford, England: Basil Blackwell.

Bellack, A. A., Kliebard, H. M., Hyman, R. T., & Smith, F. L. (1966). *The language of the classroom*. New York: Teachers College Press.

Bredo, E., & McDermott, R. P. (1992). Teaching, relating, and learning. *Educational Researcher, 21*(5), 31–35.

Bruner, J. S. (1983). *Child's talk: Learning to use language*. Oxford, England: Oxford University Press.

Bruner, J. S. (1984). Vygotsky's zone of proximal development: The hidden agenda. In B. Rogoff & J. V. Wertsch (Eds.), *Children's learning in the zone of proximal development* (pp. 93–97). San Francisco: Jossey-Bass.

Bruner, J. S. (1985). Vygotsky: A historical and conceptual perspective. In J. V. Wertsch (Ed.), *Culture, communication, and cognition* (pp. 21–34). Cambridge, MA: Cambridge University Press.

Bruner, J. S. (1990). *Acts of meaning. The Jerusalem-Harvard lectures*. Cambridge, MA: Harvard University Press.

Bruner, J., & Haste, H. (Eds.). (1987). *Making sense—The child's construction of the world*. London: Methuen.

Campione, J. C., Brown, A. L., & Connell, M. L. (1990). Metacognition: On the importance of understanding what you are doing. In R. I. Charles & E. A. Silver (Eds.), *The teaching and assessing of mathematical problem solving* (pp. 93–114). Hillsdale, NJ: Lawrence Erlbaum Associates.

Cazden, C. (1986). Classroom discourse. In M. C. Wittrock (Ed.), *Handbook of research on teaching* (pp. 432–462). New York: Macmillan.
Coulmas, F. (1992). *Die Wirtschaft mit der Sprache—Eine sprachsoziologische Studie* [The mess with language: A sociolinguistic study]. Frankfurt: Suhrkamp.
Davidson, D. (1990). A coherence theory of truth and knowledge. In A. Malachowsky (Ed.), *Reading Rorty* (pp. 120–138). London: Basil Blackwell.
Davis, H. G., & Taylor, T. J. (Eds.). (1990). *Redefining linguistics*. London: Routledge.
Davis, P. J., & Hersh, R. (1980). *The mathematical experience*. Boston: Birkhäuser.
Durkin, K., & Shire, B. (Eds.). (1991). *Language in mathematics education—Research and practice*. Milton Keynes, England: Open University Press.
Erickson, F. (1986). Quantitative methods in research on teaching. In M. C. Wittrock (Ed.), *Handbook of research on teaching* (pp. 119–161). New York: Macmillan.
Ernest, P. (1993). Constructivism, the psychology of learning, and the nature of mathematics: Some critical issues. *Science & Education, 2*(1), 87–93.
Fann, K. T. (1971). *Die Philosophie Ludwig Wittgensteins* [Ludwig Wittgenstein's philosophy]. Munich: List Verlag.
Fischer, H. R. (Ed.). (1991). *Autopoiesis—eine theorie im brennpunkt der kritik* [Autopoiesis—a theory in the focus of critique]. Heidelberg: Carl Auer.
Garfinkel, H. (1967). *Studies in ethnomethodology*. Englewood Cliffs, NJ: Prentice-Hall.
Garton, A. F. (1992). *Social interaction and the development of language and cognition*. Hillsdale, NJ: Lawrence Erlbaum Associates.
Goffman, E. (1974). *Frame analysis—An essay on the organization of experience*. Cambridge, MA: Harvard University Press.
Greeno, J. G. (1991). Number sense as situated knowing in a conceptual domain. *Journal for Research in Mathematics Education, 22*(3), 170–218.
Griffin, P., & Mehan, H. (1981). Sense and ritual in classroom discourse. In F. Coulmas (Ed.), *Conversational routine* (pp. 187–213). The Hague: Mouton.
Heritage, J. (1984). *Garfinkel and ethnomethodology*. Cambridge, England: Polity Press.
Hörmann, H. (1977). *Psychologie der Sprache* [Psychology of language]. Berlin: Springer.
Hoetker, J., & Ahlbrand, W. P. (1969). The persistence of the recitation. *American Educational Research Journal, 6*(2), 145–167.
Hofstadter, D. R. (1980). *Gödel, Escher, Bach: An eternal golden braid*. New York: Vintage Books.
Johnson, D. K. (1991). Reclaiming reality: A critique of Maturana's ontology of the observer. *Methologica, 5*(9), 7–31.
Kozulin, A. (1990). *Vygotsky's psychology: A biography of ideas*. London: Harvester Wheatsheaf.
Krummheuer, G. (1992). *Lernen mit "Format"—Elemente einer interaktionistischen Lerntheorie* [Learning with format—Elements of an interactionist theory for learning]. Weinheim, Germany: Deutscher Studien Verlag.
Lave, J., & Wenger, E. (1991). *Situated learning: Legitimate peripheral participation*. Cambridge, England: Cambridge University Press.
Leary, D. E. (1990). *Metaphors in the history of psychology*. Cambridge, MA: Cambridge University Press.
Lektorsky, V.A. (1993). Remarks on some philosophical problems of activity theory. *Multidisciplinary Newsletter for Activity Theory, 2*(1), 48–50.
Lektorsky, V. A., & Engeström, Y. (1990). *Activity: The theories, methodology & problems*. Orlando, FL: Paul M. Deutsch.
Lerman, S. (1992). The function of language in radical constructivism: A Vygotskyan perspective. In W. Geeslin & K. Graham (Eds.), *Proceedings of the Sixteenth PME Conference* (Vol. II, pp. 40–47). Durham: University of New Hampshire.
Luhmann, N. (1984). *Soziale Systeme—Grundriß einer allgemeinen Theorie* [Social systems—outline of a general theory]. Frankfurt: Suhrkamp.

Luria, A. R. (1979). *The making of the mind: A personal account of Soviet psychology*. Cambridge, MA: Harvard University Press.

Maier, H., & Voigt, J. (1991). Interpretative Unterrichtsforschung. *Untersuchungen zum Mathematikunterricht* [Interpretive research on teaching and learning: Investigations in the mathematics classroom] (Vol. 17). Cologne: Aulis Verlag Deubner.

Maturana, H. R., & Varela, F. J. (1980). *Autopoiesis and cognition: The realization of the living*. Dordrecht, Netherlands: Reidel.

Mehan, H., & Wood, H. (1975). *The reality of ethnomethodology*. New York: Wiley.

Moll, L. C. (1991). *Vygotsky and education: Instructional implications and applications of sociohistorical psychology*. New York: Cambridge University Press.

Norman, D. (Ed.). (1993). Situated action. *Cognitive Science* [Special issue], *17*(1).

Picht, G. (1987). *Aristoteles "De Anima."* Stuttgart: Klett-Cotta.

Pimm, D. (1987). *Speaking mathematically: Communication in mathematics classrooms*. London: Routledge & Kegan Paul.

Pintrich, P. R., Marx, R. W., & Boyle, R. A. (1993). Beyond cold conceptual change: The role of motivational beliefs and classroom contextual factors in the process of conceptual change. *Review of Educational Research, 63*(2), 167–199.

Pollio, H. R., Smith, M. K., & Pollio, M. R. (1990). Figurative language and cognitive psychology. *Language and Cognitive Processes, 5*(2), 141–167.

Putnam, R. T., Lampert, M., & Peterson, P. L. (1990). Alternative perspectives on knowing mathematics in elementary schools. In C. B. Cazden (Ed.), *Review of research in education* (Vol. 16, pp. 57–150). Washington, DC: American Educational Research Association.

Radatz, H. (1990). Was können sich Schüler unter Rechenoperationen vorstellen [Students' imaginations in arithmetical operations]. *Mathematische Unterrrichtspraxis, 11*(1), 3–8.

Resnick, L. B. (1989). *Knowing, learning, and instruction*. Hillsdale, NJ: Lawrence Erlbaum Associates.

Resnick, L., Levine, J. M., & Teasley, S. D. (1991). *Perspectives on shared cognition*. Washington, DC: American Psychological Association.

Rorty, R. (1989). *Contingency, irony, and solidarity*. Cambridge, MA: Cambridge University Press.

Roth, G. (1992). Kognition: Die Entstehung von Bedeutung im Gehirn [Cognition: The genesis of meaning in the brain]. In W. Krohn & G. Küppers (Eds.), *Die Entstehung von Ordnung, Organisation und Bedeutung* [The genesis of order, organization and meaning] (pp. 104–133). Frankfurt: Suhrkamp.

Schoenfeld, A. H. (Ed.). (1987). *Cognitive science and mathematics education*. Hillsdale, NJ: Lawrence Erlbaum Associates.

Schütz, A. (1975). *Collected papers, Vol. 3: Studies in phenomenological philosophy*. The Hague, Netherlands: Martinus Nijhoff.

Siegal, M. (1991). *Knowing children—Experiments in conversation and cognition*. Hove, England: Lawrence Erlbaum Associates.

Steinbring, H. (1991). Eine andere Epistemologie der Schulmathematik—Kann der Lehrer von seinen Schülern lernen [Another epistemology of school mathematics: Can teachers learn from their students?] *Mathematica Didactica, 14*(2/3), 69–99.

Struve, R., & Voigt, J. (1988). Die Unterrichtsszene im "Menon"-Dialog—Analyse und Kritik auf dem Hintergrund von Interaktionsanalysen des heutigen Mathematikunterrichts [The teaching episode in Plato's dialogue "Menon"—Analysis and critique based on analyses of the interaction in recent mathematics classrooms]. *Journal für Mathematikdidaktik, 9,* 259–285.

Tomasello, M., Kruger, A. C., & Ratner, H. H. (1993). Cultural learning. *Behavioral & Brain Sciences, 16*(3), 495–552.

van der Veer, R., & Valsiner, J. (1991). *Understanding Vygotsky: A quest for synthesis*. Oxford, England: Basil Blackwell.

Varela, F. J. (1990). *Kognitionswissenschaft—Kognitionstechnik* [Cognitive science—Cognitive psychology]. Frankfurt: Suhrkamp.

Voigt, J. (1984). *Interaktionsmuster und Routinen im Mathematikunterricht* [Patterns of interaction and routines in mathematics classrooms]. Weinheim, Germany: Beltz Verlag.

von Glasersfeld, E. (1991). *Radical constructivism in mathematics education.* Dordrecht, Netherlands: Kluwer Academic.

Walkerdine, V. (1988). *The magic of reason.* London: Routledge.

Weinert, F. E., & Kluwe, R. H. (1986). *Metacognition, motivation, and understanding.* Hillsdale, NJ: Lawrence Erlbaum Associates.

Wittgenstein, L. (1972). *Tractatus logico-philosophicus.* London: Routledge & Kegan Paul. (Original work published 1921)

Wittgenstein, L. (1974). *Philosophical investigations.* Oxford, England: Basil Blackwell. (Original work published 1945)

Wolff, C. (1962). *Philosophia prima sive ontologia.* Hildesheim, Germany: Olms. (Original work published 1728)

Wood, D. (1988). *How children think and learn.* Oxford, England: Basil Blackwell.

GLOSSARY

Abstract units of one: A cognitive construct developed to account for young children's arithmetic activity. Students' ability to enumerate the units created while counting by putting up fingers indicates that the units are abstract in quality. See also *Ten as a numerical composite, Ten as an abstract composite unit.*

Accountability: An underlying assumption of the interactionist approach maintaining that each participant of an interaction tries to demonstrate the rationality of his or her contribution. In the process of acting, the individual is trying to indicate how his or her action is accountable. See also *Argumentation.*

Accounting practices: Techniques and methods used to help demonstrate the rationality of an action or statement while acting or speaking. See also *Argumentation.*

Argumentation: (a) A primarily social process in which cooperating individuals try to adjust their interpretations and interactions by verbally presenting rationales for their actions. (b) The techniques or methods used to establish the validity or claim of a statement. During an argumentation, if one participant explains a solution, the implicit message is that the claim is valid. A successful argumentation refurbishes a challenged claim into a consensurable or acceptable one for all participants. See also *Collective argumentation.*

Authority: The relationship constituted between individuals establishing a power imbalance such that one member is viewed by all participants as more competent mathematically or socially. Mathematical and social authority do not necessarily coincide. As learners collaborate, the type of power imbalance affects the realm of developmental possibilities. See also *Mathematical authority, Power imbalance, Realm of developmental possibilities, Social authority.*

Balance task: A format used to present arithmetical tasks in the project classroom that depicts a pan balance with several boxes. The task for the children is to fill in the empty boxes so that the sums on each side balance. In this book, the symbol "/" is used to separate the two sides of the balance. Thus, the first two tasks in Fig. 2.1 (p. 20) are denoted as "8, 9/_ ," and "10, _ /17."

Basis for mathematical communication: A synonym for consensual domain. See *Consensual domain.*

Collaborative argumentation: Argumentation that is interactively constituted by several members of a group. The development of the theoretical constructs and empirical analyses reported in this book can be called *collective argumentation.*

Collection-based solution: A cognitive construct used to account for solution strategies in which the implicit metaphor appears to be that of mentally manipulating imagined collections. Collection-based solutions typically involve partitioning two-digit numbers into a "tens part" and a "ones part." See also *Counting-based solutions.*

Collective argumentation: A process of argumentation involving several people who are conjointly engaged. See also *Argumentation.*

Collectivism: A theoretical position that emphasizes the influence of participation in social interaction and culturally organized activities on an individual's psychological development. The collectivist position is exemplified by theories developed both in the Vygotskian tradition and in the sociolinguistic tradition. In both cases, mathematical learning is viewed as primarily a process of enculturation. See also *Individualism.*

Consensual domain (Maturana & Varela, 1980): The range of understandings that are taken as shared in the sense that each member of the social group can produce an acceptable interpretation or act acceptably.

Constructivism (von Glasersfeld, 1987): A theoretical perspective that characterizes students as active developers of their own ways of mathematical knowing. Learning is a process in which students organize their activity as they strive to achieve their goals.

Counting-based solution: A cognitive construct used to characterize solutions that appear to involve the curtailment of counting by ones. See also *Collection-based solution.*

Culture: The observer's description of a social system, or the participants' implicit knowledge of social practices within the culture. Learning is mainly seen as the cognitive process that leads to competent membership in the culture.

Direct collaboration: A sociological construct used to characterize an interaction in which small-group partners explicitly coordinate their attempts to solve a task. In general, interactions involving direct collaboration only occur when the children make taken-as-shared interpretations of a task and each other's mathematical activity. See also *Indirect collaboration.*

Elicitation pattern (Bauersfeld, 1988): A pattern of classroom discourse in which the teacher uses questions to discern if students are following and to lead them to a desired solution method. In general, the process can be characterized as following three phases: teacher asks open-ended questions and then begins to give clues, students begin to guess what the teacher is looking for (and teacher may oblige by giving partial answers), and teacher reviews what students should have learned.

Ethnomethodology (Mehan & Wood, 1975): A study of the patterns of interaction that emerge in a classroom, such as the famous triadic "elicitation-reaction-evaluation" pattern. This approach is related to the interactionist perspective, which focuses on the roles, obligations, and expectations of the participants and how they are implicitly negotiated.

Explanation: A rationale that is offered voluntarily. Although this concept is frequently used as a synonym for *argumentation,* it is used here in contrast to *justification,* which refers to the attempt to explicate one's thinking when challenged by other participants.

Framing: The schematized interpretation of a social event or object. Framing emerges by processes of cognitive routinization and interactive standardization. The concept refers to the cognitive constitution of meaning whereby an individual tries to create a meaning for a situation.

Habitus: A generative mechanism enabling a teacher or student to act adequately as a member of the classroom culture, even in situations never before experienced. The concept of habitus links the individual to the social dimensions of the classroom culture. The individual's habitus is formed by his or her participation in social practice. The in-

GLOSSARY

dividual can learn to participate adequately and to act in accord with certain regularities without being explicitly taught and without knowing what the regularities are. See also *Social norms.*

Hidden curriculum: The ulterior or implicit goal of a teacher or instructional designer. Examples in the experimental classroom included the development of intellectual and social autonomy (Kamii, 1985), relational beliefs about mathematics (Skemp, 1976), and task involvement rather than ego involvement as a form of motivation (Nicholls, 1983).

Hundreds board: A 10-by-10 grid labeled 1–10 on the top row, proceeding through 91–100 on the bottom row.

Illusion of competence (Gregg, 1992): A situation in which a teacher and student develop a solution together, with the student following the teacher's directives. See also *Elicitation pattern, Realm of developmental possibilities.*

Image-independent solution: A cognitive construct used to characterize solutions that do not appear to involve a reliance on situation-specific imagery. See *Image-supported solution.*

Image-supported solution: A cognitive construct used to characterize solution strategies that appear to involve a reliance on situation-specific imagery (such as referring to dimes in a money problem). See also *Image-independent solution.*

Indirect collaboration: A sociological construct used to characterize a small-group interaction in which one or both children think aloud while solving a task independently. Although neither child is obliged to listen to the other, the way in which they frequently capitalize on the other's comments indicates that to some extent they are monitoring what the partner is saying and doing. See also *Direct collaboration.*

Individualism: A theoretical position that focuses on the activity of the individual, autonomous learner as he or she participates in social interaction. This position is exemplified by neo-Piagetian theories that view social interaction as a source of cognitive conflicts that facilitate otherwise autonomous cognitive development. See also *Collectivism.*

Inquiry mathematics microculture: A classroom culture in which explanations and justifications carry the significance of acting on mathematical objects. See also *School mathematics microculture.*

Intersubjectivity: A mutual or taken-as-shared understanding of an object or event achieved through the social process of negotiation (as opposed to solely individual acts).

Language games (Wittgenstein, 1974): Linguistic processes that are special to each classroom. When the teacher and students negotiate taken-as-shared meanings and signs, they are immersed in language games. Through language games, the individuals produce objects whose existence depends on the signifying language game.

Mathematical authority: A sociological construct describing the role of one small-group partner who has been judged by both partners to be the more mathematically competent one in a working relationship. See also *Univocal explanation.*

Multivocal explanation: A sociological construct that characterizes small-group interactions in which a conflict in interpretations has become apparent and both children insist that their own reasoning is correct. In general, multivocal explanations are constituted when both children attempt to advance their perspectives by explicating their own thinking and challenging that of the partner. See also *Univocal explanation.*

Negotiation of meaning: The interactive accomplishment of intersubjectivity. In principle, objects of the classroom discourse are plurisemantic, and it is typical of the teaching–learning situation that the teacher constructs meanings for the objects that differ from those constructed by the students. Therefore, the participants have to negotiate meaning in order to arrive at a taken-to-be-shared meaning. From this perspective, mathematical meaning is not taken as existing independently from the acting individuals and from their interaction, but is viewed as accomplished in the course of social interaction. See also *Taken as shared.*

Norms: Criteria of values that are interactively constituted in classroom interactions. Based on the participants' claims and goals, they emerge through interactions and become taken as shared.

Partitioning algorithm: A computational strategy in which a number is separated into unitary conceptual entities. For example, partitioning 39 into 30 and 9 instead of 3 and 9. In this way, 30 is viewed as an unstructured composite of ones. See also *Collection-based solution.*

Patterns of interaction: A regularity that is interactively constituted by the teacher and the students. It is constituted turn by turn without necessarily being intended or realized by the participants. When the participants constitute a regularity that the observer describes as a pattern of interaction, they are stabilizing a fragile process of meaning negotiation. The individual's habitus enables him or her to participate smoothly in the constitution of a pattern of interaction. Depending on specific interests, the observer can reconstruct different patterns. For example, the format of argumentation is a pattern of interaction that describes the accomplishment of a conjointly acceptable argumentation under the condition of a framing difference. See also *Thematic pattern of interaction.*

Realm of developmental possibilities: A construct that delineates the situated conceptual advances that a child makes when participating in an interaction such as that in which an adult intervenes to support his or her mathematical activity. In contrast to Vygotsky's zone of proximal development (which is concerned with what a child can do with adult support), a child's realm of developmental possibilities addresses the way in which a child's conceptions and interpretations evolve as he or she is interacting with the adult.

Reflexive (Leiter, 1980): A property of social relationships in which accounts and settings mutually elaborate each other. Two things or matters are reflexively related if either's existence depends on the other. Reflexive relationships mentioned in this book include: the students' mathematical activity and the social relationships they established, the relation between the quality of a student's explanation and the social situation in which it is developed, the relation between mathematical themes and individual contributions, and engaging in learning and argumentation.

School mathematics microculture: A classroom culture in which explanations involve specifying instructions for manipulating symbols that do not necessarily signify anything beyond themselves. See also *Inquiry mathematics microculture.*

Situation for explanation: A discussion that is interactively constituted by the participants such that someone is required to interpret an explanation that is offered.

Social authority: A social construct describing the role of one small-group partner in a relationship in which this child generally controls the way in which the partners interact. See also *Mathematical authority.*

Social interaction: The mutually dependent actions of participants in a situation. This process enables taken-as-shared meanings to be negotiated. See also *Patterns of interaction.*

Social norms: Classroom norms that involve conventions describing how to collaborate with others, and obligations describing how to react socially to a mistake.

Sociomathematical norms: Classroom norms that involve the evaluation of insightful solutions or mathematically elegant explanations and argumentations. Although the teacher can be viewed as representing the institution of school, educational claims, and the discipline of mathematics, the constituted norms depend on the students' understanding, attitudes, willingness, and so on.

Strips-and-squares task: A format used to present arithmetic tasks with pictures of single squares and strips containing 10 squares. In this book, the symbol "I" is used to symbolize a pictured strip, and "." to symbolize a pictured square.

Symbolic interactionism (Blumer, 1969): The theoretical position that meaning develops out of interaction and interpretation between members of a culture.

Taken as shared (Streeck & Sandwich, 1979): Understood (at least to the extent of common usage) by all members of a culture (e.g., a classroom) through mutual negotiation. Through negotiation of meaning, the participants constitute taken-as-shared meanings, although they do not necessarily "share knowledge." The point is that the individual conceptions have become compatible so that the individuals interact as if they ascribe the same meanings to objects, even if the observer can reconstruct different subjective meanings. Sometimes called *taken to be shared.* See also *Negotiation of mathematical meaning.*

Ten as a numerical composite: A cognitive construct used to characterize a student's view of ten as a composite of 10 ones, as opposed to one single entity or unit of ten.

Ten as an abstract composite unit: A cognitive construct used to characterize a student's view of ten as a single entity or unit composed of 10 ones.

Thematic pattern of interaction: A pattern of interaction in which participants interactively constitute a theme as a matter of routine. See also *Patterns of interaction.*

Theme: The situational relationship between meanings that are taken to be shared by the teacher and students. The accomplishment of a theme gains stability and offers a situational context for understanding individual contributions. Based on a working consensus, the theme is a product of the negotiation of meaning.

Univocal explanation: A sociological construct used to characterize small-group interactions between a pair of students in which one child judges that the partner either does not understand or has made a mistake, and proceeds to explain his or her solution while the partner remains a passive listener. The term is used to emphasize that one child's perspective becomes dominant. See also *Mathematical authority, Multivocal explanation.*

Working consensus: A tentative agreement accomplished by negotiation in social interaction. It is seen as *modus vivendi* rather than a content-related congruence of meaning. In the teaching–learning process, a working consensus is extremely provisional and fragile. The tentative agreement appears as a working interim between more or less overt differences in the interpretation of the situation and/or objects.

REFERENCES

Bauersfeld, H. (1988). Interaction, construction, and knowledge—Alternative perspectives for mathematics education. In D. A. Grouws & T. J. Cooney (Eds.), *Perspectives on research on effective mathematics teaching: Research agenda for mathematics education* (Vol. 1, pp. 27–46). Reston, VA: National Council of Teachers of Mathematics and Lawrence Erlbaum Associates.

Blumer, H. (1969). *Symbolic interactionism: Perspectives and method.* Englewood Cliffs, NJ: Prentice-Hall.

Gregg, J. (1992). *The acculturation of a beginning high school mathematics teacher into the school mathematics tradition.* Unpublished doctoral dissertation, Perdue University, West Lafayette, IN.

Kamii, C. (1985). *Young children reinvent arithmetic: Implications of Piaget's theory.* New York: Teacher's College Press.

Leiter, K. (1980). *A primer on ethnomethodology.* New York: Oxford University Press.

Maturana, H. R., & Varela, F. J. (1980). *Autopoiesis and cognition: The realization of the living.* Dordrecht, Netherlands: Reidel.

Mehan, H., & Wood, H. (1975). *The reality of ethnomethodology.* New York: Wiley.

Nicholls, J. G. (1983). Conceptions of ability and achievement motivation: A theory and its implications for education. In S. G. Paris, G. M. Olson, & W. H. Stevenson (Eds.), *Learning and motivation in the classroom* (pp. 211–237). Hillsdale, NJ: Lawrence Erlbaum Associates.

Skemp, R. R. (1976). Relational understanding and instrumental understanding. *Mathematical Teaching, 77,* 1–7.

Streeck, J., & Sandwich, L. (1979). Good for you. Zur pragmatischen and konventionellen analyse von Bewertungen im institutionellen Diskurs der Schule. In J. Dittman (Ed.), *Arbeiten zur konversations Analyse* (pp. 235–257). Tübingen, Germany: Niemeyer.

von Glasersfeld, E. (1987). Learning as a constructive activity. In C. Janvier (Ed.), *Problems of representation in the teaching and learning of mathematics* (pp. 3–18). Hillsdale, NJ: Lawrence Erlbaum Associates.

Wittgenstein, L. (1974). *Über Gewißheit* [On certainty]. Oxford, England: Basil Blackwell.

Author Index

A

Abrahamsen, A., 275, *288*
Ahlbrand, W. P., 276, *289*
Ainley, J., 219, *226*
Anderson, J. R., 280, *288*
Atkinson, P., 36, *127*
Attewell, P., 237, 238, *268*
Axel, E., 5, *14*

B

Bakhurst, D., 4, *14*
Balacheff, N., 2, *14*, 109, *127*
Ball, D. L., 131, *161*, 204, *226*
Barnes, D., 26, *127*, 131, *161*
Barthes, R., 283, 287, *288*
Bateson, G., 135, *161*, 198, *199*
Bauersfeld, H., 1, 2, 8, 9, *14*, 27, 110, 112, *127*, 131, 165, 166, *199*, 213, *226*, 236, 238, 261, 263, 267, *268*, 272, 276, 278, 279, 280, 282, 285, 286, *288*, 294, *297*
Bechtel, W., 275, *288*
Bell, N., 6, *15*
Bellack, A. A., 272, *288*
Bermudez, T., 7, *16*
Bernstein, R. J., 8, *14*
Billig, M., 36, 110, 125, *127*, 231, *268*
Bishop, A., 2, *14*
Blumer, H., 2, *14*, 134, *161*, 166, 174, *199*, 297, *297*
Bogdan, R., 36, *129*
Boyle, R. A., 276, *290*
Bredo, E., 282, *288*

Brousseau, G., 2, *14*
Brown, A., 225, *227*, 280, *288*
Brown, J. S., 2, 4, *14*
Bruner, J. S., 13, *14*, 122, *127*, 147, *161*, 165, *199*, 230, 250, 254, *268*, 271, 273, 274, 277, 281, 284, 285, 286, 287, *288*
Bussi, M. B., 2, *14*

C

Campione, J. C., 280, *288*
Carpenter, T. P., 2, *14*, 19, *23*, 204, *226*
Carraher, D., 2, 4, *14* , *15*
Carraher, T., 2, 4, *14*
Cazden, C. B., 25, *127*, 271, *289*
Chiang, C. P., 2, *14*, 204, *226*
Cicourel, A., 173, *199*
Clark, C. M., 125, *127*
Cobb, P., 2, 3, 6, 9, 10, *14*, 17, 18, 19, 20, 22, 23, *24*, 26, 28, 29, 30, 31, 38, 40, 103, 124, *127*, *129*, 131, 133, 134, 135, 139, 144, 146, 147, 148, 150, 151, 152, 156, *161*, *162*, 165, 167, 168, 169, 172, 177, 188, 189, 190, 193, 194, 195, 196, *199*, *201*, 203, 204, 208, 211, 213, 216, *226*, *227*, 229, 230, 232, 233, 234, 236, 240, 241, 245, 249, *269*, 281
Cole, M., 2, 4, *15*, 28, 29, 111, *127*, *128*, 204, *227*
Collins, A., 2, 4, *14*
Confrey, J., 2, *14*, 18, *24*, 26, *129*, 207, *226*
Connell, M. L., 280, *288*
Coulmas, F., 283, *289*

AUTHOR INDEX

Coulthard, R., 206, *227*

D

Davidson, D., 276, 277, *289*
Davidson, N., 26, *127*
Davis, H. G., 271, 280, *289*
Davis, P. J., 3, 6, *14*, 278, *289*
Davydov, V. V., 4, *15*
Delamont, S., 36, *127*
Denzin. N. K., 167, *199*
Dillon, D., 204, *227*
Doise, W., 6, *15*, 165, *199*
Dörfler, W., 40, *127*
duBois-Reymond, M., 184, *199*
Duguid, P., 2, 4, *14*
Durkin, K., 271, *289*

E

Edwards, D., 165, *199*
Eisenhart, M. A., 2, *15*
Elster, J., 229, *268*
Engeström, Y., 284, *289*
Erickson, F., 37, *127*, 166, *199*, 271, *289*
Ernest, P., 280, *289*

F

Falk, G., 178, *199*
Fann, K. T., 274, *289*
Fennema, E., 2, *14*, 204, *226*
Fischer, H. R., 275, 277, *289*
Flores, F., 121, *129*
Follesdal, D., 229, *268*
Forman, E. A., 4, *15*, 25, 111, *127*

G

Gale, J., 37, *128*
Garfinkel, H., 166, 173, *200*, 237, 238, *268*, 272, *289*
Garton, A. F., 284, *289*
Gearhart, M., 249, *269*
Glaser, B. G., 36, *128*
Goffman, E., 169, 172, *200*, 249, 252, *268*, 272, *289*
Good, T. L., 26, *128*
Goodlad, J., 208, *226*
Gooya, Z., 27, *128*
Greeno, J. G., 2, *15*, 35, 283, *289*
Gregg, J., 2, *15*, 155, *161*, 295, *298*
Griffin, P., 2, 4, 9, *15*, 28, 111, *128*, 204, *227*, 280, *289*
Guberman, S. R., 249, *269*

H

Habermas, J., 239, 263, *268*
Halliday, M. A., 110, *128*
Hammersley, M., 36, *127*
Hanks, W. F., 4, *15*
Hasan, R., 110, *128*
Haste, H., 274, 284, *288*
Hawkins, D., 3, *15*
Heidegger, M., 224, *226*
Heritage, J., 272, *289*
Hersh, R., 3, 6, *14*, 278, *289*
Hiebert, J., 204, *226*
Hoetker, J., 276, *289*
Hofstadter, D. R., 280, *289*
Hörmann, H., 272, *289*
Human, P. G., 2, *15*
Hyman, R. T., 272, *288*

I

Inhelder, B., 237, *268*

J

Johnson, D. K., 279, *289*
Johnson, M., 40, 67, 121, *128*
Jungwirth, H., 166, *200*

K

Kamii, C, 19, 22, *24*, 131, *161*, 295, *298*
Kaput, J. J., 121, *128*
Keil, F. C., 237, *268*
Klein, W., 237, 238, *268*
Kliebard, H. M., 272, *288*
Kluwe, R. H., 280, *291*
Koch, L. C., 131, *161*
Kopperschmidt, J., 232, 234, 239, *268*
Kozulin, A., 123, *128*, 284, *289*
Kruger, A. C., 283, *290*
Krummheuer, G., 1, 5, 8, *14*, *15*, 27, 105, 108, 131, 166, 169, 172, 176, 177, *199*, *200*, 208, 232, 249, 250, 252, 261, 262, 263, *268*, *269*, 271, 272, 282, *289*

L

Labinowicz, E., 207, *226*
Lakatos, I., 35, *128*, 165, *200*
Lampert, M., 2, *15*, 35, *128*, 131, *161*, 204, *226*, 283, *290*
Lave, J., 2, 4, 5, *15*, 29, 105, *128*, 281, *289*
Leary, D. E., 277, 283, *289*
Lehmann, B. E., 237, 238, *269*

AUTHOR INDEX

Leiter, K., 135, *161*, 168, 173, 177, 192, *200*, 238, *269*, 296, *298*
Lektorsky, V. A., 281, 284, *289*
Leont'ev, A. N., 4, 9, *15*, 111, *128*
Lerman, S., 272, *289*
Levine, J. M., 280, *290*
Lin, G., 27, *128*
Linn, M. C., 103, *128*
Loaf, M., 2, *14*, 204, *226*
Luckman, T., 173, 250, *269*
Luhmann, N., 176, *200*, 275, 281, *289*
Luria, A. R., 273, 285, *290*

M

Maier, H., 178, *200*, 271, *290*
Marx, R. W., 276, *290*
Maturana, H. R., 6, *15*, 275, 277, *290*, 294, *298*
McCaslin, M., 26, *128*
McDermott, R. P., 282, *288*
McNeal, B., 2, 3, 6, *14*, *15*, 131, 133, 150, 151, *161*, 166, *200*
McPhail, J., 25, *127*
Mead, G. H., 166, *200*
Mehan, H., 2, *15*, 122, *128*, 166, *200*, 238, *269*, 272, 280, *289*, *290*, 294, *298*
Mercer, N., 165, *199*
Merkel, G., 19, *24*, 225, *227*
Miller, M., 232, 262, *269*
Minick, N., 4, *15*
Mishler, E. G., 28, *128*
Moll, L. C., 284, *290*
Moser, J. M., 19, *23*
Much, N. C., 2, *15*
Mugny, G., 6, *15*, 165, *199*
Müller, K., 250, *269*
Mulryan, C., 26, *128*
Murray, H., 125, 126, *128*, 131, *161*
Murray, J. C., 2, *15*

N

Neth, A., 187, *200*
Neuman, D., 35, *128*
Newfield, N., 37, *128*
Newman, D., 2, 4, 9, *15*, 28, 111, *128*, 204, *227*
Nicholls, J. G., 22, *24*, 295, *298*
Noddings, N., 26, *128*, 204, 207, *227*
Norman, D., 280, *290*
Nunes, T., 2, *15*

O

Olbrechts-Tyteca, L., 231, 236, 237, 259, *269*
Olivier, A. I., 2, *15*

P

Palincsar, A., 225, *227*
Paschen, H., 234, *269*
Perelman, C., 231, 236, 237, 259, *269*
Perret, J. F., 6, *15*
Perret-Clermont, A. N., 6, *15*
Peterson, P. L., 2, *14*, 204, *226*, 283, *290*
Piaget, J., 108, *128*, 237, *268*, *269*
Picht, G., 273, *290*
Pimm, D., 271, *290*
Pintrich, R. P., 276, *290*
Pollio, H. R., 282, 286, *290*
Pollio, M. R., 282, 286, *290*
Prawat, R., 225, *227*
Presmeg, N. C., 40, *128*
Putnam, H., 8, *15*
Putnam, R. T., 283, *290*

R

Radatz, H., 285, *290*
Ratner, H. H., 283, *290*
Resnick, L. B., 280, 283, *290*
Richards, J., 3, *16*, 17, 19, *24*, 28, 30, 29, 38, *128*, *129*, 131, *161*, 249, *269*
Rogoff, B., 4, *16*, 108, 111, 112, *128*
Rorty, R., 8, *16*, 272, 276, 277, 278, 287, *290*
Rosenshine, B., 207, *227*
Roth, G., 274, 277, *290*

S

Sandwich, L., 205, *227*, 297, *298*
Saxe, G. B., 2, 4, 7, *16*, 121, *128*, 196, *200*, 249, *269*
Schank, R. C., 250, *269*
Schliemann, A. D., 2, 4, *15*
Schoenfeld, A. H., 2, 6, *16*, 283, *290*
Schroeder, T., 27, *128*
Schütz, A., 166, 173, 177, *200*, 250, *269*, 284, *290*
Scribner, S., 4, *16*
Sfard, A., 40, *129*
Shimizu, Y., 26, *129*
Shire, B., 271, *289*
Shweder, R. A., 2, *15*
Siegal, M., 237, *269*, 286, *290*
Silver, E. A., 18, *24*
Sinclair, J., 206, *227*
Skemp, R. R., 22, *24*, 295, *298*
Smith, E., 26, *129*
Smith, F. L., 272, *288*
Smith, M. K., 282, 286, *290*
Söll, B., 184, *199*
Solomon, Y., 5, 7, *16*, 165, *200*

Steffe, L. P., 17, 19, 20, *24*, 29, 30, 38, 124, 125, *129*, 249, *269*
Steinbring, H., 166, 169, *200*, 287, *290*
Steinert, H., 178, *199*
Stevens, R., 207, 227
Strauss, A. L., 36, *128*
Streeck, J., 205, *227*, 297, *298*
Struve, H., 236, *269*
Struve, R., 279, *290*
Stubbs, M., 223, 227

T

Taylor, S. J., 36, *129*
Taylor, T. J., 271, 280, *289*
Teasley, S. D., 280, *290*
Thompson, P., 2, *16*, 18, *24*, 40, *129*
Todd, F., 26, *127*, 131, *161*
Tomasello, M., 283, *290*
Toulmin, S., 232, 235, 236, 239, 241, 242, 243, 244, 247, 249, 252, 263, *269*
Treffers, A., 19, *24*
Trevarthen, C., 205, 227
Tymoczko, T., 164, *200*

V

Valsiner, J., 111, 122, 123, *129*, 284, *290*
van der Veer, R., 111, 122, 123, *129*, 284, *290*
Varela, F. J., 275, 277, *290*, 294, *298*
Voigt, J., 1, 3, 8, 9, *14*, *16*, 28, 109, *129*, 134, 166, 178, 184, 187, 192, *199*, *200*, 206, 207, 213, *227*, 232, 238, 252, 258, 263, 267, *268*, 271, 272, 278, 279, 281, *290*, *291*
Völzing, P. L., 237, *269*
von Glasersfeld, E., 1, 8, *16*, 17, 18, 19, *24*, 29, 30, 38, *129*, 135, *161*, 207, *227*, 249, *269*, 281, *291*, 294, *298*
Vygotsky, L. S., 4, *16*, 111, 122, *129*

W

Walkerdine, V., 5, *16*, 280, *291*

Walloe, L., 229, *268*
Wearne, D., 204, *226*
Webb, N. M., 26, 106, *129*
Weber, R., 206, 227
Weinert, F. E., 280, *291*
Wenger, E., 4, 5, *15*, 29, 105, *128*, 281, *289*
Wertsch, J. V., 123, *129*
Wheatley, G., 17, *24*, 31, 40, *127*, 169, *199*
Wiegel, H. G., 124, 125, *129*
Wigger, L., 234, *269*
Wilson, T. P., 166, *201*
Winograd, T., 121, *129*
Wirtz, R. W., 19, *24*
Wittgenstein, L., 165, *201*, 231, 236, 253, 262, *269*, 272, 274, 275, 277, 279, 287, 288, *291*, 295, *298*
Wolff, C., 273, *291*
Wood, D., 284, *291*
Wood, H., 2, *15*, 122, *128*, 238, *269*, 272, 280, *290*, 294, *298*
Wood, T., 2, 3, 6, 10, *14*, 18, 22, 23, *24*, 26, 27, 103, 104, 122, 125, *127*, *129*, 131, 133, 134, 139, 147, 148, 150, 151, 156, *161*, *162*, 173, 181, 193, 196, 198, *201*, 203, 204, 208, 213, 216, 220, 225, *226*, *227*, 229, 230, 233, 255, 282

Y

Yackel, E., 2, 3, 6, 10, *14*, 18, 22, 23, *24*, 26, 27, 29, 103, 104, 107, 108, 110, 125, *127*, *129*, 131, 133, 134, 139, 147, 148, 150, 151, 156, *161*, *162*, 170, 173, 177, 188, 194, 196, *201*, 203, 204, 208, 211, 213, 216, *226*, *227*, 229, 230, 233, 234, 238, 240, 261, 263, *269*, 282
Yang, M. T.-L., 2, *16*

Z

Zawadowski, W., 282, *288*

Subject Index

A

Abstract composite units of ten, 31, 38–41
Abstract units of one, 19, 30, 38–39
Accomplishments, 280
Account
    accountable, 237–239
    accounting practice, 237, 258
Activity theory, *see* Theoretical perspectives
Algorithms
    addition, 213–214
    nonstandard algorithm, 214
    standard algorithm, 214
Ambiguity, 169, 224
Appropriation, 9
Argument, 232, 234–235, 247, 258, 282; *see also* Framing dependency analytic, 244
    backing, 240, 243–244, 247, 251, 261–262
    conclusion, 240, 243–244, 247
    core of, 240, 251, 253, 255–256, 258, 261, 264–267
    data, 240–241, 243, 247, 266
    explanatory relevance of, 262
    field dependency, 249–251
    finite regress, 252
    force of, 262
    formal validation of, 266
    formally valid, 253, 263, 267
    framing specific relevance, 262
    functional analysis of, 266
    inference, 240
    layout of, 239, 264
    substantial, 262
    warrant, 240–244, 247, 251, 262

Argumentation, 12–13, 66, 108, 110, 229–232, 234–235, 238–240, 245–247, 252–253, 256, 258, 261
    analysis of, 229
    analytic, 235–236
    argumentizing, 13, 267
    collaborative, 10
    collective, 232, 245, 249, 253–254, 261–262, 264, 267
    culture of, 267
    ethnography of, 230
    formal logic of, 246–247
    format of, 32, 248, 253–255, 257, 259, 261–262, 266
    functional analysis of, 239, 247, 264
    image dependent, 251
    image independent, 251
    informal logic of, 237, 240, 247
    layout of, 247
    mathematical, 11
    social interaction, 232
    substantial, 235–237, 239–240, 262
    theory of, 231
    topos, *see* Topos

B

Beliefs
    instrumental, 17
    mathematical, 78–79
    students' beliefs about roles, 22

C

Case studies
    viability of, 36

SUBJECT INDEX

Class discussion
　typical patterns of interaction, *see* Patterns of interaction
　teacher's role, 216–217
Classroom
　learning, 204
Cognitive constructs, 37–41; *see also* Abstract composite units of ten, Abstract units of one, Collection-based solution, Counting-based solution, Image-independent units, Image-supported units, Numerical composite units of ten, Part–whole relations
Collaboration
　direct, 107–108
　indirect, 108, 126
　with true peers, 126
Collaborative dialogue, 131–132
Collection-based solution, 40–41, 189, 193
Collectivism, 3–6, 273
Communication, 205, *see also* Taken as shared
Conceptual restructuring, 118–120, 123, 148–150
Consensual domain, 26, 277, 282; *see also* Consensus
Consensus
　working interim, 252–253, 255, 262–263, 266–267
Constructivism, *see* Theoretical perspectives
Context, 135
Counting-based solution, 40–41, 189
Cultural tools, 120–124
Culture, 252, 280–281, 283–284
　epoché of natural attitude, 284
　of mathematics classroom, 281
Cybernetics, 8

D

Discourse
　academic, 111
　everyday, 111
　field of, 110
　mode of, 110
　rational, 239
　tenor of, 110

E

Emergent perspective, *see* Theoretical perspectives
Ethnomethodology, *see* Theoretical perspectives

F

Folk psychology, *see* Psychology

Format, 254, 258, 266; *see also* Argumentation
Framing, 248–252, 255, 258, 263–264, 267, 272; *see also* Argument
　dependent, 261–262, 264, 266–267
　differences, 251–253, 255, 261, 263, 265–267
　frame, 249
　independent, 262
　script, 249
　working consensus, *see* Consensus

G

Goals
　student, 196
　teaching, 216

H

Hidden curriculum, 22

I

Illusion of competence, 155
Image-independent units, 39–41, 188
Image-supported units, 39–41, 188
Imagery, 40, 193
Individualism, 3, 6–7
Inquiry mathematics, 104–106
Interaction
　analysis of, 240
　classroom, 232, 262
　format of, 264
　patterns of, *see* Patterns of interaction
Interactionism, *see* Theoretical perspectives
Intersubjectivity, 8, 11–12, 109, 111, 163, 238
Interviews
　small-group sessions, 105–106
　as social events, 28–29
　tasks of, 29–32
Intuitionism, *see* Theoretical perspectives

K

Knowledge, 8
　empirical-theoretical status of, 236

L

Language games, 13, 274, 277–279
Languaging, 238
Learning, 195, 258
　classroom 204
　mathematical, 261, 264, 267
　opportunities for, 44, 148–150, 195
　situated, 280

SUBJECT INDEX 305

theory of learning mathematics, 255
Lebenswelt (Life world), 252
Locus, *see* Topos

M

Mathematical authority, 43
Meaning, 134–135; *see also* Negotiation of meaning
Metacommunication, 232
Metaphor, *see* Taken as shared
Microculture
    accomplishment of, 176–178
Micro-ethnography, 229
Multivocal explanation, 42, 107

N

Negotiation of meaning, 1, 134, 266, 282–283, 286
    ambiguity and interpretation, 167–169
    mathematical, 166–169
    taken-as-shared meanings, 172–174; *see also* Taken as shared
Neo-Piagetian perspective, *see* Theoretical perspectives
Norms
    classroom mathematical practices, 2, 124, 196
    pair collaboration norms, 216
    renegotiation of, 208
    routines, 206
    small-group norms, 104
    social norms, 22–23, 108–109, 126, 133–134, 205, 211
    sociomathematical norms, 12, 134, 196, 198–199
Notational regularities, 121
Numerical composite units of ten, 19, 38–41
Numerical significance, 144, 151, 153, 188

O

Objects, 273, 285
    arithmetical, 21, 29, 81, 104–105
    experientially real, 81
Orienting, 275

P

Pair collaboration
    social norms for, 216
    teaching goals, 216
Part–whole relations, 38–41
Patterns of interaction, 12, 32, 166, 178–184, 205, 208–213, 253, 272

direct mathematization pattern, 185–187
discussion pattern, 181–184
elicitation pattern, 178–181, 184
funnel pattern, 213, 219, 276
thematic patterns of, 32, 176–178, 184–192
typical patterns, 210
Piagetian perspective, *see* Theoretical perspectives
Platonism, *see* Theoretical perspectives
Power, 43, 101; *see also* Taken as shared
Psychology, *see also* Theoretical perspectives
    folk psychology, 13, 230, 236, 261, 267

R

Radical constructivism, *see* Theoretical perspectives
Rational discourse, *see* Discourse, rational
Rationality, 237–238, 246
Reality illusion, 13, 273–276
Realm of developmental possibilities, 29, 189, 194
Reflexive discussions, 126
Reflexive relationships
    argumentizing and learning, 264, 267
    context and actions, 136, 238, 281
    context and meaning, 177
    individual and social perspectives, 261
    learning and interaction, 192
    math activity and classroom microculture, 9–10
    math activity and social relationships, 43–44
    numerical patterns and students' interpretations, 121–122
    psychological and sociological processes, 111, 258
Reflexivity, 135–136, 266
Representation, 276, 284–285
Rhetoric, 13, 231, 249

S

Scaffolding, 287
Situation for explanation, 150–151
Social authority, 43
Social interaction, 169–172
Social norms, *see* Norms
Sociocultural perspective, *see* Theoretical perspectives
Sociohistorical perspective, *see* Theoretical perspectives
Sociolinguistics, *see* Theoretical perspectives
Sociological constructs, 41–43; *see also* Mathematical authority, Multivocal explanation, Social authority, Univocal explanation

Subtraction, 213
Symbolic interactionism, *see* Theoretical perspectives
Symbolic numerical practice, 121

T

Taken-as-shared
    basis for communication, 11, 26–27, 89, 103, 107–109, 205
    experiences, 279
    interpretation, 29, 42, 67–68, 80, 285
    issues, 282
    knowledge, 283
    mathematical practice, 113, 281
    mathematical reality, 104, 250
    meaning, 13, 276, 287
    metaphor, 86, 98, 121
    nature of linguistic act, 110
    norms, 199
    power interpretation, 43
    solution strategy, 35, 79, 101
Teaching, 203
    emerging practice of, 12, 203
    focusing pattern in, 219–220
    funnel pattern in, 213, 219, *see also* Patterns of interaction
    interactive collective activity, 204
    questioning, 210, 219
    tension in, 12, 125, 211
    transformation of, 203
Theme, 174
    evolution of, 192–196
Theoretical perspectives
    activity theory, 4–5, 280
    constructivism, 13, 18, 123–124, 264, 266, 273–274, 281
    emergent, 122–124
    ethnomethodology, 2, 231–232, 237–239, 246, 250, 266, 272
    interactionism, 164–167, 231, 250, 252, 272, 274, 281
    intuitionism, 164
    neo-Piagetian, 6–7
    Piagetian, 108–110, 112
    Platonism, 164
    psychological, 7–10
    radical constructivism, 272, 279
    sociocultural, 8, 111–112, 122–123
    sociohistorical, 9
    sociolinguistic, 5–6
    sociological, 7–10
    symbolic interactionism, 2, 8, 13, 123–124
    Vygotskian, 4–5, 111–124
Throwness, 224
Topoi, *see* Topos
Topos, 249, 255, 258–261, 264, 266

U

Univocal explanation, 42, 106–107

V

Verdinglichung, 280
Vygotskian perspective, *see* Theoretical perspectives

W

Working interim, *see* Consensus

Z

Zone of proximal development, 4, 29